Perennial Biomass Crops for a Resource-Constrained World

Susanne Barth • Donal Murphy-Bokern
Olena Kalinina • Gail Taylor • Michael Jones
Editors

Perennial Biomass Crops for a Resource-Constrained World

Springer

Editors
Susanne Barth
CELUP Oak Park Research Centre
Teagasc
Oak Park, Carlow, Ireland

Olena Kalinina
Fg. Nachwachsende Rohstoffe
und Bioenerg
Universität Hohenheim
Stuttgart, Baden-Württemberg, Germany

Michael Jones
School of Natural Sciences
Trinity College Dublin
Dublin, Ireland

Donal Murphy-Bokern
Lohne, Niedersachsen, Germany

Gail Taylor
Centre for Biosciences
University of Southampton
Southampton, Hampshire, UK

ISBN 978-3-319-44529-8 ISBN 978-3-319-44530-4 (eBook)
DOI 10.1007/978-3-319-44530-4

Library of Congress Control Number: 2016954627

Printed on acid-free paper

This Springer imprint is published by Springer Nature
The registered company is Springer International Publishing AG
The registered company address is: Gewerbestrasse 11, 6330 Cham, Switzerland

Preface

This book is a product of a conference held at the University of Hohenheim in September 2015, which brought together five European Framework Programme 7 research consortia to report the results of their work over the previous 5 years. The research was initiated in recognition of Europe's need for sustainably produced biomass to support European strategic objectives for the bioeconomy.

The conference title was 'Perennial Biomass Crops for a Resource-Constrained World' and we have retained this for the title of this book. The book contains 24 chapters which are largely the written accounts of presentations made in Hohenheim. The chapters were solicited by the editors and have been peer reviewed and placed under five broad topic areas relating to the use of perennial biomass crops in Europe. These are 'Bioenergy Resources from Perennial Crops in Europe', 'European Regional Examples for the Use of Perennial Crops for Bioenergy', 'Genotypic Selection of Perennial Biomass Crops for Crop Improvement', 'Ecophysiology of Perennial Biomass Crops' and 'Examples of End Use of Perennial Biomass Crops'.

We focus on two major issues relating to the future use of biomass energy: the identification of the most suitable second generation biomass crops, and the need to utilise land not used for intensive agricultural production (broadly referred to as 'marginal' land) so that we avoid the potential conflict between food and fuel production. Perennial biomass crops (PBCs) are crops that are established only once in a plantation's lifetime which can be harvested regularly over a lifespan of at least 20–30 years. The two main categories of plants that fit this description are perennial rhizomatous grasses and trees that can be coppiced, although there are others, such a fibre nettle (*Urtica dioica*).

In Part I: 'Bioenergy Resources from Perennial Crops in Europe', the overarching questions are related to the suitability of perennial crops for feedstocks for a European bioeconomy and in particular the need to exploit environments for biomass crops which do not compete with food crops. Bioenergy is the subject of a wide range of national and European policy measures, and the development of public policy based on an examination of bioenergy policy in Germany, the UK and Ireland is examined. New developments in the use of perennial grasses to produce protein for animal feed, the potential for long-term yields and soil carbon sequestration from the PBC, *Miscanthus* are presented. It is also shown that *Miscanthus* can complement grassland as a bioenergy source on marginal land in Europe.

In Part II: 'European Regional Examples for the Use of Perennial Crops for Bioenergy', the chapters review the use and development of PCBs across Europe. A range of PBCs has been shown to suit the different climatic and soil conditions of Europe, particularly on marginal land. *Miscanthus* has been shown to out-yield other PBCs in trials in central Europe and the Far East. Other PBCs are also shown to have high potential including *Arundo donax* in Italy, fibre nettle (*Urtica dioica*) in Lithuania and perennial grasses in maritime climates such as Ireland.

In Part III: 'Genotypic Selection of Perennial Biomass Crops for Crop Improvement', the chapters highlight how breeding work on PBCs is in its infancy compared to annual food crops and progress is slower than for annual crops because of the long selection cycles. Breeding still often starts with germplasm collection from places where these species grow wild and progresses through the screening of germplasm and the making of novel crosses and evaluation. However there is also the development of novel variation by polyploidization and mutagenesis to enlarge variation in the gene pools. Important breeding objectives for the perennial rhizomatous grass, *Miscanthus*, are frost and chilling stress where significant progress has yet to be achieved. Significant breeding efforts are also directed towards coppiced trees and giant reed for traits like heavy metal uptake. For traditional C_3 grass species improved drying rates harvested for biomass are of prime interest.

In Part IV: 'Ecophysiology of Perennial Biomass Crops', the chapters illustrate how an understanding of the ecophysiology of different PBC species and genotypes is essential to draw conclusions and forecasts on the adaptation to, as well as survival and growth on, marginal land. Marginal land by definition severely limits the productivity of crops because of a range of abiotic stresses, including shortage/excess of soil water, low nutrient availability, salinity and high and low temperatures, and it is shown that the use of mycorrhiza-based biofertilizer to provide nutrients for the growth of *Arundo donax*, tall fescue and reed canary grass on nutrient-depleted soils improves adaptation. The selection of suitable species for particularly challenging situations is of utmost importance to ensure adaption and the best biomass production potential on marginal land sites, but a systems approach can be used to guide future PBC development on marginal land.

Finally in Part V: 'Examples of End Use of Perennial Biomass Crops', the chapters illustrate how end uses of biomass can be novel and lead to new applications and products like using specific plant fragments for biorefining, as well as using *Miscanthus* as the aggregate base in concrete masonry blocks and even as horse bedding.

Oak Park, Carlow, Ireland — Susanne Barth
Lohne, Niedersachsen, Germany — Donal Murphy-Bokern
Stuttgart, Baden-Württemberg, Germany — Olena Kalinina
Southampton, Hampshire, UK — Gail Taylor
Dublin, Ireland — Michael Jones

Acknowledgments to International Peer-Review Board

We are indebted to the following specialist expert referees who reviewed the chapters for this book so thoroughly:

Christel Baum, University of Rostock, Germany
Stephen Byrne, Teagasc, Ireland
John Carroll, Teagasc, Ireland
Michael Casler, USDA, USA
John Clifton-Brown, Aberystwyth University, UK
Kerrie Farrar, Aberystwyth University, UK
John Finnan, Teagasc, Ireland
Trevor R. Hodkinson, Trinity College, Dublin, Ireland
Yasir Iqbal, University of Hohenheim, Germany
Mike Jones, Trinity College Dublin, Ireland
Gary Lanigan, Teagasc, Ireland
Michal Mos, Aberystwyth University, UK
Cora Münnich, TINPLANT Biotechnik und Pflanzenvermehrung GmbH, Germany
Brian Murphy, Trinity College Dublin, Ireland
Hilde Muylle ILVO, Belgium
Bruce Osborne, University College Dublin, Ireland
Piotr Przytarski, Greenwood Resources, Poland
Tiina Pursula, Gaia, Finland
Moritz Reckling, ZALF, Germany
Danilo Scordia, University of Catania, Italy
Hazel Smith, Southampton University, UK
Daniele Trebbi, Geneticlab, Italy
Ulrich Thumm, University of Hohenheim, Germany
Charles Warren, St Andrews University, UK

Contents

Contributors

Andrea Acampora Consiglio per la ricerca in agricoltura e l'analisi dell'economia agraria, Unità di ricerca per l'ingegneria agraria (CREA-ING), Rome, Italy

Vincenzo Alfano Consiglio per la ricerca in agricoltura e l'analisi dell'economia agraria, Unità di ricerca per l'ingegneria agraria (CREA-ING), Rome, Italy

Mike R. Allwright Centre for Biological Sciences, University of Southampton, Southampton, UK

Elena Baraza Balearic Islands University, Palma, Baleares, Spain

Bruno Barbosa MEtRiCS, Departamento de Ciências e Tecnologia da Biomassa, Faculdade de Ciências e Tecnologia, FCT, Universidade Nova de Lisboa, Caparica, Portugal

Universidade Federal do Oeste da Bahia, Barreiras, Brazil

Vicky Bartels Tinplant Biotechnik und Pflanzenvermehrung GmbH, Klein Wanzleben, Germany

Susanne Barth Teagasc, Crops, Environment and Land Use Programme, Oak Park Crops Research Centre, Carlow, Ireland

Pietro Bartocci Department of Engineering, University of Perugia, Perugia, Italy

Rupali Bhalerao Department of Forest Genetics and Plant Physiology, Umeå Plant Science Center, Swedish University of Agricultural Sciences, Umeå, Sweden

Gianni Bidini Department of Engineering, University of Perugia, Perugia, Italy

Alla Bryanskaya The Federal Research Center Institute of Cytology and Genetics, The Siberian Branch of the Russian Academy of Sciences, Novosibirsk, Russia

Brendan Burke Crops Research Department, Crops, Environment and Land Use Programme, Teagasc Oak Park, Carlow, Ireland

Natalya Burmakina The Federal Research Center Institute of Cytology and Genetics, The Siberian Branch of the Russian Academy of Sciences, Novosibirsk, Russia

John Carroll Crops Research Department, Crops, Environment and Land Use Programme, Teagasc Oak Park, Carlow, Ireland

Joanna Cerazy Institute of Plant Genetics, Polish Academy of Sciences, Poznań, Poland

Tian-Gen Chang CAS-MPG Partner Institute for Computational Biology, Shanghai Institutes for Biological Sciences, Chinese Academy of Sciences, Shanghai, China

Josep Cifre Balearic Islands University, Palma, Baleares, Spain

John Clifton-Brown Institute of Biological, Environmental and Rural Sciences, University of Aberystwyth, Aberystwyth, UK

Salvatore Luciano Cosentino Dipartimento di Agricoltura, Alimentazione e Ambiente—Di3A, University of Catania, Catania, Italy

Jorge Costa MEtRiCS, Departamento de Ciências e Tecnologia da Biomassa, Faculdade de Ciências e Tecnologia, FCT, Universidade Nova de Lisboa, Caparica, Portugal

Franco Cotana Department of Engineering, University of Perugia, Perugia, Italy

Franz Dahm University of Luxembourg, Esch-sur-Alzette, Luxembourg

Charel Faltz University of Luxembourg, Esch-sur-Alzette, Luxembourg

Francesco Fantozzi Department of Engineering, University of Perugia, Perugia, Italy

Ana Luisa Fernando MEtRiCS, Departamento de Ciências e Tecnologia da Biomassa, Faculdade de Ciências e Tecnologia, FCT, Universidade Nova de Lisboa, Caparica, Portugal

John Finnan Teagasc, Crops, Environment and Land Use Programme, Oak Park Crops Research Centre, Carlow, Ireland

Simon Fonteyne Plant Sciences Unit, Institute for Agricultural and Fisheries Research (ILVO), Melle, Belgium

Department of Plant Production, Ghent University (UGent), Ghent, Belgium

Christine Gebhardt Tinplant Biotechnik und Pflanzenvermehrung GmbH, Klein Wanzleben, Germany

Angelo Del Giudice Consiglio per la ricerca in agricoltura e l'analisi dell'economia agraria, Unità di ricerca per l'ingegneria agraria (CREA-ING), Rome, Italy

Eugenia Golubeva The Federal Research Center Institute of Cytology and Genetics, The Siberian Branch of the Russian Academy of Sciences, Novosibirsk, Russia

Tatiana Goryachkovskaya The Federal Research Center Institute of Cytology and Genetics, The Siberian Branch of the Russian Academy of Sciences, Novosibirsk, Russia

Elvyra Gruzdevienė Upytė Experimental Station of the Lithuanian Research Centre for Agriculture and Forestry, Upytė, Lithuania

Javier Gulias Balearic Islands University, Palma, Baleares, Spain

Magnus Hertzberg SweTree Technologies AB, Umeå, Sweden

Trevor R. Hodkinson Department of Botany, School of Natural Sciences, Trinity College Dublin, The University of Dublin, Dublin, Ireland

Yasir Iqbal Department of Biobased Products and Energy Crops (340b), Institute of Crop Science, University of Hohenheim, Stuttgart, Germany

Zofija Jankauskienė Upytė Experimental Station of the Lithuanian Research Centre for Agriculture and Forestry, Upytė, Lithuania

Marta Jankowska Teagasc, Crops Environment and Land Use Programme, Oak Park Crops Research Centre, Carlow, Ireland

Stanisław Jeżowski Institute of Plant Genetics, Polish Academy of Sciences, Poznań, Poland

Michael B. Jones School of Natural Sciences, Trinity College Dublin, The University of Dublin, Dublin, Ireland

Uffe Jørgensen Department of Agroecology, Aarhus University, Tjele, Denmark

Zygmunt Kaczmarek Institute of Plant Genetics, Polish Academy of Sciences, Poznań, Poland

Olena Kalinina Department of Biobased Products and Energy Crops (340b), Institute of Crop Science, University of Hohenheim, Stuttgart, Germany

JJB (Joost) Keurentjes Laboratory of Genetics, Wageningen University, Wageningen, The Netherlands

Jens Bonderup Kjeldsen Department of Agroecology, Aarhus University, Tjele, Denmark

Manfred Klaas Teagasc, Crops, Environment and Land Use Programme, Oak Park Crops Research Centre, Carlow, Ireland

Nikolay Kolchanov The Federal Research Center Institute of Cytology and Genetics, The Siberian Branch of the Russian Academy of Sciences, Novosibirsk, Russia

Kirsten Kørup Department of Agroecology, Aarhus University, Tjele, Denmark

Poul Erik Lærke Department of Agroecology, Aarhus University, Tjele, Denmark

Søren Ugilt Larsen Department of Agroecology, Aarhus University, Tjele, Denmark

Division AgroTech, Danish Technological Institute, Aarhus, Denmark

Iris Lewandowski Department of Biobased Products and Energy Crops (340b), Institute of Crop Science, University of Hohenheim, Stuttgart, Germany

Peter Lootens Plant Sciences Unit, Institute for Agricultural and Fisheries Research (ILVO), Melle, Belgium

Linda-Maria Mårtensson Department of Biosystems and Technology, Swedish University of Agricultural Sciences, Alnarp, Sweden

Irina Meshcheryakova The Federal Research Center Institute of Cytology and Genetics, The Siberian Branch of the Russian Academy of Sciences, Novosibirsk, Russia

Michele Morgante Università Di Udine, Istituto die Genomica Applicata, Udine, Italy

Cora Münnich Tinplant Biotechnik und Pflanzenvermehrung GmbH, Klein Wanzleben, Germany

Donal Murphy-Bokern Lohne, Niedersachsen, Germany

Hilde Muylle Plant Sciences Unit, Institute for Agricultural and Fisheries Research (ILVO), Melle, Belgium

Nikolay Nechiporenko The Federal Research Center Institute of Cytology and Genetics, The Siberian Branch of the Russian Academy of Sciences, Novosibirsk, Russia

Szymon Ornatowski Institute of Plant Genetics, Polish Academy of Sciences, Poznań, Poland

Luigi Pari Consiglio per la ricerca in agricoltura e l'analisi dell'economia agraria, Unità di ricerca per l'ingegneria agraria (CREA-ING), Rome, Italy

Sergey E. Peltek Institute of Cytology & Genetics SB RAS, Novosibirsk, Russia

The Federal Research Center Institute of Cytology and Genetics, The Siberian Branch of the Russian Academy of Sciences, Novosibirsk, Russia

Elena Petrunenko Botanical Garden-Institute FEB RAS, Vladivostok, Russia

Andrea Polle Department of Forest Botany and Tree Physiology, Büsgen Institute, Georg-August-University, Göttingen, Germany

Benjamin Rauscher Department of Biobased Products and Energy Crops (340b), Institute of Crop Science, University of Hohenheim, Stuttgart, Germany

Dirk Reheul Department of Plant Production, Ghent University (UGent), Ghent, Belgium

Isabel Roldán-Ruiz Plant Sciences Unit, Institute for Agricultural and Fisheries Research (ILVO), Melle, Belgium

Antonia Romero-Munar Balearic Islands University, Palma, Baleares, Spain

Enrico Santangelo Consiglio per la ricerca in agricoltura e l'analisi dell'economia agraria, Unità di ricerca per l'ingegneria agraria (CREA-ING), Rome, Italy

Emanuele Sanzone Consiglio per la ricerca in agricoltura e l'analisi dell'economia agraria, Unita di ricerca per l'ingegneria agraria (CREA-ING), Rome, Italy

Davide Scaglione IGA Technology Services, Udine, Italy

Simone Scalabrin IGA Technology Services, Udine, Italy

Antonio Scarfone Consiglio per la ricerca in agricoltura e l'analisi dell'economia agraria, Unità di ricerca per l'ingegneria agraria (CREA-ING), Rome, Italy

Sergey V. Shekhovtsov The Federal Research Center Institute of Cytology and Genetics, The Siberian Branch of the Russian Academy of Sciences, Novosibirsk, Russia

Vladimir Shumny The Federal Research Center Institute of Cytology and Genetics, The Siberian Branch of the Russian Academy of Sciences, Novosibirsk, Russia

Sarah Sidella Dipartimento di Agricoltura, Alimentazione e Ambiente—Di3A, University of Catania, Catania, Italy

Nikolay Slynko The Federal Research Center Institute of Cytology and Genetics, The Siberian Branch of the Russian Academy of Sciences, Novosibirsk, Russia

Hazel K. Smith Centre for Biological Sciences, University of Southampton, Southampton, UK

Qing-Feng Song CAS-MPG Partner Institute for Computational Biology, Shanghai Institutes for Biological Sciences, Chinese Academy of Sciences, Shanghai, China

Konstantin Starostin The Federal Research Center Institute of Cytology and Genetics, The Siberian Branch of the Russian Academy of Sciences, Novosibirsk, Russia

Alessandro Suardi Consiglio per la ricerca in agricoltura e l'analisi dell'economia agraria, Unità di ricerca per l'ingegneria agraria (CREA-ING), Rome, Italy

Tom De Swaef Plant Sciences Unit, Institute for Agricultural and Fisheries Research (ILVO), Melle, Belgium

Maria Tauler Balearic Islands University, Palma, Baleares, Spain

Gail Taylor Centre for Biological Sciences, University of Southampton, Southampton, Hampshire, UK

Vislojit Thapa Laboratory for Solid Structures, University of Luxembourg, Esch-sur-Alzette, Luxembourg

Ulrich Thumm Department of Biobased Products and Energy Crops (340b), Institute of Crop Science, University of Hohenheim, Stuttgart, Germany

Sergey Veprev The Federal Research Center Institute of Cytology and Genetics, The Siberian Branch of the Russian Academy of Sciences, Novosibirsk, Russia

Daniele Waldmann Laboratory for Solid Structures, University of Luxembourg, Esch-sur-Alzette, Luxembourg

Henning Wildhagen Department of Forest Botany and Tree Physiology, Büsgen Institute, Georg-August-University, Göttingen, Germany

Xin-Guang Zhu CAS-MPG Partner Institute for Computational Biology, Shanghai Institutes for Biological Sciences, Chinese Academy of Sciences, Shanghai, China

Jesko Zimmermann School of Natural Sciences, Trinity College Dublin, The University of Dublin, Dublin, Ireland

The original version of this book was revised. An erratum to this book can be found at DOI 10.1007/978-3-319-44530-4_25

Part I
Introduction into Bioenergy Resources from Perennial Crops

Chapter 1
The Role of Perennial Biomass Crops in a Growing Bioeconomy

I. Lewandowski

The Increasing Demand for Biomass in a Growing Bioeconomy

The German Bioeconomy Council (2015) defines bioeconomy as "the knowledge-based production and utilization of biological resources to provide products, processes and services in all sectors of trade and industry within the framework of a sustainable economic system." The growth of a bioeconomy involves the replacement of fossil by biogenic resources (or biomass) derived from plants, animals, or microorganisms. In the envisioned "ideal" bioeconomy, biomass production will take ecological, social, and health aspects into consideration and be internationally competitive (Staffas et al. 2013). From the definitions and ambitions of the bioeconomy, it can be concluded that its growth will require a sufficient supply of sustainably produced biomass. It is hard to predict how much biomass will be required to fulfill the increasing demands for food, biomaterial, and bioenergy simultaneously and sustainably.

An estimated 136 million tons of agricultural biomass will be required for energetic use in 2020, compared to the 72 million tons used in 2012 (Table 1.1; Scarlat et al. 2015). There are no estimates available for the amount of biomass required to produce biomaterials and biochemicals. But it is expected that more than 80 million tons of bioproducts will be produced in 2020 (Scarlat et al. 2015) and potentially about 150 million tons of fossil-based products and chemicals could be replaced by biobased products. This indicates a substantial demand for additional biomass to fulfill the EU's bioeconomy goals.

I. Lewandowski (✉)
Department of Biobased Products and Energy Crops (340b), Institute of Crop Science,
University of Hohenheim, Fruwirthstr. 23, Stuttgart 70599, Germany
e-mail: iris_lewandowski@uni-hohenheim.de

S. Barth et al. (eds.), *Perennial Biomass Crops for a Resource-Constrained World*, DOI 10.1007/978-3-319-44530-4_1

Table 1.1 Increase in biomass demand in the EU bioeconomy for selected sectors that use agricultural biomass. (**a**) For bioenergy: estimates of biomass feedstock requirements are available; (**b**) for biomaterials: estimates are only available for end products

(**a**) Bioenergy	Use in 2012	Predicted use in 2020 (according to NREAPs*)
	Biomass in million tons	Biomass in million tons
Biofuels	40	63 food-crop based 15 lignocellulosic
Heat+Electricity	280 • 178 from forestry • 72 from agriculture – 40 from energy crops – 32 from residues • 30 from waste	420 • 224 from forestry • 136 from agriculture – 84 from lignocellulosic crops – 52 from residues • 60 from waste

(**b**) Biomaterials	Use in 2012	Predicted use in 2020
	Product in million tons Biogenic (Fossil)	Product in million tons
Biochemicals	8.6 (90)	≈30
Bioplastics	0.3 (57)	≈50
Biolubricants	0.14	0.4–0.4
Biosolvents	0.6	1.1
Others: biosurfactants, enzymes, biopharmaceuticals and new biogenic products	?	?

Source: Data from Scarlat et al. (2015)
[a]*NREAPs* National Renewable Energy Action Plans of Member States

According to Scarlat et al. (2015), the potential risks arising from increased biomass production and supply in Europe are:

- The move toward a bioeconomy based on natural resources from land and sea would lead to a large increase in the demand for biomass undermining the sustainability of a biobased economy.
- Additional land use could lead to negative impacts from land use change, such as biodiversity, soil carbon and soil fertility losses.
- The need to increase crop productivity could lead to increased use of fertilizers and pesticides with additional problems related to water and soil pollution.
- Additional pressure on water resources.
- Increasing competition for resources between food supply and nonfood biomass.

The objective of this paper is to discuss the potential role that perennial biomass crops (PBC) could play in fulfilling the future demand for sustainably produced biomass in a growing bioeconomy in the EU. Here, PBC are defined as either rhizomatous grasses or trees grown as short rotation coppice (SRC).

How Can Perennial Biomass Crops (PBC) Address the Challenges Around Biomass Production and Supply in Europe?

After more than a decade of strong political support for the introduction of bioenergy, especially liquid transport biofuels, it has become obvious that resource limits are a major obstacle to the acceptance and further expansion of bioenergy (Lewandowski 2015). Here, biomass is defined more generally as all biogenic resources derived from plants, animals, or microorganisms, including edible products, such as starch, sugar and oils, and nonedible lignocellulosic biomass.

Several studies have found that the technical potential is large, but the actual supply of low-cost and sustainably produced biomass is currently still limited (for an overview, see Dornburg et al. 2010). The main reason for this is that these assessments of potential took neither infrastructural and logistical costs and constraints, nor "real" production conditions on the ground into consideration (Lewandowski 2015). These conditions include farmers' ability or willingness to produce the biomass required. Other constraints include doubts about the environmental performance of biofuel production. Political support for bioenergy was justified by promises of greenhouse gas (GHG) reductions and increased employment in rural areas. Whereas support measures resulted in increased income in rural areas, several studies revealed low GHG emission reduction potentials of liquid biofuels (Fargione et al. 2008; Searchinger et al. 2008). This is especially the case for biofuels produced from intensively managed oil-, sugar-, or starch-delivering crops requiring high amounts of nitrogen (N) fertilization. Additionally, these so-called *first generation* biofuels require biomass that is also used for food or feed. This has had a strong negative impact on their acceptance and resulted in the "food-versus-fuel" debate. The feedstock demand of first generation biofuels encroaches directly on food markets. This has an impact on food prices along with other factors, such as speculation, storage capabilities, and poor harvests. One possible way forward is the production of bioenergy from nonedible biomass, such as organic wastes and lignocellulosic biomass. Another possibility could be the use of land unsuitable for food-crop production, also referred to as "marginal land." Agricultural land is marginalized for either economic or biophysical reasons. Biomass production on economically marginalized land often delivers low economic margins, mainly due to low yields. Biophysical marginalization can be the result of low soil quality, contamination, insufficient water supply, or steep slopes, and can only be overcome by stress-tolerant crops. Good margins may be achieved on biophysically marginal land if productive use of marginal land can be made by these crops and if land use costs are low.

Perennial biomass crop production can address these issues in several ways (for an overview, see, e.g., McCalmont et al. 2015):

- **PBC have low input requirements and their cultivation is associated with very low GHG emissions.**

Due to their perennial growth and ability to recycle and store nutrients over winter in underground roots and rhizomes, PBC have comparatively low fertilization requirements. Recycling is especially efficient for N, which is associated with the highest proportion of GHG emissions in crop production. Under some circumstances, the productive cultivation of PBC without application of N fertilizers is possible. Also, apart from herbicides during the first establishment years, most PBC require little, if any, pesticide application for healthy growth. Tillage is required in the year of establishment only. Altogether this allows for biomass production with low inputs and low GHG emissions.

- **PBC cultivation does not require annual ploughing which leads to improved soil fertility, carbon sequestration, and biodiversity.**

The productive period of PBC ranges from 10 to 25 years, depending on the crop. Long-term soil rest is ensured as no intensive soil cultivation, which is required when the crop is established, is done after the year of planting. Together with the increased return of organic matter this leads to sequestration of soil carbon, improvement in soil fertility, and increase in soil biodiversity. Many PBC, such as rhizomatous grasses, are harvested in early spring and others, mainly short rotation coppice, are only harvested once every 3–4 years. Therefore, PBC fields provide shelter for mammals and birds over winter and harvesting is not performed during the breeding season.

- **PBC are water and land-use efficient.**

Miscanthus and switchgrass belong to the group of so-called *C4* crops. These have a more efficient photosynthetic pathway than *C3* crops, such as wheat or oilseed rape. The theoretical maximal photosynthetic energy conversion efficiency is 4.6 % for C3 and 6 % for C4 plants (Zhu et al. 2010). This higher efficiency of C4 crops is due to better water- and nutrient-use efficiency and supports increased land-use efficiency. However, the deep rooting systems of these crops can also lead to exploitation of groundwater resources; therefore, careful choice of the production site needs to be made.

- **PBC are stress-tolerant and can be produced under marginal conditions.**

Once established, PBC are more stress-tolerant than annual crops because they root deeply and, from the second year on, are not dependent on optimal establishment conditions, such as sufficient precipitation or soil workability. This deep rooting and long-term soil rest mean that established plantations provide protection from erosion, for example on slopes. Droughts can be overcome more easily once deep roots have developed. Tolerance of many types of stress, such as droughts, salinity, cold, and contaminations, has been reported, e.g., in various miscanthus genotypes (see results of EU project OPTIMISC in Lewandowski et al. 2015).

- **PBC biomass can be produced at low costs.**

The major crop production costs are crop establishment, agrochemicals, and harvesting. Because PBC only need to be established once in a plantation cycle of 10–25 years, establishment costs can be lower than for annual crops once cost-efficient establishment methods have been developed. Low demand for agrochemicals leads to low costs for agricultural inputs and low operational costs.

Once established successfully, the operational procedures in PBC production are reduced to fertilization and harvesting. This makes PBC cultivation also feasible in remote areas.

- **PBC produce lignocellulosic biomass which does not impact food markets.**

The main product of PBC is lignocellulosic, and thus nonedible biomass. This biomass does not directly interfere with the food market and has no direct impact on food prices. New markets need to be developed for this biomass. However, it must also be seen that the use of lignocellulosic biomass is limited and that presently it only has limited applications in the bioeconomy. The attempt to use lignocellulosic biomass in so-called *second generation* conversion technologies, such as enzymatic hydrolysis and pyrolysis, has not resulted in production at a relevant scale, mainly due to high investment and input costs and low processing efficiency.

In summary, the cultivation of PBC can contribute to the avoidance, or at least addressing, of many of the issues related to the expected increase in biomass production in the EU mentioned in this chapter.

What Role Does PBC Production Presently Play in the EU?

There are no actual statistics about production of PBC in Europe, but a survey among the Biomass 2015 conference participants revealed that the PBC with the largest cultivation area in Europe is probably miscanthus. There are approximately 20,000 ha of miscanthus in Europe: about 10,000 ha in the UK, 4000 ha in Germany, 4000 ha in France, and 500 ha each in Switzerland and Poland. However, it has been reported that the area of miscanthus cultivation in Europe depends on public intervention and declines when support is withdrawn, as has been recently observed in Ireland. The area of willow cultivation in Sweden also decreased from 14,000 ha in 2014 to 10,000 ha in 2015. Presently, the only PBC with rapidly increasing area is short rotation poplar grown on about 10,000 ha in Poland.

In the UK, the main use of miscanthus biomass is co-firing with coal for electricity generation. In Germany, use in heating is prevalent. Material uses include building materials and bio-composites. Apart from several hectares of *M. sinensis* used for thatching in Denmark, only one genotype—*M. x giganteus*—is cultivated commercially in Europe. The world's largest miscanthus production is found in China, where about 100,000 ha of *M. lutarioriparius* growing wild in the coastal area of the Dongting Lake are used for paper making. Here, miscanthus biomass is also used as building material and for food. Young miscanthus shoots are harvested in spring to be made into pickles. Currently, about 5300 t of miscanthus pickles are produced annually.

In Sweden, willow short rotation coppice (SRC) is mainly grown for heat or combined heat and power production. There are also an estimated 4000 ha of willow SRC cultivated for heat production in Germany. Several thousand hectares of reed canary grass are grown for heat and pulp and paper production in Sweden and Finland.

In summary, the cultivation area of PBC in the EU is probably less than 60,000 ha and it can be concluded that, despite all potential environmental and economic benefits, PBC currently do not play a significant role in the EU.

What Are the Reasons for the Low Production of PBC and How Can Obstacles to an Increased Production be Overcome?

According to the results presented at the Biomass 2015 conference, the reasons for the low production of PBC in the EU can be categorized into three major points:

- **Biomass production costs are still too high and yields can vary from year to year.**

High biomass production costs for PBC result from insufficient development of cultivars and agricultural production technology, along with high costs for agricultural inputs, land and labor for a relatively low-value biomass. Although they are amortized over a production period of 10–25 years, initial establishment costs for PBC are high and need to be reduced in the future. Research is currently looking at ways of reducing these costs through improved mechanization for those PBC which can be established vegetatively via rhizomes or twigs, such as miscanthus and SRC trees. Further research is necessary to develop safe establishment methods for perennial grasses via seeds.

In Europe, few PBC are available and for some, such as miscanthus, there is only one commercial genotype. The characteristics are often still very close to those of wild species as little breeding and selection has yet taken place. A major bottleneck to the breeding of PBC varieties is the high costs involved and the long breeding periods, necessary because most yield-relevant parameters cannot be assessed until after the establishment phase of 2–3 years.

There is little information on long-term yield developments in PBC. Research projects on PBC often run for only 2–4 years, which only allows the establishment period to be covered. Those reports on long-term yields which are available indicate an entire range of developments: from stable yields over long periods; through year-to-year variations; to yield decreases after an early peak.

Low production costs can only be achieved though economies of scale and efficient production technologies. The development of specialized agricultural machinery, such as planting or harvesting equipment, will remain insufficient so long as there is no larger market for such machinery. Today, farmers often seek help in self-made equipment, such as adapting potato harvesting machines for the harvest of miscanthus rhizomes. Also the development of service units, such as machinery cooperatives, will only develop once PCB production reaches a significant scale.

- **More stable markets and high-value uses are required.**

The main use of lignocellulosic PBC biomass is for heat and power production—a comparatively low-value use determined ultimately by the price for fossil fuels. Generally, in Europe, with the notable exception of wood from forests and forest industry by-products, bioenergy products require subsidy to compete in retail energy markets. Therefore, higher-value applications should be developed for PBC biomass to provide attractive market options.

As shown above, PBC cultivation can perform several ecological services, such as carbon sequestration. In principle, these ecosystem services provide a basis non-market remuneration generating additional income from PBC production.

- **Farmers' interest and acceptance is low.**

PBC are new to farmers and many have neither the knowledge nor the technical equipment to produce them. Therefore, the introduction of PBC cultivation requires the development of efficient production technology and machinery cooperatives or services.

Farmers hesitate to produce PBC because it involves dedicating their fields to long-term biomass production. They will only be willing to do this once biomass markets are stable or if long-term contracts are granted.

The integration of PBC biomass production into a farm business should be performed in such a way that a higher proportion of the value generated remains on the farm. This is the case for example with biogas production, where farmers sell bioelectricity instead of biomass. The development of on-farm biorefinery concepts which allow decentralized biomass densification and valorization can help to involve farmers in the biobased value chain. This provides them with better income opportunities than biomass production alone.

Potentials for PBC Production in a Growing European Bioeconomy

This section discusses different situations in which potentials for PBC production are seen:

- **Production of PBC on land that is marginal in biophysical or economic terms for food production or that can be improved by PBC production.**

Allen et al. (2014) estimated that 1,350,000 ha of land are potentially available for the production of PBC in Europe. This includes 50,000 ha of contaminated sites, 200,000 ha of crop land recently (<5 years) taken out of agriculture, 600,000 ha of recently abandoned grassland, 200,000 ha of fallow land, and 300,000 of underutilized land. They performed their analysis on the assumption that PBC should not replace forest, grasslands, or semi-natural habitats and should not be cultivated on land presently used as arable land. This assessment did not include areas on which

PBC could be grown as buffer strips, for example along waterways or between intensively used agricultural land and ecological sensitive areas. The estimate of 200,000 ha of crop land that is no longer farmed is only a rough approximation. Because explanatory factors for loss of agricultural land include urban expansion (68 %) and the development of scrubby vegetation communities (24 %), it is rather unlikely that a larger part of this land can be brought back to crop production (Allen et al. 2014).

If contaminated land is taken into account, the potentials for PCB cultivation may be larger. A survey by Lado et al. (2008) revealed that large areas of Europe are contaminated by heavy metals. These areas are not available for food production for health reasons. PCB crops could be produced on this land because the biomass is not used for food purposes. In those cases where PBC are able to remediate heavy metal contaminated soils they can even help revert this land into land suitable for food production. For example, *Miscanthus* x *giganteus* grown on contaminated soils can accumulate up to 5 mg Cd kg^{-1}, 150 mg Pb kg^{-1}, and 700 mg Zn kg^{-1}, indicating it as a promising crop for the phytoremediation of Pb and Zn (Pogrzeba et al. 2011, 2013). Similar results have been reported for willow, especially for Cd remediation. However, depending on the kind and severity of heavy metal contamination, the remediation process can take decades to centuries (Lewandowski et al. 2006).

In summary, PCB show good potential to make use of land which is marginal or difficult to manage, or for land amelioration. However, it has to be seen that due to low yields, marginal production conditions can also result in low profit margins (van Dam et al. 2005). Also, crop production under marginal conditions is more challenging than on good agricultural land. Therefore, crop management systems that ensure safe establishment and optimal management in these conditions need to be developed and stress-tolerant genotypes are required.

- **Integration of PBC into farming systems.**

As discussed above, it is unlikely that large areas of good agricultural land will be dedicated to PBC production. But farms can benefit from the incorporation of PBC production in many ways. An integrated approach to PBC production can be seen in France where small miscanthus fields are established in a way that improves the landscape. In the UK, arable farmers working with the company Terravesta found that using the 10 % poorer quality land for PBC bioenergy production resulted in practical examples of sustainable intensification[1] with the better 90 %, delivering increased yields and overall better economic return at farm level. Farms can benefit from the integration of PBC through the following aspects:

1. Making better use of the least suitable fields or areas of the farm to generate an income from biomass production.
2. Using PBC to improve the environmental performance of agriculture by integrating PBC in a way that fulfills ecological functions, such as carbon sequestra-

[1] Pretty et al. (2011) defined sustainable agricultural intensification as follows: "producing more output from the same area of land while reducing the negative environmental impacts and at the same time increasing contributions to natural capital and the flow of environmental services."

tion, prevention of erosion, flooding or nitrate leaching, and providing habitats for wildlife.
3. Improving soil quality and thus farm productivity by increasing the humus and nutrient content, improving soil structure, or remediation of contaminated soils.
4. Provision of biomass for on-farm use, e.g., heating purposes or animal bedding.

The following are practical examples of on-farm use of biomass:

- Some farms in France, Germany, and Austria heat their premises using miscanthus biomass. One French farm that rents out holiday flats even uses the biomass heater as an attraction in the market for sustainable holidays. In general, farms can save on heating costs by producing their own fuel.
- The French cooperative "Coopédom" produces feed pellets from dried alfalfa. The energy required is generated from the combustion of miscanthus produced on 450 ha of the cooperating farms, all of which provide a certain proportion of their fields for miscanthus production.
- A French turkey producer found that miscanthus bedding has several advantages over conventional straw bedding. Deep litter using miscanthus has the advantage of better water and ammonia adsorption avoiding the need for litter renewal during the turkey growing period. Overall, turkeys are healthier on miscanthus than on straw bedding. He grows miscanthus on a piece of marginal land on his farm to produce the required bedding material. The manure is composted and sold as valuable fertilizer.

- **Replacement of less sustainably produced biomass.**

Today, about six million ha of agricultural land in the EU are dedicated to the production of so-called *first generation* energy and industry crops. These are crops that deliver edible biomass but are used for energetic or material purposes. Rapeseed and maize are the most prominent examples. The replacement of these mostly intensively managed crops by PCB reduces nitrate leaching, erosion, and the demand for agrochemicals and increases soil carbon sequestration, soil fertility, and biodiversity. A replacement of annual crops by PBC would lead to more environmentally benign crop production. However, the quality of lignocellulosic biomass is not adequate to replace edible feedstocks because markets require either easily degradable carbohydrates, such as sugar and starch, or oils for liquid biofuels. Most material applications that are expected to become major growth areas in the bioeconomy (see Table 1.1), such as biochemicals, are also based on sugars and oils as feedstocks. Therefore, we suggest either shaping the PBC biomass toward the requirements of bioeconomic uses through breeding and crop management, for example by:

- Breeding lignocellulosic crops with lower lignin content and lower cell wall recalcitrance;
- Early harvest of perennial grasses, e.g., miscanthus, to provide green biomass to replace maize as feedstock for biogas production;
  - Or integrating decentralized processing to shape the feedstock toward the requirements by for example:

- Ensiling of early-harvested PBC biomass to support its digestion into easily accessible carbohydrates;
- On-farm pyrolysis plants producing pyrolysis oils from PBC biomass;
- On-farm biorefinery systems producing cellulose, sugars, protein, chlorophyll, and other products from lignocellullosic biomass; and
- On-farm production units providing biomaterials, such as building material.

Many of these approaches, especially on-farm biorefining, help to increase the addition of value at farm level. Instead of only supplying "cheap" biomass to the often volatile market, high-value products can be created.

Further options for increasing sustainability are created by replacing imported biomass with domestically grown biomass. A small-scale example is reed used for thatching in Denmark, which was originally imported from the Danube region but has been replaced by locally produced *Miscanthus sinensis*. A much larger market can be served if PCB biomass is harvested "green" and proteins can be extracted before it is used for material or energetic purposes. Perennial C3 grasses, which can be grown in mixtures with N-fixing leguminous crops, also have the potential to contribute to an increased domestic protein production.

- **Development of new products making optimal use of the properties of PCB biomass.**

The bioeconomy is not only about the replacement of fossil by biogenic resources, but also about developing completely new biobased products and improving product properties through biogenic components. Examples are specialized products, such as packaging material for transport of liquids or animal bedding, which make use of the high water absorption capacity of miscanthus; and building materials and other products such as planting pots which make use of the low susceptibility of miscanthus to rotting.

There are many other examples from the area of biomaterial use of PCB biomass that will lead to new markets and income opportunities from sustainably produced biomass and biobased products.

Conclusions

- PBC have a significant potential to contribute to an improved and sustainable biomass supply in the EU and to a growing European Bioeconomy.
- PBC can replace biomass in present applications and at the same time offer opportunities for the development of new and improved biobased materials.
- R&D work on PBC has generated important knowledge and materials. However, development of PRB is a long-term task that should be continued by building on current achievements. This includes the development of stress-tolerant genotypes in breeding programmes, efficient crop management procedures, and low-cost agricultural technology, on-farm biorefinery concepts and high-value applications for lignocellulosic biomass.
- To extend the production of PBC in European agriculture ecosystem service provision should be acknowledged and remunerated.

References

Allen B, Kretschmer B, Baldock D, Menadue H, Nanni S, Tucker G (2014) Space for energy crops—assessing the potential contribution to Europe's energy future. Report produced for BirdLife Europe, European Environmental Bureau and Transport & Environment. IEEP, London

Dornburg V, van Vuuren D, van de Ven G et al. (2010) Bioenergy revisited: Key factors in global potentials of bioenergy. Energy & Environmental Science, 3:258–267

Fargione J, Hill J, Tilamn D, Polasky S, Hawthorne P (2008) Land clearing and the biofuel carbon dept. Science 319:1235–1238

Lado LR, Hengl T, Reuter HI (2008) Heavy metals in European soils: a geostatistical analysis of the FOREGS geochemical database. Geoderma 148(2008):189–199

Lewandowski I (2015) Securing a sustainable biomass supply in a growing bioeconomy. Global Food Security 6:34–42

Lewandowski I, Schmidt U, Londo M, Faaij A (2006) The economic value of the phytoremediation function. Agr Syst 89(1):68–89

Lewandowski I, Kalinina O, Kiesel A, Clifton-Brown J et al (2015) OPTIMISC—developing Miscanthus production systems for marginal lands. European Biomass Conference and Exhibition Proceedings, Vienna 2015, 6–8

McCalmont JP, Hastings A, McNamara NP, Richter GM, Robson P, Donnison IS, Clifton-Brown J (2015) Environmental costs and benefits of growing Miscanthus for bioenergy in the UK. GCB-Bioenergy. doi: 10.1111/gcbb.12294

Pretty J, Toulmin C, Williams S (2011) Sustainable intensification in African agriculture. Int J Agric Sustain 9(1):5–24

Pogrzeba M, Krzyżak J, Sas-Nowosielska A, Majtkowski W, Małkowski E, Kita A (2011) A heavy metal environmental threat resulting from combustion of biofuels of plant origin. In: L.I. Simeonov, et al. (Eds.) Environmental Heavy Metal Pollution And Effects On Child Mental Development: Risk Assessment And Prevention Strategies, Springer Science+Business Media B.V., 213–225

Pogrzeba M, Krzyżak J, Sas-Nowosielska A (2013) Environmental hazards related to Miscanthus x giganteus cultivation on heavy metal contaminated soil. E3S Web of Conferences 1, 29006

Scarlat N, Dallemand JF, Monforti-Ferario F, Nita V (2015) The role of biomass and bioenergy in a future bioeconomy: policies and facts. Environ Dev 15:3–34

Searchinger T, Heimlich R, Houghton RA, Dong F, Elobeid A, Fabiosa J, Tokgoz S, Hayes D, Yu T (2008) Use of US croplands for biofuels increases greenhouse gases through emissions from land-use change. Science 319:1238–1240

Staffas L, Gustavsson M, McCormick K (2013) Strategies and policies for the bioeconomy and bio-Based economy: an analysis of official national approaches. Sustainability, 5:2751–2769

van Dam JEG, de Klerk-Engels B, Struik PC, Rabbinge R (2005) Securing renewable resource supplies for changing market demands in a bio-based economy. Ind Crop Prod 21(1):129–144

Zhu X-G, Long SP, Ort DR (2010) Improving photosynthetic efficiency for greater yield. Annu Rev Plant Biol 61:235–261

Chapter 2
Pointers for Bioenergy Policy in a Resource-Constrained World

D. Murphy-Bokern

Introduction

Policy is about turning political vision into change in the real world. In revealing how the real world works, science plays a strong role in the development of policy. Furthermore, science-based innovation and technology play a key role in determining policy outcomes. This chapter is about the role of public policy in the development of bioenergy. It draws on experience from three European countries: the United Kingdom, Germany and Ireland to explore the development and impact of bioenergy policy in Europe. Bioenergy is relevant to a range of policy areas: energy, climate change, environment, agricultural and industrial policy. In addition to the resulting complex policy environment, the bioenergy sector is characterised by long lead-in times to change based on perennial biomass crops.

European Policy Background

Current bioenergy policy in Europe can be traced back to the early 1980s. Several decades of increasing agricultural production led to domestic production exceeding consumption for some key food commodities in Europe. European policy on non-food crop production which was first developed in the 1980s is thus rooted in what appeared to some to be a surfeit of resources, even though it was clear that this 'surplus' related to only a few commodities (Mildon 1986). Concepts for non-food uses of agricultural crops emerged out of a 1984 report 'Cereal Crops for Industrial Use in Europe' (Rexen and Munck 1984). Arising from this, the first European

D. Murphy-Bokern (✉)
Lohne, Niedersachsen, Germany
e-mail: donal@murphy-bokern.com

S. Barth et al. (eds.), *Perennial Biomass Crops for a Resource-Constrained World*, DOI 10.1007/978-3-319-44530-4_2

research programme committed to the development of non-food uses of agricultural resources was the ECLAIR Programme that ran from 1988 to 1993 (Coombs 1997). The objectives were remarkably similar to research programmes today with efforts to encourage collaboration between research providers and developers. Like today's research policy, ECLAIR also sought to build on life sciences and biotechnology. 'Alternative crops' and non-food uses were developed. Much of today's bioenergy research activities are based on work initiated in the 1980s when, for example, short rotation coppice research was started in Ireland, investigations into the potential of miscanthus were initiated in Germany, and concepts such as biorefining were developed in Denmark in the European-funded BIORAF project that ran from 1991 to 1994. Researchers' pioneering activities combined with continued political interest in finding alternatives to food production and emerging policies on sustainable development led to national research and development programmes for non-food crops. The establishment of the German Agency for Renewable Materials (*Fachagentur Nachwachsende Rohstoffe*) in 1993 is a prominent example of the growth in political interest in non-food uses of agricultural resources in this period. Similar programmes emerged in France, Italy, the UK, and Demark and it was estimated that total research spending in Europe in this area in the decade 1988–1998 was about one billion Euro (Rexen 1997).

Compared with these research activities, the wider European policy framework developed relatively late. The Treaty of Amsterdam in 1997 was a starting point. It included provision for promoting 'a harmonious, balanced and sustainable development of economic activities' with a 'high level of protection and improvement of the quality of the environment'. This provided the political foundation for a series of EU policy activities starting with the first Sustainable Development Strategy in 2001 and the Directive on Electricity Production from Renewable Energy Sources (RES). This was followed in 2003 by the Directive on the promotion of the use of biofuels and other renewable fuels in transport (Biofuels Directive).

The RES Directive of 2001 set indicative targets for member states for the proportion of electricity from renewable energy sources with the overall target of 12 % by 2010 and 20 % by 2020. At member state level for 2010 these were 12.5 % for Germany, 10.0 % for the UK, and 13.2 % for Ireland.

The 2003 Biofuels Directive set out measures to replace 5.75 % of transport fossil fuels on an energy basis by 2010, to be raised to 10 % by 2020. This resulted in the mandating of the inclusion of biofuels in conventional fuels. This market-based approach replaced a number of national direct subsidies such as the duty derogation for biodiesel that operated in Germany. The mandate stimulated the global trade in vegetable oils for non-food use, including palm oil. This led quickly to the realisation that first generation biofuels (i.e., biofuels derived from vegetable oils or ethanol from easily fermented starch and sugar) have consequences for agricultural food commodity markets. This coincided with the large increase in commodity prices in 2007 and 2008. Even though the earliest studies in Germany indicated that first generation biofuels are ineffective in terms of greenhouse gas mitigation (e.g., Wintzer et al. 1993; Kaltschmitt et al. 1997), it was not until about 2008 that consensus about the unwanted consequences of using food crops for biofuel emerged, most notably from the Gallagher Review of the indirect effects of biofuel production commissioned by the UK government in 2008 (Renewable Fuels 2008). That report stated that a slowdown in the growth of biofuels was needed and that a biofuels sector must avoid using agricultural land that would otherwise be used for food production. It was clear in stating that development must target under-utilised and marginal land.

Current relevant EU policy is largely framed by the 2009 Renewable Energy Directive (RED) which requires that 20 % of primary energy used, including at least 10 % of all energy in road transport fuels, be produced from renewable sources by 2020. These primary energy targets range from 10 % for Malta to 49 % for Sweden. The targets for the UK, Ireland and Germany are 15 %, 16 % and 18 % respectively. Of these, the UK is furthest away from its target while Germany aims to exceed its 18 % target to achieve 20 %. The biofuels provision in the RED reinforced the earlier Biofuels Directive. However, the Fuel Quality Directive (FQD) was amended to require that the whole road transport fuel mix in the EU be 6 % less carbon intensive than a fossil fuel-based baseline. These Directives specified sustainability criteria including a 35 % life-cycle GHG reduction. Crucially, while not addressing indirect land-use emissions directly, these Directives required the Commission to address indirect land use change by the end of 2010. After a great deal of scientific and policy debate, this led to a further amendment of the RED in 2015 to address the concerns that GHG emissions from first generation biofuels are high, mainly due to indirect land-use change. This resulted in a limit of 7 % for the contribution of first generation biofuels combined with stronger incentives to use other fuels in the transport mix, including renewable electricity for railways which will make up the remaining 3 %, thereby holding to the 10 % target for transport.

German Bioenergy Policy

Non-food agricultural crops accounted for 2.5 million ha of agricultural land in Germany in 2015. This is 15 % and 20 % of the agricultural and arable area respectively. Of this non-food area, 89 % (2.2 million ha) was used for energy (FNR 2015). Biomass crops for biogas, primarily for electricity, accounted for 1.4 million ha or 56 % of the non-food crop area; oilseed rape and cereals for biofuels account for 0.8 million ha (FNR 2015). Combined with wood from 11.4 million ha of forest, this allocation of land makes Germany a European leader in bioenergy. Despite this scale, analysis of several data sources, such as the total primary energy coming from biomass (957 PJ, 7.5 % in 2013) and the contribution of different types of bioenergy to end-energy consumption (BMWE 2016), indicates that bioenergy from agricultural crops accounts for only about 2 % of German primary energy use.

For bioenergy, German policy instruments relate to the use of biofuels for transport, biomass for heat, and measures supporting renewable electricity. As referred to above, until the mandating of biofuel inclusion, biodiesel use was promoted by a fuel duty derogation of about 47 Euro cents/l. This was expensive in relation to direct emission savings (Zimmer et al. 2008) costing German tax payers more than 2,000 million Euros in 2006 (Rauch and Thöne 2012). The switch from direct subsidy to mandating the use of biofuels under the Biofuels Directive simplified the system. It also impacted heavily on the German biodiesel sector in particular as small producers' niche markets for pure biodiesel disappeared.

For heat, German measures comprising support grants and loans for the installation of wood-fuelled boilers and reduced value-added tax (7 %) are largely indirect. As these instruments are not directly related to perennial biomass crops or heat and electricity, they are not discussed further here.

The development of 'arable biomass' for electricity was largely driven initially by the Electricity Feed-in Act (1990), followed in 2000 by the Act on Granting Priority to Renewable Energy Sources (EEG). Because of their far-reaching and pioneering effects, the history and political impact of these is described here.

Biogas warrants explanation. Biogas is about 60 % methane. It arises naturally from the degradation of organic matter in the absence of oxygen. Marsh gas is biogas. The English term 'anaerobic digestion' for the biogas process appropriately reflects the connection to food production and use as in practice only 'digestible' organic matter can be used effectively. In nature, it complements aerobic digestion (composting), enabling decomposition of organic matter in anaerobic environments. Anaerobic digestion is an effective way of generating a high-quality energy carrier from wet co-products and wastes ('substrate') from the food system. Anaerobic digestion also conserves mineral nutrients. The residue (digestate) is an effective fertiliser and the process helps close nutrient cycles in the food system. The biogas in Europe is generally used directly to drive gas engines or upgraded to near pure methane injected into the gas network or used as a transport fuel.

The Electricity Feed-in Act and the Act on Granting Priority to Renewable Energy Sources

The Electricity Feed-in Act (*Stromeinspreisungsgesetz*) of 1990 required network operators to purchase all electricity offered from renewable generation with minimum feed-in tariffs for generation units smaller than 5 MW. The feed-in tariffs were determined by a proportion of the retail price of electricity in the previous 2 years, ranging from 75 to 90 %. Since the full retail price then was well below the cost of generating electricity using solar voltaic, this measure failed to stimulate a significant solar sector. The tariffs were high enough though to stimulate investment in land-fill biogas and wind on favourable sites. For wind in particular, this stimulated development of the very large wind turbines we have today. This was a significant technical achievement. While ground-breaking in terms of showing how an electricity network can be opened up to small-scale generation, the mechanism in general was regarded as too blunt to promote anything other than the lower cost and more mature renewable generation technologies. However, it pioneered the way forward in and beyond Germany.

Based on a local initiative in 1992, the City of Aachen developed a local scheme for variable feed-in tariffs for renewable electricity. This set up a local levy on electricity bills to differentially support renewable generation within the local municipal network. As a result, different feed-in tariffs for different renewable energy technologies were set up so that in effect a local subsidy scheme was targeted to stimulate new high-cost technologies, notably solar voltaic. This principle was called the Aachen Model. It is significant because it was adopted nationally in the Act on Granting Priority to Renewable Energy Sources (*Erneuerbare Energie* Gesetz, (*EEG*)) which replaced the Electricity Feed-in Act in 2000.

The EEG now regulates feed-in tariffs and is at the core of German renewable energy policy. This legislation was revised in 2004, 2009, 2012 and 2014. As described in detail later, these revisions adjusted the feed-in tariffs and related conditions directly for electricity and indirectly for biogas fed into the gas grid. Tariffs and other conditions are guaranteed for 20 years after the generation unit goes onto the grid. Rooted in the Aachen model with different payments for different technologies, the EEG has had a profound effect on the development of renewable electricity in Germany and beyond.

About half of Germany's primary energy use is for heating, mostly using oil, gas and wood. To address the risk of imbalance in efforts to support renewable energy, the Renewable Heat Law (Erneuerbare-Energie-Wärmegesetz (EEWärmeG)) was introduced in 2009. This legislation mandates the use of renewable heat or other mitigating technologies such as combined heat and power or the use of energy-saving technologies (e.g., insulation). The initial outlay for renewable heat technology is generally higher compared to standard oil or gas heating and so the legislation also makes provision for grants for installing heating based on solar or biomass. The long-term aim is to achieve a carbon neutral built environment by 2050.

The Impact of the EEG: A Case Study in Impacts of Strong Policy Intervention

Because the EEG in particular has had a major impact on the development of bioenergy directly in Germany and indirectly in other countries, the detail of its development and impact is relevant and set out here. The main features evolved through the four revisions (2004, 2009, 2012, and 2014 (BMWE 2016)).

On its introduction in 2000 following its predecessor, the EEG set out a relatively simple feed-in tariff system with feed-in tariffs ranging from 6 to 10 Euro cent/kWh for all technologies except solar voltaic, which received 48.1–50.6 Euro cents. The feed-in tariff for biomass was about 10 Euro cents/kWh for units up to 500 kW, with slightly lower payments for larger units up to 5 MW. The 2004 revision picked up on the principles of the Aachen model by giving a bonus on top of the base payment for electricity from 'renewable biomass'. For generation units going onto the network in 2004, electricity generated from biomass crops received about 15.9 Euro cents (up to 500 kWh) compared to base of 9.9 Euro cents for electricity from other biomass resources, for example food waste (all guaranteed for 20 years).

This differentiation was the foundation of the distorting effect of the EEG rooted in the Aachen model principle of linking subsidy to private costs rather than to public benefits.

The total non-food cropping area increased from 1.4 million ha in 2004 to 2.0 million ha in 2007, almost all of this growth being due to the increase in the area of maize for biogas. By 2006 and 2007, agricultural economists in particular (e.g., Isermeyer and Zimmer (2006) and Görmann *et al.* (2007)) were expressing concern that the biogas boom was both unsustainable and counterproductive in terms of the stated aims of the EEG. Even the German government's own Scientific Advisory Council for Agriculture issued clear warnings of fundamental flaws in the policy (BMEL 2007). However, by this time an influential biogas lobby had developed representing interests connected to maize production inputs and the biogas equipment sector which had expanded rapidly. Corporate activity on finance markets also contributed to the growing public unease about the sector. At the height of the boom in 2006 and 2007, initial public offerings (IPO, 'going public') were followed by profit warnings causing a stock market crash across the sector. Substantial loss of investors' capital and other negative financial events in this highly subsidised sector contributed to growing public concern.

By 2009, concerns grew beyond agricultural economists and the negative consequences were becoming clear to the wider public leading to a wide range of local citizens' initiatives to oppose further expansion (e.g., as reflected by Stechmesser, n.d.). In particular, negative impacts on land use caused protest within the agricultural community and widespread local conflicts over land resources. In Lower Saxony, the expansion of maize production of about 100,000 ha between 1999 and 2007 was associated with corresponding decline in the grassland area. Analysis showed that building biogas plants raised local land rents by about 200 Euro/ha/year on average in areas already characterised by high land rents due to intensive livestock production (Theuvsen et al. 2011) indicating that land owners are the ultimate beneficiaries. This posed profound risks to farm businesses producing food. It also drove counterproductive environmental effects with land-use change (NLWKN 2010). Local examples of the conversion of permanent grass to energy maize production on peatland were reported underscoring evidence of serious unintended consequences and increased greenhouse gas emissions. The increase in maize production led to what was commonly referred to as '*vermaisung*' (a new word in the German language meaning to cover the landscape with maize monoculture) with adverse impacts on biodiversity. In addition, there were several counter-intuitive impacts due to difficult-to-foresee effects from related legislation, particularly the regulation of fertiliser and manure use. Largely because digestate was treated as an inorganic source of nitrogen and therefore not directly regulated under the Nitrates Directive, the growth in the energy maize area was associated with local increases in nitrate levels in ground water. In effect, from a regulatory perspective, biogas production converted regulated organic nitrogen into less regulated inorganic nitrogen causing an overall increase in reactive nitrogen inputs into cropping systems. This caused serious additional concerns about impacts on ground water quality in livestock production areas.

Despite the evidence of resource constraints and serious market distortions, the revision of the EEG in 2009 reinforced growth in the sector. A more complex payment system to steer development was introduced. Reflecting the influence of lobbying and following the principle of the Aachen Model, payments were increased to compensate for increased costs of the crops grown (substrate) in effect supporting a self-reinforcing destructive cycle of high subsidy followed by inflation in substrate costs. For generation units between 150 kW and 500 kW, the basic payment was set at 9.2 Euro cents/kWh. Bonuses were added to this for various technical features associated with the use of biomass crops (mostly maize). These included a bonus for using biomass crops (7 Euro cents), a bonus for adding slurry (1 Euro cent), a bonus for using heat (3 Euro cents), and several bonuses for additional technical features. The result was that in practice the payments for electricity from biomass crops (maize) were increased to up to about 20 Euro cents/kWh for the typical farm-based unit with CHP (BMU 2009). Because the bonuses for slurry and other sustainable substrates co-fermented with maize were paid on all gas produced rather than just the gas from these more sustainable materials, the 'slurry bonus' increased the incentive to produce gas from maize.

The number of biogas plants grew from about 500 in 2000 when the EEG was introduced to about 7,500 in 2012 with an installed capacity in excess of 3 GW. In 2011, the Scientific Advisory Council for Agriculture within the Federal Ministry for food and Agriculture (BMEL) updated its advice on bioenergy emphasising again the damaging effects of the expansion of biogas. This high-level scientific advisory body went as far as to remind the policy community of the distorting effects of biogas industry lobbying on public debate. By 2012, concerns about the adverse impacts of the EEG had reached the highest levels. The President of Germany, Joachim Gauck, warned of central economic planning ('Planwirtschaft') playing a damaging role in energy policy (Gauck 2012). Frequent news media items discussed the issue (e.g., Klawitter 2012). The rising levy within the retail electricity price was also impacting on consumers, including businesses. It reached 5 Euro cents per kWh in 2013 with a total transfer of 20,000 million Euros from consumers to renewable electricity generators (BMWE 2016). Bioenergy accounted for one-fifth of this (4,000 million Euro). Even if only direct emissions from biomass production were considered, the GHG mitigation costs are high. In addition, more high-profile commercial failures illustrated that even with high subsidy the sector was under economic strain. The sector was in effect self-destructing due to resource constraints.

The 2012 revision of the EEG was a relatively mild intervention to address these concerns. For the typical agricultural biogas plant, the base tariff for electricity feed-in between 150 kW to 500 kW was increased to 12.3 Euro cents but the bonus for biomass crop material was reduced to 6 Euro cents. A higher bonus (8 Euro cents) was paid for other types of biomass such as residues from land maintenance. Critically, other bonuses were eliminated and the payment of the resulting 18–19 Euro cents tariff for bioenergy from biomass crops was conditional on the inclusion of slurry and the use of the heat. In practice, compared with the 2009 EEG, tariffs for the typical agricultural unit of less than 500 kW were reduced by only 1–2 Euro cents and a relatively high level of subsidy in terms of direct emission savings and output remained in place. However, conditions were placed on developers, such as the requirement to market heat from CHP and use slurry as part of the substrate.

Reflecting the stressed state of the relevant resource markets and local political resistance, these changes were sufficient to significantly slow the expansion of biogas from energy crops.

In contrast to the UK's market-based approach, the German EEG discriminates against large-scale generation even though there is substantial evidence that direct replacement of coal with biomass is one of the more effective mitigation measures (e.g., Wintzer et al. 1993; Zimmer et al. 2008). This might be rooted in a tendency to support particular technical solutions using micro-managing interventions. Central planning thinking is referred to by Isermeyer et al. (2012) as a feature of German policy discourse in this area. Debate was characterised by positive appraisal of the technology by technologists rather than assessment of the impact of technology on natural resources. This was associated with the quest for certain types of socio-political outcomes with terms such as 'local', 'decentral', 'self-sufficiency', 'energy autarky', 'energy regions' and 'energy villages'. Added to this was the biogas sector's tendency to confuse the effect of replacing a fossil energy carrier with the effects of the generation technology used, especially with respect to combined heat and power. The co-generation of heat was widely presented as a special feature of biogas. The overall result was increased complexity in interventions followed by complex regulation to address the unintended consequences. Complex interventions such as different bonuses for different classes of substrates, different scales of generation and subsidy bonuses for using heat also motivated ineffective investment and increased the risk of subsidy abuse and fraud.

The EEG was revised again in 2014 to reinforce the trend set in 2012. This revision was a radical re-orientation onto market principles. There is an increasing requirement for generators to market the electricity they produce. At the outset in 2015, a single payment of up to 11.8 Euro cents/kWh was established for up to 500 kW units using biomass crops. In contrast to the early versions of the EEG, tariffs for biogas from organic waste are now higher than those for biomass crops at up to 15.3 Euro cents/kWh, and higher still at 23.7 cent for units smaller than 75 kW using at least 80 % slurry. In effect, the EEG has moved away from the Aachen Model and now aligns payment more to greenhouse gas mitigation effects (i.e., public benefits) instead of costs. These payments are further reduced at 3 month intervals and the expansion of the sector is capped at 100 MW per year through a register of new units. Operators can top-up income through marketing and obtain premia for generation at times of peak demand and from the sale of heat. In practice, EEG 2014 means that the main option now for expanding biogas in agriculture is the development of small units (less than 75 kW with more than 80 % slurry in the substrate on a weight basis).

Wider Consequences of the Effect of the EEG on Bioenergy

The EEG has had a major impact on the public perception of bioenergy in Germany. The word 'bioenergy' is for many Germans living in rural areas now synonymous with biogas from maize, a connection reinforced by official government publications that promoted bioenergy in the earlier years of the EEG. This is even though the greater proportion of bioenergy in primary energy terms comes from wood used mostly for heat. Influenced by the German biogas sector, several other countries have introduced similar incentives, notably Italy. However, other than those short-lived efforts, the over-riding effect has been policy lessons from a real-world example of the negative consequences of high subsidy and direct intervention. The lessons learnt are now seen in the more recent policy development in Ireland and the United Kingdom.

Policy Developments in Ireland

The Renewable Energy Directive commits Ireland to 16 % renewable energy by 2020 made up of 40 % of electricity, 12 % of heat and 10 % of transport energy. Ireland is a relative newcomer to bioenergy policy even though research on bioenergy cropping, for example at Teagasc Oakpark, can be traced back to the 1970s. Current policy is based on a 2007 White Paper 'Developing a sustainable energy future for Ireland'. A second White Paper in 2015 (DCENR 2015) 'Ireland's transition to a low carbon energy future 2015-2030' describes a wide range of measures to support renewable energy generation and to support energy efficiency and saving. The proportion of renewables in the energy mix has risen steadily since 2007, largely due to wind energy. By 2015, Ireland was half way to the RED targets for 2020. However, this means that even though there has been steady growth in renewable energy output, the growth rate needs to increase further if the targets are to be met. The greatest challenge is in the renewable heat area.

Growth has been driven largely by the Renewable Energy Feed-in Tariff (REFIT) for renewable electricity. Like the German feed-in tariffs, these are funded by levies on electricity bills. However, feed-in tariffs have been generally lower than in Germany and the amount of aided generation is capped. Payments range from 10 Euro cents/kWh for biogas without CHP above 500 kW to 15 Euro cents for biogas with CHP below 500 kW. Straight combustion of biomass crops (mostly co-firing with peat) receives 9.5 Euro cents.

The Council of European Energy Regulators found that Ireland's approach to supporting renewable energy is one of the most cost effective in the EU (DCENR 2015). The 2015 White Paper emphasises achieving cost-effective GHG emissions mitigation and cost-effective energy. It also emphasises stability for investors. Policy documents generally (e.g., Clancy 2015) draw on economic analysis to cost policy options. These indicate that using biomass for renewable heat is potentially a most cost-effective option which in addition would directly address the gap between achievements and targets in the heat area. A renewable heat initiative is being developed. In addition to a state-owned electricity network, there are large areas of public land under forestry and the state-owned peat development company (Bord na Mona) is increasingly involved in renewable energy. A joint-venture between the state-owned forestry company and Bord na Mona is planned for the development of bioenergy.

Policy Developments in the United Kingdom

The Renewable Energy Directive commits the UK to 15 % renewable energy by 2020 with renewable energy accounting for 30 % of electricity, 12 % of heat and 10 % of transport. Policy to reach these targets is set out in the National Renewable Energy Action Plan (NREAP).

The 2012 Bioenergy Strategy sets out the framework principles that underpin policies (DECC 2012). There is emphasis on genuine carbon reductions, cost-effective mitigation and flexible responsive policy. Perhaps reflecting the German experience, the strategy explicitly acknowledges the risk of unintended consequences and resource constraints and sets out low-risk deployment pathways. It highlights energy from waste, biomass for heat, biofuels and biomass for electricity as a transitional technology. Policy makers draw on advice from the Committee on Climate Change which says that by 2050 bioenergy penetration levels of around 12 % are feasible without jeopardising sustainability but that policy must address the challenge of indirect land-use change (CCC 2011). In contrast to the German policy, the UK Strategy gives priority to the replacement of coal. The strategy clearly acknowledges policy risks but concludes that uncertainty is not so great as to justify inaction. It states that the pre-eminent concern of the UK Government in bioenergy policy is that bioenergy offers a genuine reduction in greenhouse gas emissions, that this reduction must be cost effective, and that the biomass must be produced sustainably. It says that sustainability standards need to be 'more stringent'.

Mainly through the 2008 Energy Act, UK bioenergy policy is delivered through four major instruments: renewable obligation, contracts for difference, feed-in tariffs for electricity and feed-in tariffs for heat.

The market-based Renewable Obligation (RO) has been the main mechanism for renewable electricity, including from biomass. Renewable obligation certificates (ROCs) are issued to generators who supply renewable electricity into the grid. Grid operators need to show these certificates in relation to the amount of electricity they sell to consumers. The certificates are traded between generators and retailers, the price for which is in effect a subsidy for generation above the base wholesale price for electricity. This price is set by a buy-out price that is nearly 5 pence/kWh (about 6.5 Euro cents). The obligation in terms of the proportion of renewable electricity in the system is set by the renewable obligation rate which has risen steadily and is currently 0.29 ROC per MWh. The award of ROCs for renewable generation is banded ranging from 0.5 ROC/MWh for electricity from biomass co-fired with coal to 1.9 ROC/MWh for anaerobic digestion (biogas) and dedicated biomass with CHP. The rate of award is fixed for 20 years.

Despite its modest cost compared to the EEG in Germany, the RO is regarded as costly. It will be discontinued in 2017 and is being replaced by contracts for difference. These are contracts between generators and public authorities which top-up generators' electricity prices to an agreed 'strike price'. The top-up between the market price for electricity and the strike price is in effect a subsidy. The level of subsidy varies depending on the difference between the normal generators' market price and the strike price set for the target generating technology. These contracts are awarded through a bidding process so a price-finding mechanism influences the level of subsidy. Through a government-owned company, these top-ups are funded by levies on consumers. In the biomass area, strike prices generally range from about 10 to 15 Euro cents/kWh.

Contracts for difference apply to large schemes. This mechanism is therefore complemented by feed-in tariffs for smaller-scale generation (up to 5 MW). In the biomass area, only anaerobic digestion qualifies. The development of biogas was signalled in 2011 (DECC 2012) with the implied intention to base AD on waste and co-products. Tariffs for electricity from biogas vary slightly around 18 Euro cents/kWh depending on the scale of the system but are substrate-neutral. It is now clear that this feed-in tariff is stimulating the production of arable (food) crops for biogas production, with consequences similar to the German experience predicted (Monbiot 2014). The total amount of electricity aided is subject to cap and the tariffs are regularly revised without further primary legislation.

The schemes mentioned above do not provide support for small and medium-sized use of woody biomass. This gap is now filled with the world's first feed-in tariff scheme for renewable heat. This is a fixed payment for the renewable heat generated. The payment comes directly from public finances. Practically, all generators of renewable heat are eligible. The tariff for heat from biomass is now about 7 Euro cents/kWh. Initial reports indicate that the scheme has found favour with applicants.

Despite the relatively modest level of subsidy provided by the RO, it has stimulated a very significant use of wood pellets for electricity generation, most notably at the Drax power station in Yorkshire. Drax alone burns 5 million tonnes of wood pellets per year. There are other power stations in the UK also burning wood pellets. The import of wood pellets to the UK from the USA has reached five million tonnes or one-third of internationally traded wood pellets. The real effect of this on carbon emissions is hotly debated (e.g., Rose 2014; Pearce 2015).

Policy Approaches Supporting Bioenergy

The IPCC (2014) estimates that agriculture, forestry and land-use are responsible for 24 % of greenhouse gas emissions. Therefore, bioenergy activities that increase agricultural activities risk increasing these agricultural emissions to such an extent that emission savings elsewhere (e.g., in the energy sector) are cancelled out. This has been overlooked in policy development and recent developments in Germany and United Kingdom show that the non-food uses of biomass can counter their public purpose. Bioenergy is therefore a particularly challenging policy area. The developments in the three countries set out above provide the basis of pointers to policy development.

Prioritising Climate Protection

The UK Government's Review on the Economics of Climate Change (The Stern Review (Stern 2006) identifies climate change as the greatest and widest-ranging market failure ever seen (Stern 2006) necessitating substantial intervention in markets. The 2015 United Nations Climate Change Conference in Paris (COP21) moves such interventions towards global collective action. This means that seeking genuine reductions in global greenhouse gas emissions at reasonable cost and with minimal negative side-effects is likely to be the foundation of policy affecting bioenergy. It follows that such interventions would prioritise cost-effective and substantial mitigation above supplementary considerations such as energy security or national efforts to promote certain energy sectors and technologies. The COP21 agreement also explicitly protects food production capacity.

Market-Based Intervention

German bioenergy policy has been characterised by complex interventions that promoted the development of specific technologies and caused distortions and unintended consequences. In contrast, UK policy in particular is characterised by market-based approaches aimed at minimising mitigation costs. There is now convergence between these approaches. The UK has adopted feed-in tariffs to stimulate specific technologies, albeit at lower levels of subsidy. Analogous to the UK approach behind contracts for difference, for the next revision of the EEG German policy-makers are introducing a bidding process to identify renewable energy projects most worthy of public support.

Timing

Early policy action and foresight is very important in this area where there are long lead-in times to outcomes. In Ireland, a substantial renewable energy policy emerged late in response to the Renewable Energy Directive of 2001. Even though Ireland has a very substantial renewable energy resource and a relatively low population density, this standing start means that Ireland is unlikely to meet targets for 2020. Progress has been hindered by weaknesses in related policy areas, for example planning regulations that have permitted a dispersed pattern of rural settlement that is now hindering the development of renewable energy (DCENR 2015).

The Level of Intervention in Markets

For anaerobic digestion, practical experience shows that feed-in tariffs or equivalent support in excess of about 16 Euro cents/kWh will trigger the use of crops normally grown for food, particularly maize where climate allows. The German EEG has not impacted significantly on woody biomass even though wood-based generation is eligible. This observation combined with the experience of very significant large-scale use of woody biomass for electricity in the UK indicates the level of payment in the 10–18 Euro cents/kWh range has a critical effect on the energy technology and resource chosen. A price of 10–14 Euro cents/kWh is enough to mobilise woody biomass for large-scale generation but is generally not sufficient to stimulate biogas generation using annual crops. This level of support also mobilises food wastes/food industry residues etc., especially if these come with gate fees and the heat can be used effectively. Payment at about 18 Euro cents/kWh can mobilise food crops for biogas. Mobilising animal slurry for biogas presents a special challenge as experience shows that payments in excess of 20 Euro cents/kWh are required to compensate for capital costs of the small scale, the low gas yields and intermittent substrate supplies. The problem here is that this level of subsidy stimulates crop production for substrate and so Monbiot (2014) suggests that a clear distinction be made between biomass in co-products or waste and biomass from annual crops with complete withdrawal of subsidy for energy from annual arable crops.

Protecting Food Resources and Prioritising Marginal Land Use

The negative public reaction that we have seen in Germany shows that adverse impacts on food production capacity and farmland ecosystems must be avoided. Perennial woody plants grown on marginal land are regarded as key to this because they:

1. Combine high biomass yields with protection of soil carbon stocks;
2. Have low greenhouse gas emissions from primary production;
3. Have low levels of other pollution emissions and reduce soil erosion;
4. Support diversification of crop cover in most landscapes and
5. Are relatively productive on land not suited to economic food production ('marginal land').

The potential of economically marginal land in the UK is very significant (Lovett et al. 2014). Soldatos (2015) defined marginal land as land with a low opportunity cost. The market would favour the use of these marginal lands for bioenergy if the returns exceed the relatively low opportunity costs more than other uses of this land. However, the lower costs of marginal lands may not compensate for the lower yields (Soldatos 2015). This means that market-based intervention alone may not effectively stimulate production on marginal land more than on land well suited for food production. Targeting of marginal land may be required.

The configuration of the food sector as determined by consumption is also very important factor determining the potential resource available. In a study modelling the effects of consumption scenarios, Westhoek et al. (2014) showed that a 50 % reduction in the consumption/production of livestock (in line with public health recommendations) would potentially make up to 24 million ha of agricultural land in the EU available for other uses. They make the point that optimising such a scenario for climate protection requires the use of perennial biomass crops and this area could supply about 4 % of Europe's primary energy.

Research and Innovation Policy

As outlined earlier, research into the modern non-food use of biomass can be traced back to the 1980s. Investment in research is consistent with other policy areas and the product (knowledge) is usually a public good in itself. Investment targeted to increase the productivity of biomass on marginal land specifically is particularly attractive from a public policy perspective. It can dovetail with market-related policy to tilt development towards production on marginal land in these areas. European Union investment in agricultural research has grown substantially in the last 10 years. Despite differences between Framework Programmes, research into the non-food use of biomass has been substantial throughout. Following the Commission's Communication on the bioeconomy (European Commission 2012), the later stages of FP7 saw a significant shift in funding towards research supporting the non-food bio-based sector (European Commission 2014). The research arising from this funding boost was required to heavily engage SME participants. Much of it was planned following the food crisis of 2007 and 2008 and therefore emphasised either high value non-food uses of biomass or the production of biomass from perennial species grown on marginal land. The following are suggestions for research policy:

What and where is the potential land resource? The Common Agricultural Policy and several related policies are major determinants of agricultural land-use decisions. Investigations of current land-use and the related distribution of ecosystem service provision could identify land most suited to bioenergy production. There is already a large body of knowledge and advanced tools that can be used (e.g., Lovett et al. 2014). Such investigation would look at different uses in relation to a range of ecosystem services, not just the biomass yield potential of sites. From this, the potential resource can be assessed with insight into potential for multiple benefits, for example where there is scope for combining soil protection with biomass production. From this regionally tailored approaches within the CAP and related policies can be developed.

Plant breeding. Plant genetic improvement is the foundation of improving production but is subject to significant market failure (Moran et al. 2007) leading to private sector under-investment. This market failure is likely to be large for perennial biomass crops, particularly for adaptation to marginal land. Tolerance to abiotic stresses that characterise marginal land is reported in miscanthus, poplar, giant reed

(Biomass 2015). New breeding methods, such as molecular assisted breeding, can accelerate the breeding process and are particularly relevant to perennial crops with long generation cycles. Informed by insights into the available land resource and thus the potential market for improved planting material, an overview of relevant public and private breeding efforts in Europe should lead to a plan for appropriate public investment in plant breeding. In this, biomass yield, particularly under marginal conditions, is by far the most important target for research and for plant breeding. It is the foundation of the economic and environmental viability of these cropping systems.

Reducing the financial risk of crop establishment. The feasibility of public intervention to reduce the financial risks arising from high establishment costs should be investigated. It is noted though that there is less than encouraging experience from national schemes that have aided planting. The Defra Energy Crops Scheme that ran from 2000 to 2013 provided planting grants but uptake was low (Lindegaard 2013). However, these schemes tended to be targeted at normal agricultural land. A framework for integrating the benefits, for example by targeting marginal land or to reduce the risk of flooding, was not used.

Grant-aid for planting is not the only mechanism. Integrated with local land-use strategies for multiple benefits, Pillar II funds (Rural Development) could be used to generate critical technical mass at local level. Options include supporting local processing, machinery development or cooperation. This will address low farmer acceptance and make the use of perennial biomass crops to optimise agricultural land use a normal part of a more diverse farm sector. Farmers hesitate in dedicating their land to long-term biomass production. Even where economic disadvantages are fully addressed, Warren et al. (2016) reported that farmers' socio-cultural identity limits adoption of short rotation coppice. They will only be willing to do this if biomass markets are reliable or if long-term contracts are granted. Therefore, the development of biomass marketing structures should be supported. This could also be supported under Pillar II.

Conclusion

Achieving the ambitious aims of decarbonising much of our economy by 2050 requires a big commitment to a wide range of changes: very significant reductions in energy use, reductions in emissions from the food system, protection of high-carbon stock ecosystems such as forest and peatland, increased carbon sequestration in agricultural soils and the use of renewable energy. The 2015 COP21 agreement in Paris points towards increasing emphasis on reducing greenhouse gas emissions effectively. Even in a densely populated region such as Europe, bioenergy has an important role to play, as it already does. The developments over the last 20 years show clearly that there are significant resource constraints that must be recognised by the policy community if public acceptance for an increased role for bioenergy is to be sustained. This means establishing more sustainable and cost-effective market-based instruments that protect food resources from competition and which integrate biomass production from perennial species. This also means concentrating on waste and co-products of food and forestry, and biomass grown on marginal land using perennial species.

A presentation to accompany this paper is available at www.murphy-bokern.com/presentations

Acknowledgement Donal Murphy-Bokern works on the WATBIO project (Development of improved perennial non-food biomass and bioproduct crops for water-stressed environments) which is a collaborative research project funded from the European Union's Seventh Programme for research, technological development and demonstration under grant agreement No 311929.

References

Biomass (2015) Abstracts of the Conference "Perennial biomass crops for a resource-constrained world". www.biomass2015.eu./fileadmin/einrichtungen/biomass2015/Biomass2015_Abstracts.pdf

BMEL (2007) Wissenschaftlicher Beirat Agrarpolitik beim Bundesministerium für Ernährung, Landwirtschaft und Verbraucherschutz Nutzung von Biomasse zur Energiegewinnung. Empfehlungen an die Politik

BMEL (2011) Wissenschaftlicher Beirat Agrarpolitik beim Bundesministerium für Ernährung, Landwirtschaft und Verbraucherschutz. Förderung der Biogaserzeugung durch das EEG. Stellungnahme zur geplanten Novellierung zur geplanten Novellierung des Ernuebare-Energien-Gesetze

BMU (2009) Übersicht über die EEG-Vergütungsregelungen für 2009 gemäß Bundestagsbeschluss zum EEG Vergütung ct/kWh

BMWE (2014) Die EEG-Reform Biomasse. www.biogaspartner.de/fileadmin/biogas/documents/Politik_und_Recht/eeg-2014-faktenblatt-biomasse.pdf

BMWE (2016) Informationsportal Erneuebare Energien. Das Erneurbare-Energie-Gesetz. www.erneuerbare-energien.de accessed on 23.6.2016.

CCC (2011) Bioenergy review. Committee on Climate Change. www.theccc.org.uk

Clancy M (2015) Renewable heat in Ireland to 2020. Achieving Ireland's 2020 renewable heat target: Analysis of policy options

Coombs J (1997) EU R&D programmes in the non-food area. Plant based specialty products and biopolymers. Nordic Council of Ministers

DCENR (2015) Ireland's transition to a low carbon energy future 2015-2030. Department for Communications, Energy and Natural Resources

DECC (2012) UK Bioenergy Strategy. Department for Energy and Climate Change

European Commission (2012) Communication from the Commission to the European Parliament, the Council, the European Economic and Social Committee of the Regions. Innovating for sustainable growth: a bioeconomy for Europe. COM 60

European Commission (2014) An ex-post evaluation of the rationale, implementation and impacts of EU Seventh Framework Programme (2007–2013), Cooperation Theme 2: Food, agriculture and fisheries, and biotechnology. Report to the European Commission

FNR (2015) Entwicklung der Anbaufläche für nachwachsende Rohstoffe. Fachagentur Nachwachsende Rohstoffe e.V.

Görmann H, Kreins P, Breuer T (2007) Germany—the European energy-corn-belt? Agrarwirtschaft 56:263–271

Gauck J (2012) Speech of the Federal President at the opening of the Environment Week. *Eröffnung der Woche der Umwelt 2012*. www.bundespraesident.de

Isermeyer F, Zimmer Y (2006) Thesen zur Bioenergiepolitik in Deutschland: Agrarökonomen warnen vor negativen Folgen einer Fortführung der bisherigen Förderpolitik—Derzeit überhitzte Bioenergiekonjunktur birgt Gefahr von Fehlinvestitionen—Businessplan muss Grundlage für Neuausrichtung schaffen. Agra Europe (Bonn) 47(23):1–15

Isermeyer F, Bolte A, Dieter M, de Witte T, Zimmer Y (2012) Bewertung der Leopoldina-Studie 2012 zur Bioenergie. Johann Heinrich von Thünen-Institute, Braunscweig, Germany

IPCC (2014) Climate Change 2014: Synthesis Report. Contribution of Working Groups I, II and III to the Fifth Assessment Report of the Intergovernmental Panel on Climate Change

Kaltschmitt M, Reinhardt GA, Stelzer T (1997) Life cycle analysis of biofuels under different environmental aspects. Biomass Energ 12(2):121–134

Klawitter N (2012) Corn-mania: Biogas boom in Germany leads to modern-day land grab. Spiegel Online International. http://www.spiegel.de/international/germany/biogas-subsidies-in-germany-lead-to-modern-day-land-grab-a-852575.html

Lindegaard K (2013) Why we need an energy crops scheme. Position Paper. Crops for energy. http://www.crops4energy.co.uk

Lovett A, Sünnenberg G, Dockerty T (2014) The availability of land for perennial energy crops in Great Britain. GCB Bioenergy 6:99–107

Mildon R (1986) Surpluses and the common agricultural policy. In: Raymond WF, Larvor R (eds) Alternative uses for agricultural surpluses. Elsevier Applied Sciences, New York

Monbiot G (2014) How a false solution to climate change is damaging the natural world. Biomass and Bioenergy. George Monbiot's blog. www.theguardian.com

Moran D, Barnes A, McVittie A (2007) The rationale for Defra investment in R&D underpinning the genetic improvement of crops and animals. Defra report for project IF0101. SAC Commercial Ltd, Scotland, United Kingdom

NLWKN (2010) Grundwasser Band 10. Niedersächsischer Landesbetrieb fürWasserwirtschaft, Küsten- und Naturschutz

Pearce D (2015) DRAX, UK—up in flames. How biomass burning wrecks Europe's forests. Case study report. Fern

Rauch A, Thöne M (2012) Biofuels—at what cost? International Institute for Sustainable Development

Rexen F (1997) Information on Nordic and EU R&D programmes in the non-food area. Plant based specialty products and biopolymers. Nordic Council of Ministers

Rexen F, Munck L (1984) Cereal crops for industrial use in Europe. European Commission

Renewable Fuels Agency (2008) The Gallagher Review of the indirect effects of biofuels production

Rose D (2014) The bonfire of insanity http://www.dailymail.co.uk/news/article-2581887/The-bonfire-insanity-Woodland-shipped-3-800-miles-burned-Drax-power-station-It-belches-CO2-coal-huge-cost-YOU-pay-cleaner-greener-Britain.html#ixzz41DIwM8n6

Soldatos P (2015) Economic aspects of growing perennial grasses for fibre and bioenergy in South Europe. Abstracts from the International Conference on 'Perennial biomass crops for a resource-constrained world' 7–9 September 2015. Hohenheim, Germany. www.biomass2015.eu

Stechmesser G (n.d.) Initiativen mit Weitblick. www.initiativen-mit-weitblick.de

Stern N (2006) The economics of climate change (the stern review). HM Treasury, London

Theuvsen L, Plumeyer C-H, Emmann C (2011) Einfluss der Biogasproduktion auf den Landpachtmarkt in Niedersachsen. Niedersächsisches Ministerium für Ernährung, Landwirtschaft, Verbraucherschutz und Landesentwicklung

Warren CR, Burton R, Buchanan O, Birnie RV (2016) Limited adoption of short rotation coppice: the role of farmers' socio-cultural identity. J Rural Stud 45:175–183

Westhoek H, Lesschen J, Rood T, Wagner S, De Marco A, Murphy-Bokern D, Leip A, van Grinsven H, Sutton M, Oenema O (2014) Food choices, health and environment: effects of cutting Europe's meat and dairy intake. Glob Environ Chang 26:196–205

Wintzer D, Fürniss B, Klein-Vielhauer S, Leible L, Nieke E, Rösch C, Tangen H (1993) Technikfolgenabschätzung zum Thema Nachwachsende Rohstoffe. Reihe A: Angewandte Wissenschaften, Sonderheft, Schriftenreihe des BML, Landwirtschaftsverlag, Münster-Hiltrup

Zimmer Y, Berenz S, Döhler H, Isermeyer F, Leible L, Schmitz N, Schweinle J, Toews T, Tuch U, Vetter A, Witte T de (2008) Klima- und energiepolitische Analyse ausgewählter Bioenergie-Linien. Braunschweig: vTI, V, 120 p, Landbauforsch SH 318

Chapter 3
Perennial Grasses for Sustainable European Protein Production

U. Jørgensen and P.E. Lærke

European Agriculture is Challenged

Agricultural innovation has underpinned the development of modern societies by a combination of many small incremental innovations as well as a few major changes based on radical innovations (van der Veen 2010). European agriculture has been innovated over many centuries and current systems are highly efficient. To reach our aims of further sustainable intensification of the agricultural system, it will be necessary to rethink and develop radically innovated chains (Kuyper and Struik 2014). One obvious example is developing alternatives to high-input, high-emission annual grain and seed crops. In Denmark, 77 % of grain production is currently used as feed (Gylling et al. 2013).

The EU produced around 20 % of the world pig meat and 15 % of the world chicken meat in 2012. This creates a high demand for high-quality protein suitable for intensive production of monogastric animals. The EU is a net importer of high-protein crop commodities and is almost 70 % dependent on imports (Table 3.1); for soya bean products this figure is over 97 %, with an import around 30 million tonnes of soyameal annually (Parajuli et al. 2015). This dependency of a single commodity increases the vulnerability of European animal production to price volatility (EIP-AGRI 2014).

Another great challenge for Europe is to reduce the environmental impacts from agriculture as stipulated in numerous directives and policies such as the Water Framework Directive (WFD); the Nitrates Directive; and the new EU climate policy

---

The original version of this chapter was revised. An erratum to this chapter can be found at DOI 10.1007/978-3-319-44530-4_25

U. Jørgensen (✉) • P.E. Lærke
Department of Agroecology, Aarhus University, Blichers Allé 20, Postboks 50, DK-8830 Tjele, Denmark
e-mail: Uffe.jorgensen@agro.au.dk

S. Barth et al. (eds.), *Perennial Biomass Crops for a Resource-Constrained World*, DOI 10.1007/978-3-319-44530-4_3

Table 3.1 EU balance of protein-rich feeds in 2012 (EIP-AGRI 2014)

Material	EU production (Mt)		EU consumption (Mt)	
	Products	Protein	Product	Proteins
Soya beans/meals	1189	452	34,134	15,904
Rapeseed and sunflower seed/meals	27,481	5213	19,721	6329
Pulses	3045	670	2800	616
Dried forage	4056	771	3900	741
Miscellaneous plant sources	2877	654	5859	1260
Subtotal	38,648	7760	66,414	24,850
Fish-meal	398	275	599	433
Total	39,046	8035	67,013	25,283

with specific greenhouse gas (GHG) reduction goals for agriculture. These goals can be very difficult to reach in some areas with current production systems, at least in a cost-efficient way. Denmark has an especially vulnerable aquatic environment due to sandy soils, a long coastline, and high precipitation. Since the 1980s Denmark has halved nitrate leaching by implementing numerous measures on improved manure handling, fixed N-norms for all crops, statutory catch crops, etc. (Dalgaard et al. 2014). Still, fulfilling the WFD means that some areas must halve their nitrate leaching once again, and radical changes are required to reduce losses while maintaining a profitable agricultural production.

Annual Versus Perennial Crops

Compared with annual grain and seed crops, the production of perennial crops reduces significantly the losses of nutrients, the need for pesticides, and supports soil carbon build-up (Cadoux et al. 2014; Pugesgaard et al. 2015). Even though water quality will be improved by cropping systems with increased growing season, water quantity (surplus for ground water and river discharge) may be reduced due to higher annual evapotranspiration. However, new research shows that not all perennial crops increase water use (Ferchaud et al. 2015). In addition, there is an increased water infiltration capacity in pastures compared with annual crops (Franzluebbers et al. 2014), which can reduce the loss by water run-off. The potential for securing a water supply for a long growing period is highest in humid north-western Europe, even though optimizations may also be feasible in water-limited areas such as the use of winter-grown green biomass (Tsiplakou et al. 2014).

In humid temperate regions grasses and legumes can capture solar radiation more efficiently than annual grain and seed crops, in which a considerable part of the growing season is used for crop ripening, harvest, soil tillage, and sowing (Cadoux et al. 2014; Dohleman and Long 2009; Pugesgaard et al. 2014; Pugesgaard et al. 2015). Potentially, biomass yields can be doubled and the potential is expected to increase with climate change (Jørgensen et al. 2012). Most promising are grasses with C_3 photosynthesis in northern and C_4 grasses in southern regions (Cadoux et al. 2014).

Green Biorefining of Grasses Can Provide Sustainable Protein Feed and Bioenergy

Green harvested biomass does not fit well into current agricultural markets shaped for handling grain and seeds to feed monogastric animals. But an industrial separation and upgrade of green biomass hold promise for coupling the best agronomic potentials with new industrial technologies into a radically improved agro-industrial system (Kammes et al. 2011; Parajuli et al. 2015; Sanders and Bos 2013; Sharma et al. 2011). Of particular interest is the high content of protein of a good amino acid composition (high contents of lysine and methionine) in grass and legume crops (Houseman and Connell 1976; Maciejewicz-Rys and Hanczakowski 1990), and Leaf-Protein-Concentrates may be very suitable substitutes for soya products, the production of which has major environmental impacts in, e.g., Brazil (Harvey 2014). Self-supply with protein feed is especially interesting for organic production systems where issues of GMO, trusty certification, and high prices question the future sustainable development of the organic animal sector based on soya imports.

Following separation of the biomass into a press juice containing the easily soluble proteins, the fiber fraction of the press cake is a valuable cattle feed with a high digestibility (some addition of molasses may be necessary (Klop et al. 2015)) containing protected proteins that will pass the rumen un-degraded. Thus, simple decentralized separation in high-value feed fractions for both mono- and polygastric animals can create value from locally grown biomass and reduce the need for imports (Seppala et al. 2014). For fish production large amounts of soya, wheat, and sunflower are used, and this sector may be boosted with delivery of a more sustainable feed. In addition to increasing the current amount of protein feed production in Europe, the implementation of more productive cropping systems will result in by-products available for either material or energy use. Because valuable food or feed is produced, this coproduct biomass will be available without a negative ILUC effect, which in most cases makes the production of, for example, rape seed biodiesel and dedicated energy crops unsustainable (Tonini and Astrup 2012).

Leaf protein has even been proposed for direct human consumption, which would further increase resource efficiency of food production, and there are several examples of usage of, e.g., lucerne protein in food products albeit only to a limited extent (Linnemann and Dijkstra 2002). However, utilization for direct human consumption is much less straightforward than the use for animal feed (Chiesa and Gnansounou 2011). The animal production sector uses 30 % of the earth's entire land surface, mostly permanent pasture, but also 33 % of the global arable land is used to produce feed for livestock (Steinfeld et al. 2006), and the animal feed sector has a major economic importance.

A Scenario for Increased Biomass Production and Utilization for Biorefining in Denmark

Gylling et al. (2013) investigated whether it would be possible in Denmark to increase the production of biomass from agriculture and forestry by ten million tonnes without causing a reduction in food production, expansion of the farmed area, and adverse impact on the aquatic environment and biodiversity. Preservation of soil fertility and carbon content were likewise important factors. The report describes three scenarios: (1) a business-as-usual scenario with just an increased utilization of the existing agriculture and forestry resources. (2) A biomass-optimized scenario where both agriculture and forestry are adjusted to produce the maximum level of biomass. (3) An environment-optimized scenario with perennial grasses as an important measure to improve productivity and environmental performance.

The results showed that the optimized scenarios were able to deliver an additional eight to ten million tonnes of biomass for a biorefinery industry by 2020 (Fig. 3.1). This is possible without compromising the food production if 10–15 % of the green and yellow categories of biomass are turned into animal feed in the refineries. Especially, the green biomass category (grasses in the environment scenario) contributed considerably in the optimized scenarios and would be able to deliver significant amounts of protein feed. However, much depends on how efficiently the biomass is converted and how the animal fodder can be extracted, stored, and utilized efficiently in the livestock sector. The process of introducing new cropping and harvesting methods and new crops to agriculture is complex, and its

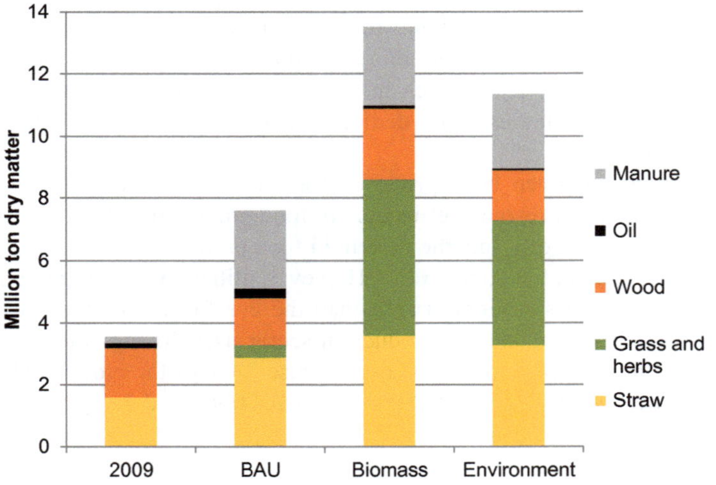

Fig. 3.1 Scenarios for increased delivery of biomass for Danish biorefineries compared with the use of biomass for energy in 2009 (Gylling et al. 2013). The scenarios are: business-as-usual (BAU), biomass-optimized, and environment-optimized

implementation will not happen automatically if farmers do not perceive advantages from it. An active collaboration between industry, farming, authorities, and research will therefore need to be established (Gylling et al. 2013; Parajuli et al. 2015).

In the environment scenario, a reduction of nitrate leaching equivalent to 23,000 tonnes N annually was estimated for Denmark (Gylling et al. 2013). This is approximately the reduction required to fulfil the WFD. Even though much more organic matter will be mobilized for biorefining, we estimated that soil carbon levels would be largely unchanged in the environment scenario (Jørgensen et al. 2013).

New results (see below) from the EU project GrassMargins (www.grassmargins.org) and from Danish projects (e.g., www.biovalue.dk) seems to confirm that the anticipated increase in total biomass yield and reduction in nitrate leaching are realistic when converting land currently used for grain crop production into grass production.

Materials and Methods

Field experiments were set up at three locations with coarse sand, sandy loam, and loamy sand soils. The grass species reed canary grass (*Phalaris arundinacea*), tall fescue (*Festuca arundinacea*), cocksfoot grass (*Dactylis glomerata*), miscanthus (*M.* x *giganteus),* festulolium (hybrid between *Festuca* and *Lolium*), and two grass-clover mixtures are compared with innovative combinations of annual species and reference traditional crop rotations or monocultures (e.g., continuous maize) with four field replications at each site. The crops are managed with fertilizer, pesticides, and irrigation at levels anticipated to assure optimal growth; however, the grass-clover mixtures were not given any N-fertilizer after establishment. Concentrations of nitrate were measured with ceramic suction cups installed at 1.25 m depth in all replicates and sampled approx. every 3 weeks.

First Year Results on Grass Productivity and Environmental Impact

The dry matter yields of selected crops at one location are given in Fig. 3.2 for the first full experimental year, 2013. Miscanthus was still in the establishment phase with a low yield. However, festulolium produced 22.5 tonnes ha^{-1} dry matter in three cuts, which was about the double of total (straw + grain) grain crop yields of spring barley and winter wheat that are the current major crops in Denmark. The low yield of the grass-clovers can be mainly attributed to the absence of nitrogen fertilization. The highest yield was obtained in beets (root + top); however, not much protein is expected in beets but they may be used for ethanol or biogas production as already practiced in some places (Börjesson et al. 2015). In 2014 grain crops

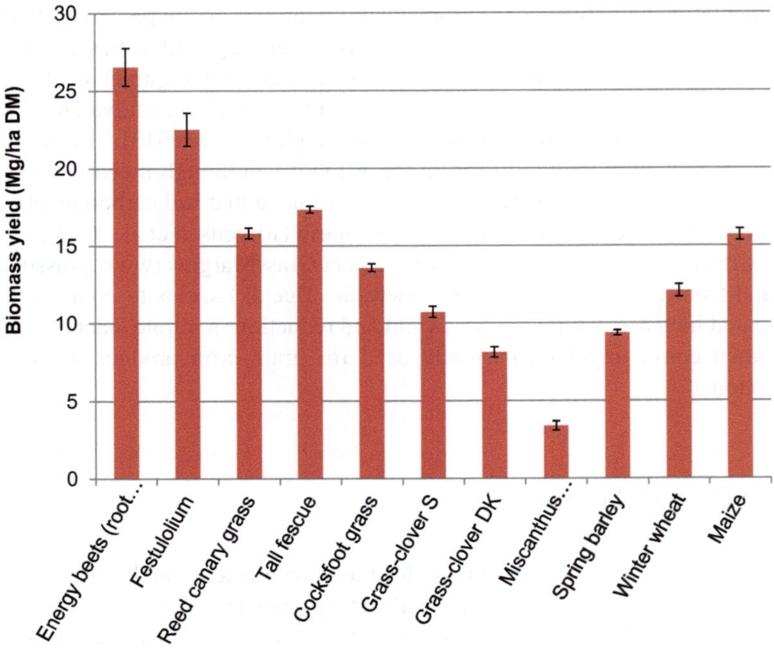

Fig. 3.2 Total dry matter yields of selected crops at a loamy sand at AU Foulum, Denmark in 2013. Bars indicate ± SE

performed better than in 2013, but all pure grass crops (still except miscanthus) produced more with approx. 20 tonnes ha^{-1} dry matter (data not shown). Under warmer conditions than in Denmark, maize is often more productive or as productive as grasses (Muylle et al. 2015) but in this study from Belgium grasses were only fertilized 249 kg N ha^{-1}, which was probably not enough to increase yields of grass in pure stand above 15–20 tonnes ha^{-1}.

Mean nitrate-N concentrations in soil water where winter wheat (fertilized 175 kg N ha^{-1}) or spring barley (130 kg N ha^{-1}) had been harvested were 17–20 mg L^{-1} during the following autumn and winter (14 October 2013 to 31 March 2014), while they were 5–10 mg L^{-1} below intensively fertilized festulolium (425 kg N ha^{-1}), unfertilized grass-clover and beet (130 kg N ha^{-1}) (Fig. 3.3). This means that nitrate leaching may be halved by shifting from grain crops to intensive grass or beet production. Such sustainable intensification seems sounder than the other possible routes to reduced environmental impact, namely unfertilized grass-clover as indicated in Fig. 3.3. However, other studies have reached higher yields in legume-based systems not applied nitrogen fertilizer (Hauggaard-Nielsen et al. 2016) that have lower emission of nitrous oxide than fertilized pure grass systems (Schmeer et al. 2014) and may provide a better overall GHG balance.

A change in cropping pattern is of course conditional on economic uses for the grass biomass, such as the above-mentioned extraction of protein for feed or food

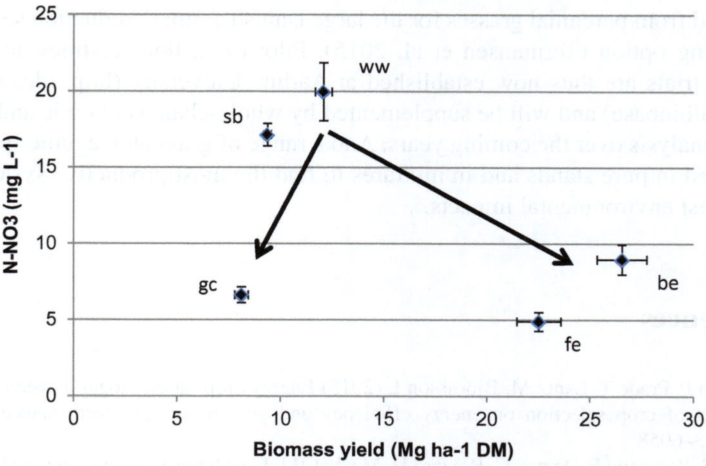

Fig. 3.3 Mean nitrate concentrations in autumn and winter after the production year 2013 versus dry matter yields of winter wheat (ww), spring barley (sb), festulolium (fe), beets (be), and grass clover (gc). Bars indicate ± SE. Possible development routes for reducing nitrate leaching from the current cereal-based production systems are indicated by arrows

purposes. Otherwise, consideration of ILUC can undermine the GHG benefit if the grasses are used solely for energy (Tonini and Astrup 2012). Additional environmental benefits by changing into grass production are achieved by the large difference in the use of pesticides in grass crops: according to the Danish statistics on pesticide use grass crops were treated with 0.04 standard pesticide applications per ha in 2013 compared with a mean of 3.76 standard applications across the whole arable land (Danish Environmental Agency 2014). Usually, grass crops will increase soil carbon content while annual crops tend to decrease it (Taghizadeh-Toosi et al. 2014).

One possible drawback for the intensive grass crop production may be emissions of nitrous oxide that can compromise the GHG budgets and simple IPCC calculations give high emissions as they are directly related to the amount of N-fertilization. The nitrous oxide emissions still need to be measured in our experiments. However, the first N-balances show that we harvest more nitrogen than we apply in fertilizer to the grasses, and there seems not to be much room for denitrification in the system. Intensive grass production may not significantly improve biodiversity but the low pesticide use and low losses of nutrients reduce the impact on adjacent natural areas.

Future Perspectives

Grasses and legumes have higher contents of protein with better quality (high lysine and methionine contents) than grain and seed crops under Danish conditions (Jørgensen et al. 2013). Thus, substituting imported soya bean protein with protein extracted from perennial grasses for the large Danish animal production sector is an interesting option (Termansen et al. 2015). Pilot extraction facilities and animal feeding trials are thus now established at Aarhus University (http://dca.au.dk/en/research/biobase) and will be supplemented by whole-chain economic and sustainability analysis over the coming years. And a range of grass and legume species are compared in pure stands and in mixtures to find the most productive systems with the lowest environmental impacts.

References

Börjesson P, Prade T, Lantz M, Björnsson L (2015) Energy crop-based biogas as vehicle fuel-the impact of crop selection on energy efficiency and greenhouse gas performance. Energies 8:6033–6058

Cadoux S, Ferchaud F, Demay C, Boizard H, Machet JM, Fourdinier E, Preudhomme M, Chabbert B, Gosse G, Mary B (2014) Implications of productivity and nutrient requirements on greenhouse gas balance of annual and perennial bioenergy crops. Glob Chg Biol Bioenergy 6:425–438

Chiesa S, Gnansounou E (2011) Protein extraction from biomass in a bioethanol refinery—possible dietary applications: use as animal feed and potential extension to human consumption. Bioresour Technol 102:427–436

Dalgaard T, Hansen B, Hasler B, Hertel O, Hutchings NJ, Jacobsen BH, Jensen LS, Kronvang B, Olesen JE, Schjorring JK, Kristensen IS, Graversgaard M, Termansen M, Vejre H (2014) Policies for agricultural nitrogen management-trends, challenges and prospects for improved efficiency in Denmark. Environmental Research Letters 9.

Danish Environmental Agency (2014) Bekæmpelsesmiddelstatistik 2013. Orientering fra Miljøstyrelsen nr. 6, 2014, 100p

Dohleman FG, Long SP (2009) More productive than maize in the Midwest: how does Miscanthus do it? Plant Physiol 150:2104–2115

EIP-AGRI (2014) Focus Group on protein crops: Final Report, 48 pp. https://ec.europa.eu/eip/agriculture/en/content/eip-agri-focus-group-protein-crops-final-report

Ferchaud F, Vitte G, Bornet F, Strullu L, Mary B (2015) Soil water uptake and root distribution of different perennial and annual bioenergy crops. Plant and Soil 388:307–322

Franzluebbers AJ, Sawchik J, Taboada MA (2014) Agronomic and environmental impacts of pasture-crop rotations in temperate North and South America. Agr Ecosyst Environ 190:18–26

Gylling M, Jørgensen U, Bentsen NS, Kristensen IT, Dalgaard T, Felby C, Johannsen VK (2013) The +10 Million Tonnes Study: increasing the sustainable production of biomass for biorefineries. University of Copenhagen, Frederiksberg, 32p

Harvey M (2014) The food-energy-climate change Trilemma: toward a socio-economic analysis. Theory, Culture & Society 31:155–182

Hauggaard-Nielsen H, Lachouani P, Knudsen MT, Ambus P, Boelt B, Gislum R (2016) Productivity and carbon footprint of perennial grass-forage legume intercropping strategies with high or low nitrogen fertilizer input. Sci Total Environ 541:1339–1347

Houseman RA, Connell J (1976) Utilization of products of green-crop fractionation by pigs and ruminants. Proc Nutr Soc 35:213–220

Jørgensen U, Olesen JE, Lærke PE, Sørensen KK, Raulund-Rasmussen K, Jensen PE, Felby C (2012) Solens energi kan omdannes mere effektivt til biomasse. In: Robust og Bæredygtig Bioenergi, 15-17, Proceedings of the Danish Strategic Research Council seminar at DONG Energy 5/12 2012

Jørgensen U, Elsgaard L, Sørensen P, Olsen P, Vinther FP, Kristensen EF, Ejrnæs R, Nygaard B, Krogh PH, Bruhn A, Rasmussen MB, Johansen A, Jensen SK, Gylling M, Bojesen M (2013) Biomasseudnyttelse i Danmark—potentielle ressourcer og bæredygtighed. DCA Rapport Nr 033, 127p

Kammes KL, Bals BD, Dale BE, Allen MS (2011) Grass leaf protein, a coproduct of cellulosic ethanol production, as a source of protein for livestock. Anim Feed Sci Technol 164:79–88

Klop A, Durksz D, Zonderland J, Koopmans B (2015) Livestock research report 790. University, Wageningen, Grasraffinage en gebruik van rasvezel in de rundveevoeding

Kuyper TW, Struik PC (2014) Epilogue: global food security, rhetoric, and the sustainable intensification debate. Curr Opin Environ Sustain 8:71–79

Linnemann AR, Dijkstra DS (2002) Toward sustainable production of protein-rich foods: appraisal of eight crops for Western Europe. PART I. Analysis of the primary links of the production chain. Crit Rev Food Sci Nutr 42:377–401

Maciejewicz-Rys J, Hanczakowski P (1990) Improvement of the nutritive value of cereals by leaf protein supplementation. J Sci Food Agric 50:99–104

Muylle H, Van Hulle S, De Vliegher A, Baert J, Van Bockstaele E, Roldán-Ruiz I (2015) Yield and energy balance of annual and perennial lignocellulosic crops for bio-refinery use: a 4-year field experiment in Belgium. Eur J Agronomy 63:62–70

Parajuli R, Dalgaard T, Jorgensen U, Adamsen AP, Knudsen MT, Birkved M, Gylling M, Schjorring JK (2015) Biorefining in the prevailing energy and materials crisis: a review of sustainable pathways for biorefinery value chains and sustainability assessment methodologies. Renew Sustain Energy Rev 43:244–263

Pugesgaard S, Olesen JE, Jorgensen U, Dalgaard T (2014) Biogas in organic agriculture- effects on productivity, energy self-sufficiency and greenhouse gas emissions. Renew Agric Food Syst 29:28–41

Pugesgaard S, Schelde K, Larsen SU, Lærke PE, Jørgensen U (2015) Comparing annual and perennial crops for bioenergy production influence on nitrate leaching and energy balance. Glob Chg Biol Bioenergy 7:1136–1149

Sanders JP, Bos HL (2013) Raw material demand and sourcing options for the development of a bio-based chemical industry in Europe. Biofuels, Bioproducts & Biorefining 7:260–272

Schmeer M, Loges R, Dittert K, Senbayram M, Horn R, Taube F (2014) Legume-based forage production systems reduce nitrous oxide emissions. Soil Tillage Res 143:17–25

Seppala A, Kyntaja S, Blasco L, Siika-Aho M, Hautala S, Byman O, Ilvesniemi H, Ojamo H, Rinne M, Harju M (2014) Grass silage extract, feed component suitable for pigs—prospects for on farm biorefinery. Proceedings of the 5th Nordic Feed Science Conference, Uppsala, Sweden, 10–11 June 2014, 163–168

Sharma HSS, Lyons G, McRoberts C (2011) Biorefining of perennial grasses: a potential sustainable option for Northern Ireland grassland production. Chem Eng Res Design 89:2309–2321

Steinfeld H, Gerber P, Wassenaar T, Castel V, de Haan C (2006) Livestock's long shadow: environmental issues and options. Food & Agriculture Organisation, Rome, Italy

Taghizadeh-Toosi A, Olesen JE, Kristensen K, Elsgaard L, Ostergaard HS, Laegdsmand M, Greve MH, Christensen BT (2014) Changes in carbon stocks of Danish agricultural mineral soils between 1986 and 2009. Eur J Soil Sci 65:730–740

Termansen M, Gylling M, Jørgensen U, Hermansen J, Hansen LB, Knudsen MT, Adamsen APS, Ambye-Jensen M, Jensen MV, Jensen SK, Andersen HE, Gyldenkærne S (2015) Grøn biomasse. DCA Rapport 68, 38 p

Tonini D, Astrup T (2012) LCA of biomass-based energy systems: a case study for Denmark. Appl Energy 99:234–246

Tsiplakou E, Economou G, Hadjigeorgiou I, Kominakis A, Zervas G (2014) Plant species mixtures for forage production for ruminant feeding under Mediterranean conditions. Exp Agric 50: 426–437

Van der Veen M (2010) Agricultural innovation: invention and adoption or change and adaptation? World Archaeol 42:1–12

Chapter 4
Long-Term Yields and Soil Carbon Sequestration from *Miscanthus*: A Review

M.B. Jones, J. Zimmermann, and J. Clifton-Brown

Introduction

Miscanthus, particularly the triploid hybrid (*Miscanthus x giganteus*), is now widely adopted as a bioenergy crop in Europe and north America and substantial areas have been planted by farmers. It is characterised by high yields in temperate climates, low moisture content at harvest, high water and nitrogen use efficiencies and a low susceptibility to disease (Jones and Walsh 2001). The acceptability of *Miscanthus* has been based on the assumption that it is a low input-high output crop that requires little energy for maintenance and gives good yields of harvestable dry matter with small amounts of fertiliser application. *Miscanthus* is a woody rhizomatous C_4 grass species which originated in south-east Asia and was initially introduced to Europe as an ornamental plant. It is a perennial plant with an estimated productive life time of up to 20 years which yields high quality lignocellulosic material for both energy and fibre (Jones and Walsh 2001). Currently, *Miscanthus* is used mainly for anaerobic digestion to produce methane or for direct combustion. *M. x giganteus* is established from rhizomes, a relatively costly process compared to establishment from seed and for a grower to make a profit, models show that yields need to be stable and high for at least 5–10 years after planting (Lewandowski et al. 2003; Christian et al. 2008; Heaton et al. 2004). Evidence to support the enthusiasm for *Miscanthus* has come from a very large number of small plot trials that were established, mainly in Europe, in the early 1990s. In their early years these trials did not, of course, provide information on the long-term sustainability of the crop and predictions of yield

M.B. Jones (✉) • J. Zimmermann
School of Natural Sciences, Trinity College Dublin, The University of Dublin, Dublin 2, Ireland
e-mail: mike.jones@tcd.ie

J. Clifton-Brown
Institute of Biological, Environmental and Rural Sciences, University of Aberystwyth, Gogerddan, Aberystwyth SY23 3FL, UK

S. Barth et al. (eds.), *Perennial Biomass Crops for a Resource-Constrained World*, DOI 10.1007/978-3-319-44530-4_4

potential have been largely based on growth models using information gathered from recently established crops (Hastings et al. 2009; Miguez et al. 2012). The limitation of these models is that they do not simulate a possible decline in productivity in later years (Lewandowski et al. 2000). Furthermore, it was necessary to move from plots to larger scale farm trials to gain more knowledge of the longer term prospects for yields. In recent years, the availability of yield data from some long-term trials exceeding 15 years has allowed a more detailed analysis of the sustainability of *Miscanthus*. Currently, the longest continuous replicated trials of *M. x giganteus* yields in Europe reported in the peer-reviewed literature have spanned 12 years (Angelini et al. 2009), 14 years (Christian et al. 2008; Gauder et al. 2012; Clifton-Brown et al. 2007) and 20 years (Larsen et al. 2014). Here, we extend the measurement period reported by Clifton-Brown et al. (2007) in Ireland to 23 years and review the outcomes of some of the analyses and the conclusions they have drawn for the long-term future of *Miscanthus* as a viable energy crop. Our aim was to answer the following questions: How long does it take to reach a ceiling yield? What is the variability of the maximum yield? Do yields decline over time and when?

Long-term studies have also indicated that *Miscanthus* accumulates and sequesters carbon in the soil which has benefits for the overall greenhouse gas balance of the system and soil quality. Knowledge about the potential for carbon sequestration in soils was fragmentary in the initial stages of the trials and was again largely based on models of carbon sequestration parameterised using data from a small number of short-term observations (Matthews and Grogan 2001). Several features of *Miscanthus* physiology and the agricultural practices associated with its cultivation suggest that there is a large potential for soil carbon sequestration in this crop. Firstly, perennial plants such as *Miscanthus* allocate high proportions of the assimilated carbon belowground as a carbon reservoir for growth in the spring (Kuzyakov and Domanski 2000), and the perennial nature also reduces tillage to initial planting and crop discontinuation, significantly reducing soil disruption and therefore increasing soil organic carbon (SOC) stability (Roberts and Chan 1990). Secondly, *Miscanthus* is typically harvested in spring after the above ground biomass has senesced to minimise moisture content at the time of harvest. The high pre-harvest losses of leaves in particular increase the input of this carbon to the soil. Thirdly, because the *Miscanthus* receives little or no N fertiliser this results in a high C:N ratios in the soil which leads to a slower decomposition of the plant residues. Stabilisation of soil organic carbon for long-term sequestration occurs through a number of processes which leads to the formation of more recalcitrant soil humic fractions (Jones and Donnelly 2004). When C_4 *Miscanthus* is grown on land that was formerly vegetated with C_3 plants it is possible to use the shift in the carbon stable isotope signal to determine the proportion of soil organic matter derived from the new C_4 *Miscanthus* source (Balesdent et al. 1987). Evidence for the amount of carbon sequestered under *Miscanthus* has come from relatively short-term measurements of total soil carbon (Kahle et al. 2001) and of the *Miscanthus* carbon isotope signal occurring in recalcitrant fractions of the soil (Dondini et al. 2009a) combined with model predictions (Dondini et al. 2009b). Recent measurements confirm that *Miscanthus* crops accumulate carbon in the soil (Zimmermann et al. 2012;

Zimmermann et al. 2013), but there are still great uncertainties about the rate at which this carbon is accumulating and whether there is a ceiling for carbon accumulation. Our aim was therefore to answer the following questions: Do soils under *Miscanthus* sequester carbon? What is the rate of carbon sequestration? Does the rate change over time and how is it influenced by former land use? Is there a ceiling for the amount of carbon that is sequestered?

Long-Term Yields

Lewandowski et al. (2003), in a review of experience in Europe with *Miscanthus* as a novel energy crop, suggested that the crop is characterised by a yield increase during the establishment phase, lasting 3–5 years, followed by a ceiling phase with stable yields. Once mature yields are reached, *Miscanthus* is expected to have a plantation lifetime of 20–25 years (Lewandowski et al. 2000). Later work that has followed the evolution of *Miscanthus* yields, by Clifton-Brown et al. (2007) for 14 years, Christian et al. (2008) for 14 years and Angelini et al. (2009) for 12 years, all observed a third growth phase characterised by a yield decline that could not be explained by unfavourable climatic conditions during the growing season, beginning after 10, 11 and 3 years respectively. Figure 4.1 shows the yield data for the site

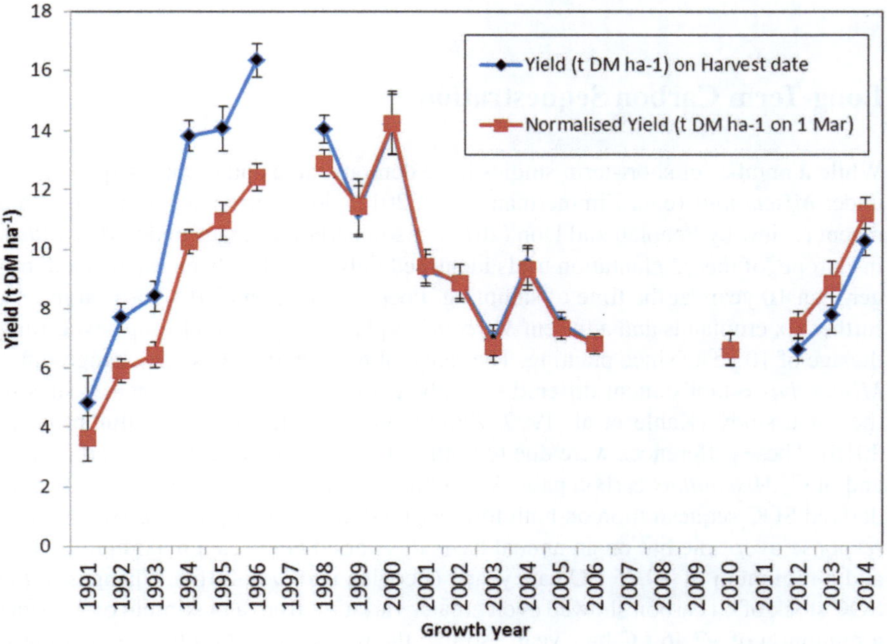

Fig. 4.1 Long-term yields (t DM ha^{-1}) of *Miscanthus* at Cashel, Ireland from 1991 to 2014. Measured yields (blue) have been 'normalised' to 1st March (red), see Clifton-Brown et al. (2007)

at Cashel, Ireland monitored by Clifton-Brown et al. (2007), extended for a further 8 years. This data suggests a recovery in yield following the initial decline. Lesur et al. (2013) have carried out an analysis of long-term yield trends of *M. x giganteus* grown in Europe, drawing on data from 37 field trials at 16 sites spanning five countries. Statistical models were used to describe yield evolution in each of the trials. The models were used to estimate three characteristics of yield trends over time viz. (1) the maximum yield reached across years, (2) the growing season when the maximum yield was reached and (3) the yield decrease associated with the period of decline. The duration of the establishment phase to maximum yield was variable and shown to be strongly determined by the planting method (Lesur et al. 2013). Maximum yields were found to be highly variable and this variability was largely explained by climatic factors, demonstrated by a strong relationship between maximum yield and latitude. Model fits to the yield data showed that yield evolution was best described by a decline over time but the decline intensity was variable so that in some trials yields remained nearly steady for up to 20 years while in others decline was severe after 5–10 years (Lesur et al. 2013). Arundale et al. (2014) have investigated the yield decline in *Miscanthus* in the Midwestern USA. Using data from several locations across Illinois over a period of 8–10 years, they found that yield initially increased until it reached a maximum during the fifth growing season and then declined to a stable but lower level in the eighth season. The possible causes of the decline were ascribed to exhaustion of the soil nutrients, especially nitrogen, soil compaction and/or pest and disease pressure.

Long-Term Carbon Sequestration

While a number of short-term studies have demonstrated soil carbon sequestration under *Miscanthus* (e.g., Zimmermann et al. 2012), long-term studies are rare. In a recent review by Poeplau and Don (2014) of soil carbon changes under *Miscanthus* in Europe, of the 22 plantation trials identified only three had been established longer than 10 years at the time of sampling. Poeplau and Don (2014) also sampled a further six croplands and adjacent *Miscanthus* plantations across Europe exceeding the age of 10 years since planting. The range of measured SOC stock changes after *Miscanthus* establishment differed strongly among studies but also within sites in the same study (Kahle et al. 1999, Zimmermann et al. 2012; McCalmont et al. 2015). These differences were due to both differences in the rate of C_3 carbon loss and of C_4 *Miscanthus* carbon gain. A compilation of literature data on *Miscanthus*-derived SOC sequestration on both former grasslands and croplands shows a linear response over time but on an annual basis this varied between a maximum of 3.07 and a minimum of −0.25 t C ha^{-1} year^{-1} (Poeplau and Don 2014). Changes in the SOC stock of C_3 carbon showed even greater variation from a maximum of 7.95 and a minimum of −7.46 t C ha^{-1} year^{-1} so that the overall effect on total SOC ranged between a maximum of 4.51 and a minimum of −6.85 t C ha^{-1} year^{-1} (Poeplau and Don 2014). When the analysis was confined to six former cropland sites Poeplau and Don (2014) measured a mean SOC accumulation in the top 0–30 cm of 0.40 ± 0.20 t C ha^{-1} $year^{-1}$, which is much lower than the mean total SOC change in all the currently published sites they reviewed (1.68 ± 0.70 t C ha^{-1} $year^{-1}$). They concluded that non-vegetation changes such as application of manure can override the effect of vegetation alone. In addition, Poeplau and Don (2014) showed that the soil fractions with which soil organic carbon is associated significantly influence its decomposability over time and therefore influence the ability of the soil to sequester carbon in the long term. Mishra et al. (2013) used combined process-based and geospatial models to estimate that the effect of growing *Miscanthus* on SOC accumulation in the croplands of the continental United States would range from 0.16 to 0.82 t C ha^{-1} $year^{-1}$. Dondini et al. (2009a, b) combined SOM fractionation with ^{13}C natural abundance analyses to trace the fate of *Miscanthus*-derived carbon in various physically protected soil fractions down to a depth of 60 cm under a 14 year old *Miscanthus* plantation, established on former arable land. The results showed that there was little evidence of sequestration of *Miscanthus*-derived carbon below 30 cm soil depth but that significant amounts of carbon storage occurred in the top 30 cm and this was largely located in the physically protected SOM fractions (Dondini et al. 2009a, b). However, Poeplau and Don (2014) at five sites they sampled, in contrast, did find *Miscanthus*-derived SOC in the 50–80 cm depth increment.

It has been predicted that the pattern of change in SOC following the planting of *Miscanthus* will be very different on conversion from grassland than from arable land (Anderson-Teixeira et al. 2009). Don et al. (2012) predicted C losses from the conversion of perennial grasslands to *Miscanthus* that could compromise the GHG mitigation potential. However, data on grassland conversions is less extensive than on arable conversions. Zatta et al. (2014) have shown that after 6 years from conversion from C_3 grassland to four novel *Miscanthus* hybrids and *M. giganteus* the amount of *Miscanthus*-derived C in total SOC was about 12 % and positively correlated to below ground biomass of the different hybrids. However, there was no significant change in total SOC as the C_4 carbon largely replaced the initial C_3 grassland carbon. Based on these observations and the use of a SOC partitioning model, RothC (Coleman and Jenkinson 1999) projected forward for 20 years, Zatta et al. (2014) conclude that planting *Miscanthus* on former grassland results in no significant change in SOC. Zatta et al. (2014) also found evidence of a priming effect where the initial SOC content decreased more in the presence of higher below ground biomass. The significance of this for total SOC content needs to be examined in more detail.

Conclusions

Yield evolution over time is important in any assessment of the economic and environmental value of *Miscanthus* as a source of bioenergy. Currently, farm-scale modelling approaches aimed at improved understanding of the potential economic perennial energy crop supply from *Miscanthus* assume a simplistic pattern of yield development which does not account for the yield decline and possible recovery (Alexander and Moran 2013). In particular, further work is needed to understand the factors inducing yield decline and to determine whether the decline can be halted or reversed. There is also a need to understand better the variability in ceiling yield and finally the reasons for differences between commercial and experimental yields.

Land conversion losses of SOC will play a major role in determining the carbon balance of biofuel crops. Clearing and tillage of grasslands for the purpose of *Miscanthus* cultivation results in carbon losses that offset SOC sequestration by the *Miscanthus* until the new carbon accumulation reaches a level that restores the initial loss following crop planting. By contrast, converting arable land to *Miscanthus* results in an immediate SOC benefit. Here, it is clear that soil carbon sequestration augments the GHG reduction associated with fossil fuel displacement. For conversion from grassland the evidence appears to be that the net effect on SOC content is small but that time to achieving this balance is variable. The evidence also suggests that significant biofuel carbon debts will not be encountered (Fargione et al. 2008). However, the potential of soil carbon sequestration to counter losses of carbon at the time of establishment may be dependent on the yield evolution of *Miscanthus*, as low yielding plantations may need to be discontinued before initial SOC losses have been replenished.

References

Alexander P, Moran D (2013) Impact of perennial energy crops income variability on the crop selection of risk averse farmers. Energy Policy 52:587–596

Anderson-Teixeira KJ, Davis SC, Masters MD, Delucia EH (2009) Changes in soil organic carbon under biofuel crops. GCB Bioenergy 1:75–96

Angelini LG, Ceccarini L, Di Nasso NN, Bonari E (2009) Comparison of *Arundo donax* L. and *Miscanthus x giganteus* in a long-term field experiment in central Italy: analysis of productive characteristics and energy balance. Biomass and Bioenergy 33:635–643

Arundale RA, Dohleman FG, Heaton EA et al. (2014) Yields of *Miscanthus x giganteus* and *Panicum virgatum* decline with stand age in the Midwestern USA. GCB Bioenergy 6:1–13

Balesdent J, Mariotti A, Guillet B (1987) Natural ^{13}C abundance as a tracer for studies of soil organic matter dynamics. Soil Biol Biochem 19:25–30

Christian DG, Richie AB, Yates NE (2008) Growth, yield and mineral content of *Miscanthus x giganteus* grown as a biofuel for 14 successive harvests. Ind Crops Prod 28:320–327

Clifton-Brown JC, Breuer J, Jones MB (2007) Carbon mitigation by the energy crop, *Miscanthus*. Glob Change Biol 13:2296–2307

Coleman K, Jenkinson D (1999) RothC-26.3. A model for the turnover of carbon in soils. Model description and window users guide. IACR Rothamsted, Herts

Don A, Osborne B, Hastings A (2012) Land-use change to bioenergy production in Europe: implications for the greenhouse gas balance and soil carbon. GCB Bioenergy 4:372–391

Dondini M, van Groenigen K-J, Del Gardo I, Jones MB (2009a) Carbon sequestration under *Miscanthus*: a study of ^{13}C distribution in soil aggregates. GCB Bioenergy 1:321–330

Dondini M, Hastings A, Saiz G et al. (2009b) The potential of *Miscanthus* to sequester carbon in soils: comparing field measurements in Carlow, Ireland to model predictions. GCB Bioenergy 1:413–425

Fargione J, Hill J, Tilman D et al. (2008) Land clearing and the biofuel carbon debt. Science 319:1235–1238

Gauder M, Graeff-Honninger S, Lewandowki I, Claupein W (2012) Long-term yield and performance of 15 different *Miscanthus* genotypes in southwest Germany. Ann Appl Biol 160:126–136

Hastings A, Clifton-Brown J, Wattenbach M et al. (2009) Future energy potential of *Miscanthus* in Europe. GCB Bioenergy 1:180–196

Heaton EA, Long SP, Voight TB et al.(2004) Miscanthus for renewable energy generation: European Union experience and projections for Illinois. Mitigation and Adaptation Strategies for Global Change 9:433–451

Jones MB, Donnelly A (2004) Carbon sequestration in temperate grassland ecosystems and the influence of management, climate and elevated CO_2. New Phytol 164:423–429

Jones MB, Walsh M (2001) Miscanthus for Energy and Fibre. James & James, London

Kahle P, Boelcke B, Zacharias S (1999) Effects of *Miscanthus x giganteus* cultivation on chemical and physical soil properties. J Plant Nutr Soil Sci 162:27–32

Kahle P, Beuch S, Boelcke B et al. (2001) Cropping of *Miscanthus* in Central Europe: biomass production and influence on nutrients and soil organic matter. Eur J Agron 15:171–184

Kuzyakov Y, Domanski G (2000) Carbon input by plants into the soil. Review. J Plant Nutr Soil Sci 163:421–431

Larsen SU, Jorgensen U, Kjeldsen JB, Laerke PE (2014) Long-term *Miscanthus* yields influenced by location, genotype, row distance, fertilisation and harvest season. Bioenergy Res 7:620–635

Lesur C, Jeuffroy M-H, Makowski D et al. (2013) Modeling long-term yield trends of *Miscanthus x giganteus* using experimental data from across Europe. Field Crop Res 140:252–260

Lewandowski I, Clifton-Brown JC, Scurlock JMO, Huisman W (2000) Miscanthus: European experience with a novel energy crop. Biomass and Bioenergy 19:209–227

Lewandowski I, Scurlock JMO, Lindvall E, Christou M (2003) The development and current status of perennial rhizomatous grasses as energy crops in the US and Europe. Biomass and Bioenergy 25:335–361

Matthews RB, Grogan P (2001) Potential C sequestration rates under short-rotation coppice willow and *Miscanthus* biomass crops: a model study. Aspects Appl Biol 65:303–312

McCalmont JP, Hastings A, McNamara NP et al. (2015) Environmental costs and benefits of growing *Miscanthus* for bioenergy in the UK. GCB Bioenergy. doi:10.1111/gcbb.12294

Miguez FE, Maughan M, Bollero GA, Long SP (2012) Modelling spatial and dynamic variation in growth, yield, and yield stability of the bioenergy crops *Miscanthus x giganteus* and *Panicum virgatum* across the conterminous United States. GCB Bioenergy 4:509–520

Mishra U, Torn MS, Fingerman K (2013) *Miscanthus* biomass productivity within US croplands and its potential impact on soil organic carbon. GCB Bioenergy 5:391–399

Poeplau C, Don A (2014) Soil carbon changes under *Miscanthus* driven by C_4 accumulation and C_3 decomposition—toward a default sequestration function. GCB Bioenergy 6:327–338

Roberts WP, Chan KY (1990) Tillage-induced increases in carbon dioxide loss from soils. Soil Tillage Res 17:143–151

Zatta A, Clifton-Brown JC, Robson P et al. (2014) Land use change from C3 grassland to C4 *Miscanthus*: effects on soil carbon content and estimated mitigation benefit after six years. GCB Bioenergy 6:360–370

Zimmermann J, Dauber J, Jones MB (2012) Soil carbon sequestration during the establishment phase of *Miscanthus x giganteus*: a regional scale study on commercial farms using 13C natural abundance. GCB Bioenergy 4:453–461

Zimmermann J, Dondini M, Jones MB (2013) Assessing the impact of the establishment of *Miscanthus* on soil organic carbon on two contrasting land-use types in Ireland. Eur J Soil Sci 64:747–756

Chapter 5
Miscanthus-Complemented Grassland in Europe: Additional Source of Biomass for Bioenergy

O. Kalinina, U. Thumm, and I. Lewandowski

Introduction

Grasslands represent one of the largest habitat types globally (Graetz 1994; Scurlock and Hall 1998). The FAO estimates that they cover 52.5 million km², or 40.5 % of the Earth's landmass (FAO 2005). Grasslands contribute to the livelihoods of over 800 million people, serving as an important feed source for livestock, providing habitat for wildlife and contributing to the storage of water and conservation of plant genetic resources. In the European Union alone, grasslands cover over 30 % of the total agriculturally utilised area (EUROSTAT). Temperate grasslands make an important contribution to the biodiversity of agricultural and recreational landscapes. They are usually characterised by soils with high organic carbon content and thus act as natural carbon sinks (Adams et al. 1990; Flanagan et al. 2002; Sala et al. 1996; Scurlock and Hall 1998; Suyker and Verma 2001).

The majority of grasslands in Europe are traditionally extensively maintained through grazing or cutting (Isselstein et al. 2005). A large part of grassland biomass is used for dairy farming. However, during the last three decades there have been notable changes in grassland use. Enhanced animal performance with increased milk and meat production has been achieved through a shift in ruminant diet composition. The increase of concentrates in feed has led to a fall in the demand for roughage. At the same time, the productivity of managed grasslands has increased. It is now estimated that about 9–15 million ha of grassland in the European Union will be surplus to the needs of food production by 2020 (Prochnow et al. 2009a,b). According to various studies (Hartmann et al. 2011; Rösch et al. 2007), the area of surplus grassland in the EU represents about 13–22 % of permanent grassland.

O. Kalinina (✉) • U. Thumm • I. Lewandowski
Department of Biobased Products and Energy Crops (340b), Institute of Crop Science, University of Hohenheim, Fruwirthstraße 23, Stuttgart 70599, Germany
e-mail: olena.kalinina@uni-hohenheim.de

S. Barth et al. (eds.), *Perennial Biomass Crops for a Resource-Constrained World*, DOI 10.1007/978-3-319-44530-4_5

Changes in agricultural practices and land-use pressure have led to a decline in grasslands over the last decades. They are either converted to other land uses, such as maize production, or cease to be used for food production. In Germany, in particular, it is estimated that approximately 25 % of all grassland will not be utilised for food production in the near future due to the low nutritive value of the forage and the high costs of harvesting it (Federal Statistical Office 2008; Prochnow et al. 2007; Wachendorf et al. 2009). However, the conversion of grassland to more economically productive arable land destroys natural grassland ecosystems and increases carbon dioxide emissions. Grasslands on marginal, nutrient-poor soils are often especially rich in biodiversity and provide habitat for many endemic species. Therefore, loss or deterioration of such areas leads to a loss of important ecological services (Aguiar 2005; Beringer et al. 2011; Young et al. 2005). Fertilisation could increase biomass yields of grasslands on nutrient-poor soil, making them more attractive for use in animal husbandry or as forage, but would negatively impact biodiversity (Aguiar 2005; Isselstein et al. 2005; Hautier et al. 2009). Thus, there is a need to develop improved grassland management practices and sustainable solutions that effectively address both biodiversity protection and grassland profitability issues.

Adler et al. (2009) investigated the introduction of C4 grasses into C3 grass communities as an approach to improving the productivity of permanent grassland and maintaining or even increasing biodiversity and carbon stocks. In their studies, the increase in ratio of C4 perennial grasses was found to correlate with the increase of aboveground biomass yield in natural grassland ecosystems (Adler et al. 2009). In Minnesota, perennial grasslands with C3 and C4 grasses managed for the conservation of soil and biodiversity produced 0.5–5.7 t ha^{-1} biomass per year. The conversion efficiency of this biomass to ethanol (450 l t^{-1}) was similar to that of some dedicated bioenergy crops such as switchgrass (Jungers et al. 2013). Due to their complementary growth patterns (Seasonal Growth Pattern of Grasses 2015), the combination of the warm-season and cold-season species could help to increase annual biomass yield in grassland.

Miscanthus is currently the most productive perennial C4 grass in temperate climates and is one of the strongest candidate energy crops (Beale et al. 1996; Dohleman and Long 2009; Jones and Walsh 2001; Naidu et al. 2003; Wang et al. 2008). With its high water- and nutrient-use efficiency, perennial growth and highly effective C4 photosynthesis, miscanthus can be grown in a wide range of climatic conditions (Beale and Long 1997; Beale et al. 1999; Clifton-Brown et al. 2001; Lewandowski and Schmidt 2006; Lewandowski et al. 2000). In addition to its outstanding biomass yields, it can also have a number of environmental benefits (Clifton-Brown et al. 2007; Clifton-Brown et al. 2004; Hansen et al. 2004; Semere and Slater 2007a,b).

Besides yield increases, biomass quality aspects may also favour the introduction of miscanthus into C3 grasslands. The potential use of biomass is dependent on its composition. *Miscanthus* biomass is more suitable for combustion than the biomass of C3 grasses. *Miscanthus* is capable of producing a much higher dry matter yield than typical grasslands in temperate climates. Its lignin content is higher, but ash and nitrogen contents are usually lower than that of grassland biomass (Arnoult and Brancourt-Hulmel 2015; Baxter et al. 2012; Heinsoo et al. 2010; Kludze et al. 2013; Lewandowski and Kicherer 1997; Prochnow et al. 2005, 2007, 2009a,b). *Miscanthus* biomass is also known to have a lower moisture content (higher dry matter) than that of semi-natural grasslands in Europe (Baxter et al. 2014; Kludze et al. 2013).

This paper discusses options for the introduction of miscanthus into C3 grass communities with the aim of increasing overall biomass productivity without jeopardising biodiversity and carbon stocks. Based on a scientific literature review and our observations, we first provide a short overview of natural miscanthus grasslands, their current performance and uses, and then examine possibilities of introducing miscanthus into European grassland for bioenergy production.

Natural *Miscanthus* Grasslands and Their Current Uses

Miscanthus species are native to Japan, the Philippines, India, China, Malaysia and Polynesia (Greef and Deuter 1993; Ohwi 1964; Watson and Dallwitz 1992). Some miscanthus species also occur naturally in the Himalayas and southern Africa (Wikberg 1990).

Miscanthus grasslands are very common in the countries of miscanthus origin, in Japan and China (Hayashi et al. 1981; Jinno and Umeno 1995; Mutoh et al. 1968; Shimada et al. 1975; Tsuyuzaki 2004). Most naturally occurring miscanthus grasslands in Japan are of the species *Miscanthus sinensis* (Itow 1962; Mutoh et al. 1985; Naito and Nakagoshi 1995; Numata 1975). Other species—*Miscanthus sacchariflorus*, *Miscanthus sinensis* and *Miscanthus floridulus*—are found on dry grassy hillsides in southern regions of Japan, along the Pacific coast (Koyama 1987).

Miscanthus is very common in Japan, often growing near active volcanoes and on poor, dry soils. The naturally occurring *Miscanthus sinensis* grasslands are usually maintained by regular mowing or grazing to prevent forest succession (e.g., at the Kawatabi Experimental Farm in northern Honshu or in the Mt. Aso area in central Kyushu). *Miscanthus sinensis* dominated areas are usually cut once or twice a year (Asami et al. 1995). There have been multiple local studies of the biomass production (Hayashi et al. 1981; Iwaki et al. 1964; Kayama et al. 1972; Mutoh et al. 1968, 1985; Shimada et al. 1975; Tsuchida and Numata 1979) and carbon and nutrient dynamics in such grasslands (Iwaki et al. 1964; Iwata 1971; Ni 2002; Toma et al. 2012; Yazaki et al. 2004; Yamane and Sato 1960). These studies reported the largest contribution of *Miscanthus sinensis* (dominant species) to the above- and belowground biomass of the grassland communities. Hayashi et al. (1981) reported a maximum aboveground dry yield of natural *Miscanthus sinensis* grassland in central Japan of 3346 g biomass m^{-2} and a net production of the miscanthus grassland stand of 959 g biomass m^{-2} per year, with over 50 % of biomass allocated to the underground plant parts. Overall, miscanthus contributes the largest proportion of the biomass yield (e.g., over 96 %, according to studies in Japan and China), thus providing the major portion of harvestable dry matter yield in miscanthus-dominated grasslands (Hayashi et al. 1981; Mutoh et al. 1985).

In Japan, *Miscanthus sinensis* has traditionally been used as thatching material for roofs for many centuries (Otaki 1997, 1999). There, it has also often been used as a forage grass and organic fertiliser (Otaki 1997; Stewart et al. 2009). Its biomass is usually harvested in autumn and baled. Extensive management of wild *Miscanthus sinensis* as a forage grass and building material possibly began in Japan over 1000 years ago (Otaki 1999; Stewart et al. 2009). Nowadays, these grasslands are primarily managed by burning or mowing (Asami et al. 1995; Otaki 1999), although it has been suggested that they can also persist for a long time without any management on abandoned land (Sakanoue 2001).

China has the largest area of miscanthus worldwide. Here, several species (*Miscanthus sinensis*, *Miscanthus lutarioparia*, *Miscanthus floridulus* and *Miscanthus sacchariflorus*) are found widespread in natural grasslands and are also genetically highly diverse (Adati and Shiotani 1962; Nie et al. 2014; Xiao et al. 2013). Natural grasslands dominated by miscanthus not only cover large areas but are also highly productive. In 1990, the natural *Miscanthus floridulus* grassland in the Chinese Anhui province alone was reported as exceeding 173,340 ha, with the fresh yield per hectare reaching 37.5 t (Nanxian and Yunfeng 1990).

Using biomass from miscanthus grasslands for bioenergy is not uncommon in the countries where the species occurs naturally, but other uses are more dominant. In China, the biomass of naturally occurring miscanthus is successfully used for paper production and in the food industry (Xi 2000). It is used for forage and as building materials in both China and Japan. One recent study in Japan demonstrated that mixed miscanthus grassland biomass can be used for combustion and bioenergy production (New Energy 2014).

In China, the biomass of naturally growing miscanthus is at present seldom used for bioenergy production. Although bioenergy is at high demand in rural areas (where over 60 % of the country's population resides) and accounts for up to 76 % of total rural energy, mostly crop residues and fuelwood are currently used for combustion purposes (Sang and Zhu 2011).

Interestingly, naturalised *Miscanthus sinensis* and *Miscanthus sacchariflorus* can also be found in many parts of the USA. These once escaped from ornamental cultivation areas and now grow in a wide range of habitats, showing the ability of this perennial grass to establish and thrive within plant communities in temperate grasslands (Basak 2015; Dougherty et al. 2014; Horton et al. 2010). *Miscanthus* was recorded as an escaped species in the USA by Hitchcock (1950) and Pohl (1978) already many decades ago. In recent years, *Miscanthus sacchariflorus* has been found in the Midwestern United States, usually persisting through its large rhizomatous root system and spreading easily among the natural plant communities of the moist roadside ditches. Escaped *Miscanthus sinensis* has also been found near Washington, D.C., where it was growing in a wooded area along the roads in a former park. *Miscanthus sinensis*, which easily spreads by seed, is known to naturalise in Western North Carolina and Pennsylvania and can be found along many roadsides, especially in disturbed, sunny, dry sites (Meyer and Tchida 1999; Meyer 2015; Quinn et al. 2010). *Miscanthus sinensis* is also known to spread in and colonise some disturbed areas in Australia (http://weedsbluemountains.org.au).

These observations have raised concerns that non-indigenous miscanthus species may become invasive (Jørgensen 2011; Meyer et al. 2010; Raghu et al. 2006). They seem to spread in two ways: via viable seeds—mostly seen in *Miscanthus sinensis*—and via creeping, horizontal shoots in *Miscanthus sacchariflorus* (Meyer and Tchida 1999; Quinn et al. 2010). The triploid sterile *Miscanthus sinensis* x *sachariflorus* hybrid, *Miscanthus x giganteus*, is believed to pose a lower risk of escape because it can neither form viable seeds, nor does it have 'running' shoots.

Biodiversity in *Miscanthus* Plantations and Natural *Miscanthus* Grasslands

The biodiversity, life form composition and productivity of naturally occurring *Miscanthus sinensis* grasslands have been extensively studied in the countries of origin of this species, Japan and China (Hayashi et al. 1981; Mutoh et al. 1985; Nie et al. 2014; Shimada et al. 1975; Shimoda and Oikawa 2006).

The species composition of native miscanthus grasslands in Japan is highly diverse and varies depending on climate, soil type, topography, prior land use and management (Anten and Hirose 1999; Stewart et al. 2009). A number of studies have investigated the botanical composition and have found high biodiversity in natural *Miscanthus sinensis* grasslands (Anten and Hirose 1999; Hayashi et al. 1981; Nie et al. 2014; Shimada et al. 1975), although miscanthus was usually the dominant species and highest biomass contributor (Hayashi et al. 1981; Mutoh et al. 1985). *Miscanthus sinensis* grasslands on steep slopes were observed to have higher plant species diversity than grasslands on the valley floor in Japan (Koyanagi et al. 2008). According to Mutoh et al. (1985), *Miscanthus sinensis* growing in 15-year old patches in semi-natural grassland in Japan was forming completely the top folia stratum (1–2 m high) of the canopy and represented as much as 96.5 % of the total standing biomass. Ninety-six other species from 42 families were identified, indicating a high biodiversity in this miscanthus-dominated grassland (Mutoh et al. 1985). Another study in Japan reported as many as 29 species per m^2 in *Miscanthus sinensis* grassland (Hayashi et al. 1981).

The effects of cultivated miscanthus on biodiversity in temperate climates still need to be investigated comprehensively. Different literature sources report mainly beneficial effects of miscanthus plantations on wildlife and plant communities. However, whether positive or negative aspects of miscanthus introduction are reported often depends on the comparator used (e.g., crop, field margins or natural grassland) and previous land use of a particular site (Fargione et al. 2009; Northrup and Wittemyer 2013; Semere and Slater 2007a,b; United States Department of Agriculture 2011).

Fig. 5.1 Weeds growing under the miscanthus canopy of a 16-year-old field trial in south-west Germany

A United States Department of Agriculture (2011) assessment found newly planted stands of *Miscanthus x giganteus* to have a neutral effect on wildlife, but wildlife and insect diversity to be lower in mature miscanthus stands than in field borders. *Miscanthus* cultivation has been found to have positive effects on farmland bird and invertebrate populations (Bellamy et al. 2009; Semere and Slater 2007a). Semere and Slater (2007b) reported that, although a lower number of small birds and mammals was observed in miscanthus stands compared to field margins, young miscanthus stands were indeed richer in weed vegetation, birds and small mammals than reed canary grass or arable fields.

Our own results (unpublished data) showed a higher weed species diversity and abundance in younger (2-year-old) miscanthus stands than in mature (16-year-old) field trials in South-West Germany. Each of these trials included 15 different miscanthus genotypes. Species richness was found to correlate negatively with the density of the miscanthus stands and to be lower in mature miscanthus plantations. However, even the 16-year-old, dense *Miscanthus x giganteus* plantations supported up to 16 different weed species per 25 m^2 plot, with weeds covering up to 12 % of the plantation (Fig. 5.1). In young plantations, *Miscanthus sacchariflorus* plots had the highest canopy cover and supported the lowest number of plant species, compared to the plots of *Miscanthus sinensis*, *Miscanthus x giganteus* and *Miscanthus sinensis x sacchariflorus* hybrids. No such difference was observed in mature miscanthus stands, where the canopy cover reached 80–100 % for most genotypes.

With regard to the maintenance of diversity in C3 grasslands, it can be concluded that the most beneficial miscanthus genotypes are those that do not outcompete other species by rapid canopy closure. Also, genotypes that spread by seed or running shoots should be avoided. This does not exclude *Miscanthus sinensis* genotypes per se but restricts their use to those genotypes that form seeds so late that ripening will not succeed. Observations in naturally occurring grasslands in Asia show that miscanthus certainly has the potential to dominate these grassland communities (Hayashi et al. 1981; Mutoh et al. 1985).

Competition between miscanthus and other grassland species can potentially be controlled by an adjusted cutting management. C3 grassland plants are usually less affected by multiple or early cuts than C4 grasses due to their earlier development and lower carbohydrate storage requirements. This means that the C3/C4 ratio can be controlled depending on the function and purpose of the biomass production system (provisioning and supporting services). To introduce C4 grasses into permanent grassland by seed, some level of disturbance is necessary (Doll et al. 2011).

Establishment of *Miscanthus* in European Grasslands: Experience in Germany

Once established into C3 grasslands, miscanthus has to compete to survive. Therefore, the challenge is to find a successful method of introduction which has no negative impact on biodiversity.

For this purpose, we established miscanthus on two semi-natural grassland sites in south-west Germany that contrasted in soil nitrogen status (low and high nitrogen availability) using a no-till method (without destroying the grassland). We monitored the survival of miscanthus and biomass yields over 4 years. Five genotypes of *Miscanthus sinensis* and *Miscanthus sacchariflorus* with different morphological properties were planted, and two different establishment regimes were tested: with no preceding herbicide application to the grassland and with herbicide application in narrow strips of 20 cm. In addition, three different mowing regimes were applied: harvesting the biomass once, twice and three times per growing season (Fig. 5.2). The results obtained showed that herbicide application before the introduction of the miscanthus had only minor and short-lived (1st season) effects on its growth. Although the increase in overall biomass yield was, for the most part, negligible in the first years after establishment, the data suggest that, with appropriate genotype selection and higher miscanthus density in grassland, the biomass quality and possibly also the yield of the grassland could be improved.

Interestingly, the miscanthus showed high establishment success, varying from 65 to 100 % for the different genotypes, even though grassland competition was not completely eliminated. This challenges the view that miscanthus is a rather weak competitor and that intensive initial weed management is necessary to allow its proper establishment. Although the yield and the growth form of the miscanthus differed (as expected) from that on open agricultural fields, it was possible to maintain miscanthus in grassland under European conditions without chemical input, with a regular mowing management.

Fig. 5.2 *Miscanthus* in semi-natural grassland in south-west Germany: the three mowing regimes are clearly visible

In our field trials, the introduction of miscanthus had no apparent effect on the species composition of the grassland community within the first 4 years after establishment. In these trials, however, the miscanthus remained at a low density and accounted for only 3–5 % of the total dry biomass. The effects on biodiversity could be different if with time it became a dominant species at a specific site or if it was planted at a higher density. The long-term effects of the introduction of miscanthus into grassland on plant communities and soil carbon content and also its potential to escape and spread outside the area of cultivation need to be carefully assessed in the future.

Conclusions

In the countries of its origin, natural miscanthus grassland is used as a biomass resource for bioenergy and materials. Natural miscanthus grasslands are known to maintain high biodiversity and serve as efficient carbon sinks. Given the higher yield and better biomass quality of miscanthus than biomass from semi-natural grasslands, the addition of miscanthus to the grassland species community could potentially improve biomass quality, e.g., for combustion purposes. The possibilities of establishing miscanthus grasslands in European climatic conditions need to be further investigated. Semi-natural, mixed miscanthus grasslands could potentially serve both as biodiversity hotspots and a bioenergy resource, on the condition that low-destructive establishment methods (for miscanthus establishment in grassland) are applied and heavy chemical applications (fertilisation, herbicides) can be avoided.

Acknowledgement The work reported here was done as a part of the OPTIMISC (Optimising miscanthus biomass production) project, an international research project funded by the European Union's Seventh Programme for research, technological development and demonstration under the grant agreement FP7-289159.

References

Adams JM, Faure H, Faure-Denard L, McGlade JM, Woodward FI (1990) Increases in terrestrial carbon storage from the Last Glacial Maximum to the present. Nature 348(6303):711–714

Adati S, Shiotani I (1962) The cytotaxonomy of the genus *Miscanthus* and its phylogenic status. Bulletin Faculty Agric Mie University 25:1–24

Adler PR, Sanderson MA, Weimer PJ, Vogel KP (2009) Plant species composition and biofuel yields of conservation grasslands. Ecol Appl 19(8):2202–2209

Aguiar MR (2005) Biodiversity in grasslands: current changes and scenarios for the future. In: Reynolds SG, Frame J (eds) Grasslands: developments opportunities perspectives. FAO and Science, pp 261–280

Anten NPR, Hirose T (1999) Interspecific differences in above-ground growth patterns result in spatial and temporal partitioning of light among species in a tall-grass meadow. J Ecol 87: 583–597

Arnoult S, Brancourt-Hulmel M (2015) A review on miscanthus biomass production and composition for bioenergy use: Genotypic and environmental variability and implications for breeding. Bioenergy Res 8(2):502–526

Asami K, Hattori T, Akamatsu H (1995) A study on management by cutting of the embankment vegetation. The Japanese Institute of Landscape Architecture 58(5):125–128

Basak S (2015) Genetic diversity of miscanthus in the US naturalized populations based on molecular marker. College of Agriculture, Environment and Nutrition Sciences Graduate Theses and Dissertations. Paper 2

Baxter XC, Darvell LI, Jones JM, Barraclough T, Yates NE, Shield I (2012) Study of *Miscanthus x giganteus* ash composition—variation with agronomy and assessment method. Fuel 95: 50–62

Baxter XC, Darvell LI, Jones JM, Barraclough T, Yates NE, Shield I (2014) *Miscanthus* combustion properties and variations with *Miscanthus* agronomy. Fuel 117:851–869

Beale CV, Bint DA, Long SP (1996) Leaf photosynthesis in the C4-grass Miscanthus × giganteus, growing in the cool temperate climate of southern England. J Exp Bot 47(2):267–273

Beale CV, Long SP (1997) Seasonal dynamics of nutrient accumulation and partitioning in the perennial C 4-grasses Miscanthus × giganteus and Spartina cynosuroides. Biomass Bioenergy 12(6):419–428

Beale CV, Morison JI, Long SP (1999) Water use efficiency of C4 perennial grasses in a temperate climate. Agric For Meteorol 96(1):103–115

Bellamy PE, Croxton PJ, Heard MS, Hinsley SA, Hulmes L, Hulmes S, Nuttall P, Pywell RF, Rothery P (2009) The impact of growing miscanthus for biomass on farmland bird populations. Biomass and Bioenergy 33(2):191–199. doi:10.1016/j.biombioe.2008.07.001

Beringer T, Lucht W, Schaphoff S (2011) Bioenergy production potential of global biomass plantations under environmental and agricultural constraints. GCB Bioenergy 3(4):299–312

Clifton-Brown JC, Lewandowski I, Andersson B, Basch G, Christian DG, Kjeldsen JB, Jørgensen U, Mortensen J, Riche AB, Schwarz KU, Tayebi K, Teixeira F (2001) Performance of 15 *Miscanthus* genotypes at five sites in Europe. Agron J 93:1013–1019

Clifton-Brown JC, Stampfl PF, Jones MB (2004) *Miscanthus* biomass production for energy in Europe and its potential contribution to decreasing fossil fuel carbon emissions. Glob Chang Biol 10(4):509–518

Clifton-Brown JC, Breuer J, Jones MB (2007) Carbon mitigation by the energy crop, *Miscanthus*. Glob Chang Biol 13(11):2296–2307

Dohleman FG, Long SP (2009) More productive than maize in the Midwest: how does Miscanthus do it? Plant Physiol 150(4):2104–2115

Doll JE, Haubensak KA, Bouressa EL, Jackson RD (2011) Testing disturbance, seeding time, and soil amendments for establishing native warm-season grasses in non-native cool-season pasture. Restoration Ecol 19(101):1–8. doi:10.1111/j.1526-100X.2010.00687.x

Dougherty RF, Quinn LD, Endres AB, Voigt TB, Barney JN (2014) Natural history survey of the ornamental grass *Miscanthus sinensis* in the introduced range. Invasive Plant Sci Manage 7(1): 113–120

FAO (2005) Grasslands of the world. In: Suttie JM, Reynolds SG, Batello C (eds). Plant production and protection series 34. http://www.fao.org/docrep/008/y8344e/y8344e00.htm. Accessed 10 Nov 2015

Fargione JE, Cooper TR, Flaspohler DJ, Hill J, Lehman C, McCoy T, McLeod S, Nelso EJ, Oberhauser KS, Tilman D (2009) Bioenergy and wildlife: threats and opportunities for grassland conservation. Bioscience 59(9):767–777

Flanagan LB, Wever LA, Carlson PJ (2002) Seasonal and interannual variation in carbon dioxide exchange and carbon balance in a northern temperate grassland. Glob Chang Biol 8(7):599–615

Federal Statistical Office (2008) Statistical Yearbook 2008 for the Federal Republic of Germany. Federal Statistical Office, Wiesbaden, Germany

Graetz D (1994) Grasslands. Changes in land use and land cover: a global perspective 125–147

Greef JM, Deuter M (1993) Syntaxonomy of *Miscanthus* x *giganteus* GREEF et DEU. Angewandte Botanik 67:87–90

Hansen EM, Christensen BT, Jensen LS, Kristensen K (2004) Carbon sequestration in soil beneath long-term *Miscanthus* plantations as determined by $_{13}C$ abundance. Biomass and Bioenergy 26(2):97–105

Hartmann S, Diepolder M, Lichti F (2011) Grünland als Biogassubstrat. Biogas Fachforum Bayern (I-12/2011)

Hautier Y, Niklaus PA, Hector A (2009) Competition for light causes plant biodiversity loss after eutrophication. Science 324(5927):636–638

Hayashi I, Hishinuma Y, Yamasawa W (1981) Structure and function of *Miscanthus sinensis* grasslands in Sugadaira, Central Japan. Vegetatio 48:17–25

Heinsoo K, Melts I, Sammul M, Holm B (2010) The potential of Estonian semi-natural grasslands for bioenergy production. Agric Ecosyst Environ 137(1–2):86–92. doi:10.1016/j.agee.2010.01.003

Hitchcock AS (1950) Manual of the grasses of the United States. USDA Misc. Publ. 200. Reprinted in 1971 by Dover Publ, New York

Horton JL, Fortner R, Goklany M (2010) Photosynthetic characteristics of the C_4 invasive exotic grass *Miscanthus sinensis* Andersson growing along gradients of light intensity in the southeastern United States. Castanea 75:52–66

Isselstein J, Jeangros B, Pavlu V (2005) Agronomic aspects of biodiversity targeted management of temperate grasslands in Europe—a review. Agron Res 3(2):139–151

Itow S (1962) Grassland vegetation in uplands of western Honshu. Japan Part I. Distribution of grassland. Jpn J Ecol 12:123–129

Iwaki H, Midorikawa B, Hogetsu K (1964) Studies on the productivity and nutrient element circulation in Kiragamine Grassland, central Japan. II. Seasonal change in standing crop. Botanical Magazine, Tokyo 77:447–457

Iwata E (1971) Ecological studies on the secondary vegetation of Kitakami Mountain Range, with special reference to grassland vegetation. Res Bulletin Faculty Agriculture, Gifu University 30:288–430

Jinno N, Umeno M (1995) Soil properties at habitats of gramineous plants (*Phragmites communis*, *P. japonica*, *Phacelurus latifolius*, *Miscanthus sacchariflorus*, *M. sinensis*) on Honmyo River water system in Isahaya City, Nagasaki Prefecture. Science Bulletin of the Faculty of Education. Nagasaki University 53:19–26

Jones MB, Walsh M (2001) *Miscanthus* for energy and fibre. James and James (Science Publishers), Berlin

Jørgensen U (2011) Benefits versus risks of growing biofuel crops: the case of *Miscanthus*. Curr Opin Environ Sustain 3(1–2):24–30

Jungers JM, Fargione JE, Sheaffer CC, Wyse DL, Lehman C (2013) Energy potential of biomass from conservation grasslands in Minnesota, USA. PLoS One 8(4), e61209. doi:10.1371/journal.pone.0061209

Kludze H, Deen B, Dutta A (2013) Impact of agronomic treatments on fuel characteristics of herbaceous biomass for combustion. Fuel Process Technol 109:96–102

Kayama R, Yano N, Fujimoto M (1972) Studies on the relationship between Miscanthus sinensis community and soil – (3) Seasonal variation of production of Miscanthus sinensis grassland and the soil conditions. Jpn J Ecol 22(4):151–161

Koyama T (1987) Grasses of Japan and its neighboring regions: an identification manual. Kodansha Ltd, Tokyo

Koyanagi T, Kusumoto Y, Yamamoto S, Takeuchi K (2008) Comparison of the relationships between seed bank and species compositions of Miscanthus sinensis grassland located on different topographical conditions. J Jpn Inst Landsc Architect 71:589–592 (in Japanese with English summary)

Lewandowski I, Kicherer A (1997) Combustion quality of biomass: practical relevance and experiments to modify the biomass quality of *Miscanthus giganteus*. Eur J Agron 6:163–177

Lewandowski I, Clifton-Brown JC, Scurlock JMO, Huisman W (2000) *Miscanthus*: European experience with a novel energy crop. Biomass and Bioenergy 19:209–227

Lewandowski I, Schmidt U (2006) Nitrogen, energy and land use efficiencies of miscanthus, reed canary grass and triticale as determined by the boundary line approach. Agric Ecosyst Environ 112(4):335–346

Meyer M (2015) Department of Horticultural Science, University of Minnesota. http://miscanthus.cfans.umn.edu/habitat. Accessed 10 Nov 2015

Meyer M, Tchida C (1999) *Miscanthus* Anderss. produces viable seed in four USDA hardiness zones. J Environ Hort 17(3):137–140

Meyer MH, Paul J, Anderson NO (2010) Competitive ability of invasive *Miscanthus* biotypes with aggressive switchgrass. Biol Invasions 12(11):3809–3816

Mutoh N, Yoshida KH, Yokoi Y, Kimura M, Hogetsu K (1968) Studies on the production processes and net production of *Miscanthus sacchariflorus* community. Jpn J Botany 20:67–92

Mutoh N, Kimura M, Oshima Y, Iwaki H (1985) Species diversity and primary productivity in *Miscanthus sinensis* grasslands. 1. Diversity in relation to stand structure and dominance. Botanical Magazine Tokyo 98:159–170

Naidu SL, Moose SP, Al-Shoaibi AK, Raines CA, Long SP (2003) Cold tolerance of C4 photosynthesis inMiscanthus×giganteus: adaptation in amounts and sequence of C4 photosynthetic enzymes. Plant Physiol 132(3):1688–1697

Naito K, Nakagoshi N (1995) The conservation ecology of *Iris rossii* Baker (*Iridaceae*), a threatened plant in rural Japan. J Plant Res 108:477–482

Nanxian Z, Yunfeng X (1990) Plant resources and its exploitation, utilization of the genus Miscanthus in Anhui province. J Wuhan Bot Res 4:010

New Energy Foundation (2009) Electric power business using grass—first in Japan. Asian Biomass Office Web. http://www.asiabiomass.jp/english/topics/090216_06.html. Accessed on 10 Apr 2014

Ni J (2002) Carbon storage in grasslands of China. J Arid Environ 50(2):205–218

Nie G, Zhang X-Q, Huang L-K, Xu W-Z, Wang J-P, Zhang Y-W, Ma X, Yan Y-H, Yan H-D (2014) Genetic variability and population structure of the potential bioenergy crop *Miscanthus sinensis* (Poaceae) in Southwest China based on SRAP markers. Molecules 19:12881–12897

Northrup JM, Wittemyer G (2013) Characterising the impacts of emerging energy development on wildlife, with an eye towards mitigation. Ecol Lett 16(1):112–125

Numata M (1975) Introduction: process and results of the Japanese grassland project. In: Numata M (ed) Ecological Studies in Japanese Grasslands with Special Reference to the IBP Area. Productivity of Terrestrial Communities, University of Tokyo Press, Tokyo, 1–7

Ohwi J (1964) Flora of Japan. Smithsonian Inst, Washington, DC

Otaki N (1997) Grassland and people's life. History of Ichinomiya Town. In: Editorial Board Committee of Ichinomiya Town (ed) Nature and Culture Anthology in Aso, Ichinomiya, Japan, 42–45

Otaki N (1999) Aso grassland for a thousand years. Burning stopped, grassland endangered. Environ Res Q 114:31–36

Pohl RW (1978) How to know the grasses. Wm. C. Brown Co. Pub., Dubuque.—Technical keys and information on grasses
Prochnow A, Heiermann M, Drenckhan A, Schelle H (2005) Seasonal pattern of biomethanisation of grass from landscape management. In: Agricultural Engineering International: the CIGR E-journal, Vol. VII, Manuscript EE 05 011. http://cigrjournal.org/index.php/Ejournal. Accessed 2 Dec 2015
Prochnow A, Heiermann M, Idler C, Linke B, Mähnert P, Plöchl M (2007) Biogas vom grünland: potenziale und erträge (Biogas from grassland: potential and yield). In: Grünlandverband D (ed) Gas aus Gras und was noch? 1/2007. Deutscher Grünlandverband, Berlin, Germany, pp 11–22
Prochnow A, Heiermann M, Plöchl A, Linke B, Idler C, Amon T, Hobbs PJ (2009a) Bioenergy from permanent grassland—a review: 2. Combust Bioresour Technol 100(21):4945–4954
Prochnow A, Heiermann M, Plöchl A, Linke B, Idler C, Amon T, Hobbs PJ (2009b) Bioenergy from permanent grassland—a review: 1. Biogas Bioresour Technol 100(21):4931–4944
Quinn L, Allen D, Stewart J (2010) Invasiveness potential of *Miscanthus sinensis*: implications for bioenergy production in the United States. GCB Bioenergy 2:310–320. doi:10.1111/j.1757-1707.2010.01062.x
Raghu S, Anderson RC, Daehler CC, Davis AS, Wiedenmann RN, Simberloff D, Mack RN (2006) Adding biofuels to the invasive species fire? Science 313:1742
Rösch C, Raab K, Skarka J, Stelzer V (2007) Energie aus dem Grünland—eine nachhaltige Entwicklung? Wissenschaftliche Berichte FZKA, 7333
Sakanoue S (2001) Long-term trends in *Miscanthus sinensis* grassland vegetation: a 20-year field observation. Grassland Sci 47:430–435
Sala OE, Lauenroth WK, Burke IC (1996) Carbon budgets of temperate grasslands and the effects of global change. Scope-Scientific Committee on problems of the environment. Int Council Sci Unions 56:101–119
Sang TAO, Zhu W (2011) China's bioenergy potential. GCB Bioenergy 3(2):79–90
Scurlock JMO, Hall DO (1998) The global carbon sink: a grassland perspective. Glob Chang Biol 4:229–233
Seasonal Growth Pattern of Grasses (2015) The UC Guide to Healthy Lawns. Agriculture and Natural resources. University of California. http://www.ipm.ucdavis.edu/TOOLS/TURF/ESTABLISH/seasongrth.html. Accessed 18 Dec 2015
Semere T, Slater FM (2007a) Invertebrate populations in miscanthus (*Miscanthus × giganteus*) and reed canary-grass (*Phalaris arundinacea*) fields. Biomass and Bioenergy 31(1):30–39
Semere T, Slater FM (2007b) Ground flora, small mammal and bird species diversity in miscanthus (*Miscanthus × giganteus*) and reed canary-grass (*Phalaris arundinacea*) fields. Biomass and Bioenergy 31(1):20–29
Shimada Y, Iwaki H, Midorikawa B, Ohga N (1975) Primary productivity of the *Miscanthus sinensis* community at the Kawatahi IBP area—Standing crop of aboveground parts. In: Numata M (ed) Ecological studies in Japanese grasslands with special reference to the IBP Area. Univ. Tokyo Press, Tokyo. JIBP Synthesis 13, 110–114
Shimoda S, Oikawa T (2006) Temporal and spatial variations of canopy temperature over a C_3-C_4 mixture grassland. Hydrol Process 20:3503–3516
Stewart JR, Toma Y, Fernández FG, Nishiwaki A, Yamada T, Bollero G (2009) The ecology and agronomy of *Miscanthus sinensis*, a species important to bioenergy crop development, in its native range in Japan: a review. GCB Bioenergy 1:126–153. doi:10.1111/j.1757-1707.2009.01010.x
Suyker AE, Verma SB (2001) Year-round observations of the net ecosystem exchange of carbon dioxide in a native tallgrass prairie. Glob Chang Biol 7(3):279–289
Toma Y, Armstrong K, Ryan Stewart J, Yamada T, Nishiwaki A, Fernández FG (2012) Carbon sequestration in soil in a semi-natural *Miscanthus sinensis* grassland and *Cryptomeria japonica* forest plantation in Aso, Kumamoto, Japan. GCB Bioenergy 4(5):566–575
Tsuchida K, Numata M (1979) Relationships between successional situation and production of *Miscanthus* sinensis communities at Kirigamine Heights, central Japan. Vegetatio 39:15–23

United States Department of Agriculture (2011) Technical note on 4 July 2011. Natural Resources Conservation Service. Plant Materials Program

Wachendorf M, Richter F, Fricke T, Graß R, Neff R (2009) Utilization of semi-natural grassland through integrated generation of solid fuel and biogas from biomass. I. Effects of hydrothermal conditioning and mechanical dehydration on mass flows of organic and mineral plant compounds, and nutrient balances. Grass Forage Sci 64(2):132–143

Wang D, Portis AR, Moose SP, Long SP (2008) Cool C4 photosynthesis: pyruvate Pi dikinase expression and activity corresponds to the exceptional cold tolerance of carbon assimilation in Miscanthus × giganteus. Plant Physiol 148(1):557–567

Watson L, Dallwitz MJ (1992) The grass genera of the world. CAB international

Wikberg S (1990) The genus Miscanthus. A summary of available literature. Projekt Agro-Fiber (Sweden)

Xi Q (2000) Investigation on the distribution and potential of giant grasses in China: *Triarrhena, Miscanthus, Arundo, Phragmites* and *Neyraudia*. Cuvillier Verlag

Xiao L, Jiang JX, Yi ZL, Ai X, Qin JP, Liu SL, Lin C (2013) A study on phenotypic diversity of *Miscanthus sinensis* natural population in Guangxi province [J]. Acta Prataculturae Sinica 4:006

Yamane I, Sato K (1960) Distribution of organic and inorganic constituents in soil profiles of mountain grassland and their investment from plant to soil. Bulletin of the Agricultural Research Center, Tohoku University 11:447–458

Yazaki Y, Mariko S, Koizumi H (2004) Carbon dynamics and budget in a *Miscanthus sinensis* grassland in Japan. Ecol Res 19(5):511–520

Young J, Watt A, Nowicki P, Alard D, Clitherow J, Henle K, Johnson R, Laczko E, McCracken D, Matouch S, Niemela J, Richards C (2005) Towards sustainable land use: identifying and managing the conflicts between human activities and biodiversity conservation in Europe. Biodivers Conserv 14(7):1641–1661. doi:10.1007/s10531-004-0536-z

Part II
European Regional Examples for the Use of Perennial Crops for Bioenergy

Chapter 6
"Soranovskii": A New Miscanthus Cultivar Developed in Russia

T. Goryachkovskaya, N. Slynko, E. Golubeva, S.V. Shekhovtsov, N. Nechiporenko, S. Veprev, I. Meshcheryakova, K. Starostin, N. Burmakina, A. Bryanskaya, N. Kolchanov, V. Shumny, and S.E. Peltek

Introduction

Wood is currently the only industrial cellulose source in Russia. It is about 40–44 % cellulose (Hames 2009). Forests accumulate biomass over decades and are ecologically important for the biosphere. The introduction of novel plant species with high biomass yields may be a promising method for producing cellulose-containing biomass and source materials for green chemistry and bioenergy. Technologies for producing bioethanol, protein concentrates, and technical fibers have already been developed. *Miscanthus × giganteus* is currently a promising crop for relevant raw materials. Miscanthus grows rapidely: while forests produce 100 t per ha of plant biomass in 30 years, miscanthus plantations typically yield 250 t per ha during that period (Shumny et al. 2010). There are three miscanthus species found in the Russian Far East: *Miscanthus sacchariflorus*, *M. purpurascens*, and *M. sinensis*. The climate of West Siberia is too cold for *Miscanthus ×* giganteus, so we developed the Soranovskii cultivar based on Far Eastern plants. It can be successfully cultivated in West Siberia (Shumny et al. 2010). However, it does not form seeds so its propagation relies on rhizomes. Mitochondrial DNA sequences of this species belong to *M. sacchariflorus*.

From the biomass of miscanthus cultivar Soranovskii we obtained experimental samples of cellulose and bioethanol. Currently, we plan to investigate two aspects of "green chemistry": to produce carbohydrate-containing substrate to be used as microbial medium, and to search for cellulosolytic microorganisms that can grow

T. Goryachkovskaya • N. Slynko • E. Golubeva • S.V. Shekhovtsov • N. Nechiporenko
S. Veprev • I. Meshcheryakova • K. Starostin • N. Burmakina • A. Bryanskaya
N. Kolchanov • V. Shumny • S.E. Peltek (✉)
The Federal Research Center Institute of Cytology and Genetics, The Siberian Branch of the Russian Academy of Sciences, Lavrentieva, 10, Novosibirsk 630090, Russia
e-mail: peltek@bionet.nsc.ru

S. Barth et al. (eds.), *Perennial Biomass Crops for a Resource-Constrained World*, DOI 10.1007/978-3-319-44530-4_6

on unhydrolyzed milled miscanthus biomass. These methods may allow us to create technologies for production of bioethanol, lactic acid, and other products of microbial processing.

Cultivating Miscanthus in West Siberia

We demonstrated that miscanthus cultivar Soranovskii can yield 10–15 t/ha of stem biomass per year in West Siberia. Two or three years are required to reach plantation maturity in terms of yield potential after which the plantation yields at this level for 12–17 years. The cultivar arises from a collection of Far Eastern miscanthus specimens propagated and tested at Novosibirsk. In this collection, we found an unusual form with very long rhizomes that we expected could produce a plantation that establishes rapidly. This observation led to the development of the cultivar "Soranovskii" that has been successfully cultivated in West Siberia for over 10 years on experimental field exceeding 5 ha.

Soranovskii does not produce seeds and so is propagated using rhizomes. Rhizomes of the Soranovskii rapidly colonize soil forming a continuous plantation, a process that takes 3 years. Soranovskii can grow on a wide variety of soils, but requires fertilizers. It reaches the productivity peak by the third year, and may be cropped for 15–20 years. Dry biomass yield was 10–16 t/ha on trial fields. Absence of sexual reproduction significantly impedes selection process and increases propagation costs.

For cultivation, rhizomes were placed into furrows with an interval of 60 cm between the rows. After 3 years, the resulting plantation is 2–2.5 m high with stem density of 200–220/m^2. Rhizomes are 5–20 deep with the total length of 60–65 m per m^2, forming a uniform net. Over a meter of rhizome containing growth buds may be formed during a season. This propagation method is simple, as rhizomes form a compact ribbon and may be easily separated from each other. Soranovskii rhizomes have a high regeneration capacity. Rhizomes can be cut into short (5–10 cm) pieces with growth buds to increase propagation coefficient. Using vegetative propagation we managed to achieve the propagation coefficient of 1:40.

Identification of the Parent Species of the Miscanthus Cultivar Soranovskii

Miscanthus cultivar Soranovskii has morphological features of both *M. sinensis* and *M. sacchariflorus* (creeping rhizomes, reddish tuft hairs). We used plastid sequences to investigate the origin of this cultivar. Based on miscanthus sequences available in GenBank, we designed primers for amplification of a 202 bp fragment that included the intergenic spacer between the plastid tRNA-Leu and tRNA-Phe genes. This short fragment contains two polymorphic positions that differ between *M. sinensis* and *M. sacchariflorus*. The DNA sequence Soranovskii (KU847994) was identical to the corresponding sequences of *M. sacchariflorus* from GenBank (LC060108-LC060114, LC060116, AJ426569, AY116246-AY116247, HQ822042, N544243-JN544248,JN642297-JN642300),as well as to *M.×giganteus* (LC060123-LC060126, AJ426567-AJ426568) and *M. lutarioriparius* (JN642302-JN642303). It is known that *M.* × *giganteus* is an allotriploid hybrid of *M. sacchariflorus* and *M. sinensis* and contains the plastid genome of *M. sacchariflorus* (Hodkinson et al. 2002a, 2015). *M. lutarioparius* is one of the synonyms of *M. sacchariflorus spp. lutarioparius*, a subspecies of *M. sacchariflorus* (Sun et al. 2010). *M. changii* is also referred to as *M. longiberbis var. changii*. We failed to find its systematic affinity to *M. sacchariflorus* and *M. sinensis*.

The sequence we obtained for Soranovskii had a T>G transition in position 60 compared to *M. sinensis* AB622625. It differed from all other *M. sinensis* sequences by two substitutions, T>G in position 60, and A>T in position 183, as well as *M. oligostachyus*, *M. junceus*, and *M. floridulus* (AB622623, AB622624, AB622626-AB622628, AJ426570-AJ426573, EU434104, GQ870006, JN544253, JN544254, JN642289-JN642291).

M. oligostachyus, *M. junceus*, and *M. floridulus* are closely related to *M. sinensis* and *M. sacchariflorus*; *M. oligostachyus* is also sometimes referred to as *Miscanthus sinensis var. purpurascens*. According to molecular phylogenetic analysis (Hodkinson et al. 2002b), representatives of this group do not form separate branches on the phylogenetic tree. In addition, systematics of this group is impeded by cases of incorrect identification.

Therefore, based on plastid DNA, miscanthus cultivar Soranovskii is either *M. sacchariflorus* or a hybrid of *M. sacchariflorus* and some other species.

Biomass Compounds

Plant biomass contains various compounds such as cellulose, hemicellulose, starch, pectins, lignin, protein, and ash. These components are found in all plants, but in various proportions (Hames 2009). Extractives include all substances found outside plant cells that do not form cell walls. They can be easily extracted using water or organic solvents (Aristidou and Penttilä 2000). Biomass of any plant can be used to obtain cellulose or technical fibers. Extraction of some other useful substances is also feasible: inulin, rubisco, enzymes, lipids and fatty acids, pigments, organic acids, amino acids, and potassium fertilizers (Fowler et al. 2003). Plants have varying composition of plant cell wall polysaccharides and extractives, and so may be suitable for food, forage, or bioenergetics. To produce ethanol from plant biomass containing no simple sugars, it is necessary to hydrolyze plant cell wall polysaccharides (Henry 2010; Wi et al. 2015).

The length of cellulose chains depends on the number of anhydroglucose monomers (polymerization level), which depends on the species and extraction process. Cellulose produced from wood has cellulose chain length about 300–1700, while

Table 6.1 Miscanthus biomass composition (%)

Component	Miscanthus "*Soranovskii*"	*Miscanthus x giganteus* (Lee and Kuan 2015)
Hemicellulose	26	23.5
Cellulose	43.8	41.1
Total lignin	22.8	21.7

that produced from cotton has about 800–10,000 depending on treatment procedure. Treatment by acids with subsequent hydrolysis with cellulases may completely process cellulose to D-glucose, while treatment with hydrochloric acid at 105 °C yields microcrystalline cellulose that is widely used in chemical, pharmaceutical, and paint industries. Glucan with polymerization level of 20–30 has all properties of cellulose.

Miscanthus biomass is often used as bedding for farm animals because it does not form much dust, it absorbs water, and it is resilient to bacterial degradation. It is also promising for the paper industry (Fowler et al. 2003) and is used a source for producing fiberboard and other construction materials (Salvadó et al. 2003).

A comparison of biomass components of miscanthus cultivar Soranovskii and *Miscanthus × giganteus* is given in Table 6.1.

Biomass Saccharification

Plant cell wall consists of cellulose fibers cross-linked by lignin bound to hemicellulose. Modern methods of fiber destruction mostly employ physicochemical methods: fine milling, steam explosion, chemical delignification, or a combination of these. These processes are expensive, inefficient, and produce large amounts of toxic substances that impede subsequent processing by microorganisms.

We assessed the perspective of using commercial cellulosolytic enzymes for saccharification of miscanthus cultivar Soranovskii, as well as other species: *Phalaris arundinacea*, *Thrachomitum lancifolium*, and *Sida hermaphrodita*. Soranovskii contains 44 % of cellulose; *P. arundinacea*, *44.2 %*; *T. lancifolium*, up to *70 %*; and *S. hermaphrodita*, *40 %*.

Biomass hydrolysis by microorganisms is a crucial stage in the global carbon cycle. Cellulose is weakly accessible to microorganisms due to its low water solubility. There are at least five mechanisms used by microorganisms to hydrolyze cellulose. Each of them is based on enzymatic depolymerization of plant cell wall polysaccharides. Extracellular localization of enzyme complexes imposes its limitations: cellulolytic proteins must be resilient to proteolysis. Products of cellulose hydrolysis are suitable substrates for other microorganisms that create very diverse communities. Efficient hydrolysis of cellulose requires combined action of at least three enzyme types: endoglucanase (1,4-β-D-glucan glucohydrolase, EC 3.2.1.4), exoglucanase (1,4-β-D-glucan cellobiohydrolase, EC 3.2.1.91), and β-glucosidase (β-D-glucosid glucohydrolase, EC3.2.1.21). Cellulolytic systems of fungi (e.g., *Trichoderma reesei*) contain all three main activities (Mukherjee et al. 2012; Dashtban et al. 2011). Endoglucanases attack amorphous zones of cellulose fibers, forming gaps for exoglucanases, which can split off cellobiose from more crystallized areas. In the end, β-glucosidases hydrolyze cellobiose.

For enzymatic hydrolysis of grass biomass we used fungal commercial enzymes: Cellolux A and Cellolux F (NPO Sibbiopharm), a xylanase from *Thermomyces lanuginosus*, a cellulase from *Aspergillus niger* (Sigma), as well as Cellobiase F10 and Cellulase B1 from *Pen. verruculosum*, kindly provided by A.P. Sinitsyn.

Biomass was milled using the MAN-30 chopper (ZAO MVM, Russia). The powder was mixed with water at 1:10 ratio (g/ml). Prior to enzymatic hydrolysis we treated the samples with alkaline peroxide (1 % $Ca(OH)_2$ and 4 % hydrogen peroxide) at 100 °C. The main goal of this stage was to dissolve hemicellulose and make cellulose more accessible to enzymes, as enzyme complexes require pretreatments for complete hydrolysis. Total amount of reducing sugars was determined using the colorimetric method using 3,5-dinitrosalycilic acid. Enzymatic hydrolysis was performed for 72 h for all specimens. The content of enzyme mixtures and the mass ratio of enzyme complexes to biomass are given in figure legends. Hydrolysis by Cellolux A and Cellobiase F10 was performed at 55 °C; by *A. niger* cellulase and *T. lanuginosus* xylanase, at 37 °C.

Percentage of biomass conversion to sugars was calculated based on holocellulose amount. Primary plant cell wall is built of cellulose fibers embedded into matrix of other polysaccharides. Lignocellolose biomass contains about 70 % of polysaccharides, including hexose (cellulose) and pentose (hemicellulose) (Aristidou and Penttilä 2000). Complete hydrolysis of these polysaccharides yields a mixture of hexoses (glucose, galactose, and mannose) and pentoses (arabinose and xylose) (Kumar et al. 2009; Schädel et al. 2010). Total content of cellulose and hemicellulose (70 %) sets the potential upper limit of biomass conversion to sugars. The main component of hemicellulose is a branched polymer consisting of D-xylopyranose molecules connected by β-1,4 bonds. Enzymatic hydrolysis of hemicellulose requires multiple enzymes, first of all endo-β-1,4-xylanases. In this study we used T. lanuginosus xylanase.

As the main cellulases we used Cellolux A (CalA) or *A. niger* cellulase (CalAn). Endoglucanases randomly cleave internal glucoside bonds, rapidly increasing the number of reducing polysaccharide chain ends. Exoglucanases split off oligosaccharides (mainly cellobiose) from reducing or nonreducing ends, which results in rapid increase in oligosaccharides concentration, but slow reduction of polymer length (Zhang et al. 2006). Cellobiose hydrolysis is required for efficient degradation of plant biomass, as cellobiose inhibits endo- and exoglucanases (Shen et al. 2008), and cellobiase F10 catalyzes this reaction.

Results of biomass hydrolysis of miscanthus cultivar Soranovskii, *P. arundinacea*, *T. lancifolium*, and *S. hermaphrodita* are presented in Fig. 6.1. From Fig. 6.1, one can see that only reedgrass (*P. arundinacea*) can be hydrolyzed by the combination of the CalA cellulase and the F10 cellobiase. Increasing the amount of xylanase

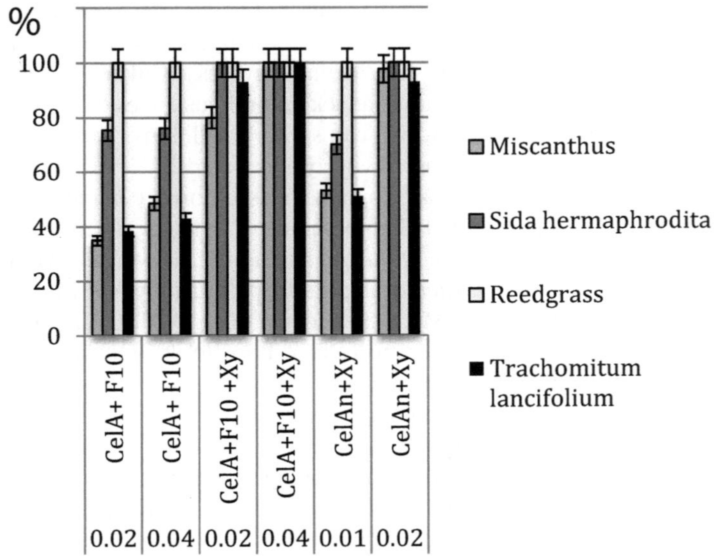

Fig. 6.1 Enzymatic hydrolysis of plant biomass by Cellolux A (CelA) together with Cellobiase F10 (F10) and/or *T. lanuginosus* xylanase (Xy), and by *T. lanuginosus* xylanase (Xy)+*A. niger* cellylase (CelAn). Mass ratio of enzyme complexes to plant biomass is shown below. The vertical axis denotes the resulting amount of hydrolyzed cellulose. Reedgrass, *P. arundinacea*; miscanthus, miscanthus cultivar Soranovskii

in reaction mixture results in complete hydrolysis of holocellulose in all samples. For *Sida* and *Phalaris*, cellobiase was not strictly required for hydrolysis in the presence of xylanase: low cellobiase activity of one of the enzymes in the reaction mixture proved to be sufficient. Cellolux A, according to the manufacturer's manual, is a complex of fungal enzymes that contains cellulases, glucanases, and xylanases. However, Cellolux A alone could not provide complete hydrolysis of Soranovskii, even at the high enzyme: biomass ratio of 0.1. Including cellobiase F10, process cellobiose to glucose, so the addition of F10 to the reaction mixture increases hydrolysis efficiency by 1.5 times (Fig. 6.2).

Phalaris biomass was the most easily hydrolized in our experiments. Complete hydrolysis of *Phalaris* requires either CelloluxA or *A. niger* cellulases with the addition of either cellobiase or xylanase at minimum doses (Fig. 6.1). Milling and pretreatment with alkaline peroxide provides optimal conditions for subsequent enzymatic hydrolysis. Thus, we managed to obtain sugar-containing substrate from Soranovskii biomass for subsequent fermentation to ethanol or organic acids by microorganisms.

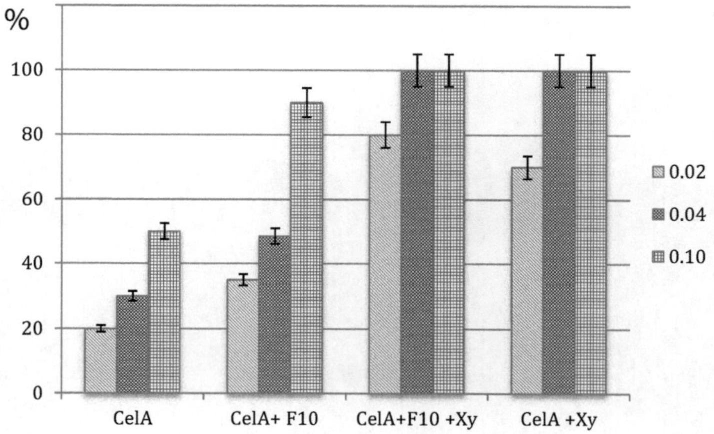

Fig. 6.2 Enzymatic hydrolysis of miscanthus cultivar Soranovskii biomass by Cellolux A (CelA) cellulase together with F10 cellobiase (F10) and/or *T. lanuginosus* xylanase (Xy). Mass ratio of the enzymatic complex to biomass is encoded by shading (right side of the picture). The vertical axis denotes the percentage of hydrolyzed holocellulose

Microorganisms Degrading Components of Miscanthus Biomass Found in Russia

In nature, plant biomass is a source of carbon in ecosystems, and all its components can be degraded to simple substances by complex microbial communities. Since cellulose is water-insoluble, bacteria and fungi mostly rely on extracellular degradation. Products of cellulose hydrolysis are a good carbon source for other microorganisms, which forms a basis for many microbial interactions. Cellulose-degrading microbial communities play a crucial role in the global carbon cycle (Leschine 1995; Wirth et al. 2012).

We obtained cellulose samples of miscanthus cultivar Soranovskii by delignification of milled biomass using chloramine B in acetic acid with subsequent removal of hemicellulose by heating with alkaline solution. The resulting cellulose had polymerization coefficient in the range of 530–610 units.

We attempted to isolate natural cellulosolytic microorganisms from natural ecosystems of the Novosibirsk oblast and the Altai krai. By incubating natural microbial cultures on various cellulose sources, we obtained 33 strains of microorganisms able to grow on miscanthus cultivar Soranovskii biomass. The medium contained 2.5 g/l $NaNO_3$, 1 g/l K_2HPO_4, 0.15 g/l $(NH_4)_2SO_4$, 0.15 g/l $MgCl_2$, 0.1 g/l NaCl, 0.1 g/l $CaCl_2$, 0.05 g/l $FeCl_3$*$6H_2O$, and 1.5 % agar. Milled miscanthus (1.5 %) was used as the sole carbon source.

Fig. 6.3 A Soranovskii field in West Siberia

We identified the obtained strains using 16S rRNA sequencing. Most of the strains belonged to the *Bacillus* genus; 3 to *Lysinibacillus*; 3 to *Achromobacter*; 1 to *Brevibacillus*; 1 to *Serratia*; five strains could not be identified.

The isolated pure cultures were transferred to the Hutchinson's medium no. 1 (Hutchinson and Clayton 1919) containing milled miscanthus (1.5 %) as a carbon source. In three days we analyzed the culture media of all 33 strains for the presence of sugars using the colorimetric approach with 3,5-dinitrosalycilic acid. No sugars were present, probably because the strains used glucose themselves, to build up their own biomass or to produce various metabolites, e.g., poly(lactate-co-3-hydroxybutyrate).

There are four main pathways of glucose processing in microorganisms:

1. $C_6H_{12}O_6$ 2 $C_2H_5OH + 2CO_2$ ethanol fermentation
2. $C_6H_{12}O_6$ 2 $CH_3CHOHCOOH$ lactic acid fermentation
3. $C_6H_{12}O_6$ 2 CH3COOH + $2CO_2$ acetic acid fermentation
4. $C_6H_{12}O_6 + 6O_2$ 6 CO_2+ H_2O glucose oxidation

To detect these metabolic pathways we searched for end products (ethanol, lactic, and acetic acids) in culture medium using an Agilent Technologies gas chromatograph. A mass spectrometer was used as a detector. A total of 33 specimens on culture medium of the studied microorganisms were analyzed. Twenty-seven of them contained up to 0.1 % of ethanol, while eight demonstrated the presence of up to 0.7 % of lactic acid. Acetic acid was not found in any of the samples.

Conclusions

In this study, we demonstrated that miscanthus may be successfully cultivated in West Siberia. *M. sacchariflorus* proved to be optimal for Siberian climate. Based on this species we obtained a novel plant cultivar, miscanthus cultivar Soranovskii (Fig. 6.3). Biomass of Soranovskii may be used as a source of cellulose or for hydrolysis with subsequent fermentation for biotechnology and green chemistry. Pretreatment with alkaline peroxide, a relatively cheap and clean process, is sufficient for effective enzymatic hydrolysis. However, enzymatic hydrolysis is still economically unfavorable, which precludes one from creating an efficient technology for obtaining sugar-containing substrates.

We also isolated cellulose-degrading microorganisms that may be used as a source of efficient enzymes for hydrolysis of miscanthus biomass. The obtained microorganism strains may be genetically modified to create economically efficient producers of ethanol and lactic acid that can grow on miscanthus biomass without preliminary hydrolysis.

Acknowledgements This study was supported by the Budget project VI.58.1.3 and by the EU FP7 Grant 150204 "Enhancing biomass production from marginal lands with perennial grasses."

Cellobiase F10 and cellulase B1 from Pen. verruculosum were kindly provided by A. P. Sinitsyn.

References

Aristidou A, Penttilä M (2000) Metabolic engineering applications to renewable resource utilization. Curr Opin Biotechnol 11(2):187–198

Dashtban M, Buchkowski R, Qin W (2011) Effect of different carbon sources on cellulase production by Hypocrea jecorina (Trichoderma reesei) strains. Int J Biochem Mol Biol 2(3):274–286

Fowler PA, McLauchlin AR, Hall LM (2003) The potential industrial uses of forage grasses including Miscanthus. University of Wales, Bangor, BioComposites Centre

Hames BR (2009) Biomass compositional analysis for energy applications. Biofuels, Methods and Protocols, pp 145–167

Henry RJ (2010) Evaluation of plant biomass resources available for replacement of fossil oil. Plant Biotechnol J 8(3):288–293

Hodkinson TR, Chase MW, Takahashi C, Leitch IJ, Bennett MD, Renvoize SA (2002a) The use of DNA sequencing (ITS and trnL-F), AFLP, and fluorescent in situ hybridization to study allopolyploid Miscanthus (Poaceae). Am J Bot 89(2):279–286

Hodkinson TR, Chase MW, Lledó DM, Salamin N, Renvoize SA (2002b) Phylogenetics of Miscanthus, Saccharum and related genera (Saccharinae, Andropogoneae, Poaceae) based on DNA sequences from ITS nuclear ribosomal DNA and plastid trnL intron and trnL-F intergenic spacers. J Plant Res 115(5):381–392

Hodkinson TR, Klaas M, Jones MB, Prickett R, Barth S (2015) *Miscanthus*: a case study for the utilization of natural genetic variation. Plant Genet Resour 13:219–237

Hutchinson HB, Clayton J (1919) On the decomposition of cellulose by an aerobic organism (Spirochaeta cytophaga, n. sp.). J Agric Sci 9(02):143–172

Kumar P, Barrett DM, Delwiche MJ, Stroeve P (2009) Methods for pretreatment of lignocellulosic biomass for efficient hydrolysis and biofuel production. Ind Eng Chem Res 48(8):3713–3729

Lee WC, Kuan WC (2015) Miscanthus as cellulosic biomass for bioethanol production. Biotechnol J 10:540–554

Leschine SB (1995) Cellulose degradation in anaerobic environments. Ann Rev Microbiol 49(1):399–426

Mukherjee PK, Horwitz BA, Kenerley CM (2012) Secondary metabolism in Trichoderma—a genomic perspective. Microbiology 158(1):35–45

Salvadó J, Velásquez JA, Ferrando F (2003) Binderless fiberboard from steam exploded Miscanthus sinensis: optimization of pressing and pretreatment conditions. Wood Sci Technol 37(3–4): 279–286

Schädel C, Blöchl A, Richter A, Hoch G (2010) Quantification and monosaccharide composition of hemicelluloses from different plant functional types. Plant Physiol Biochem 48(1):1–8

Shen Y, Zhang Y, Ma T, Bao X, Du F, Zhuang G, Qu Y (2008) Simultaneous saccharification and fermentation of acid-pretreated corncobs with a recombinant *Saccharomyces cerevisiae* expressing β-glucosidase. Bioresour Technol 99(11):5099–5103

Shumny VK, Veprev SG, Nechiporenko NN, Goryachkovskaya TN, Slynko NM, Kolchanov NA, Peltek SE (2010) A new form of Miscanthus (Chinese silver grass, Miscanthus sinensis—Andersson) as a promising source of cellulosic biomass. Adv Biosci Biotechnol 1(3): 167–170

Sun Q, Lin Q, Yi ZL, Yang ZR, Zhou FS (2010) A taxonomic revision of Miscanthus sl. (Poaceae) from China. Bot J Linnean Soc 164(2):178–220

Wi SG, Cho EJ, Lee DS, Lee SJ, Lee YJ, Bae HJ (2015) Lignocellulose conversion for biofuel: a new pretreatment greatly improves downstream biocatalytic hydrolysis of various lignocellulosic materials. Biotechnol Biofuels 24(8):228

Wirth R, Kovács E, Maróti G, Bagi Z, Rákhely G, Kovács KL (2012) Characterization of a biogas-producing microbial community by short-read next generation DNA sequencing. Biotechnol Biofuels 5(1)

Zhang YHP, Himmel ME, Mielenz JR (2006) Outlook for cellulase improvement: screening and selection strategies. Biotechnol Adv 24(5):452–481

Chapter 7
Italian Experiences on *Arundo* Harvesting: Economic and Energy Appraisal

L. Pari, A. Suardi, A. Scarfone, and E. Santangelo

Introduction

Harvesting represents one of the most energy-intensive steps within the production of energy crops (McKendry 2002; Corno et al. 2014) and its management is functional at properly addressing the feedstock towards different supply chains (Dragoni et al. 2015).

Giant reed (*Arundo donax*, L.) has attracted the interest of many researchers given its favourable agronomic traits (Angelini et al. 2009; Ceotto and Di Candilo 2010) and positive energy balance (Mantineo et al. 2009). *Arundo* is usually harvested in autumn or winter (Monti et al. 2015), but multiple harvests per year, even during summer, have been recently proposed (Dragoni et al. 2015; Ragaglini et al. 2014). The exploration of new harvesting period is aimed at increasing the versatility of use matching the feedstock qualities to specific conversion processes. The biomass harvested in winter is more suitable for combustion plants, while the anaerobic digestion requires a feedstock with high proportion of leaves and high moisture content (Dragoni et al. 2015).

Two main systems can be distinguished for the harvesting of fresh or dry *Arundo* biomass. The one-pass system involves the use of self-propelled forage harvester (SPFH) commonly employed for the harvesting of silage maize. In this case, the forager is flanked by a tractor-trailer unit receiving the chopped biomass, and then delivered to a collection point (Acampora et al. 2014; Pari et al. 2015). Another approach foresees a two-pass system where the crushed (Curt et al. 2013) or shredded (Bentini and Martelli 2013) plants are windrowed and baled.

L. Pari (✉) • A. Suardi • A. Scarfone • E. Santangelo
Consiglio per la ricerca in agricoltura e l'analisi dell'economia agraria, Unità di ricerca per l'ingegneria agraria (CREA-ING), Monterotondo, Rome, Italy
e-mail: luigi.pari@crea.gov.it

S. Barth et al. (eds.), *Perennial Biomass Crops for a Resource-Constrained World*, DOI 10.1007/978-3-319-44530-4_7

Harvesting studies have often analysed the performance of the machines leaving aside the energy and economic analysis. Angelini et al. (2005, 2009) and Mantineo et al. (2009) have made an energy assessment of the whole cultivation cycle where the harvesting phase required 3.94 GJ ha^{-1} (Angelini et al. 2009), 3.47 GJ ha^{-1} (Angelini et al. 2005) and 2.80 GJ ha^{-1} (Mantineo et al. 2009). Recently, Bentini and Martelli (2013) published an economic evaluation of a supply chain where plant shredding was followed by baling (round bale). The system was tested on a 7-year-old plantation harvested on February. The total hourly cost for shredding was 62.7 € h^{-1} and 128.8 € h^{-1} for baling.

Farmers need to carefully evaluate which system may assure the production of a high-grade feedstock with the most sustainable and low-expensive method. Thus, we deemed useful to carry out a reasoned analysis of the energy and economic impacts of some harvesting solutions examined by CREA-ING in the last 20 years. Scope of the paper was to analyse four different harvesting techniques of giant reed in the Italian conditions in order to evaluate the main factors affecting the energy and the economic costs of *Arundo* harvesting.

Material and Methods

Harvesting Systems

The harvesting systems analysed were:

1. Self-propelled forage harvester (Claas Jaguar 850 equipped with a head Orbis RU 450 XTRA) with a tractor-trailer unit (Fendt 312 Vario and Silver Car SCR 314) (Acampora et al. 2014);
2. Mower (Galfrè FR / G 190 powered by a tractor Landini Serie 5–100) plus a second step with a Claas Jaguar 850 equipped with a Pick-Up 300 HD and followed by a tractor-trailer (Fendt 312 Vario plus Silver Car SCR 314) unit (Acampora et al. 2014);
3. Self-propelled forage harvester (Claas Jaguar 850 equipped with a head Orbis RU 450 XTRA) with a tractor-trailer unit (Fendt Farmer 816 and Dumper Zaccaria ZAM 140 21) (Pari et al. 2011);
4. Shredder (Nobilil BNU 160 AD powered by a tractor Fendt Farmer 312) followed by a round-baler Welger RP 320 Farmer trailed by a Same Iron 150.7 tractor (Pari et al. 2011).

Synthesising, the analysis took into consideration four systems including one (system 1 and 3) or two (systems 2 and 4) passes. Systems 1, 2 and 3 produced chopped *Arundo*; in the system 4 the product was baled.

The Sites

The systems 1 and 2 were tested towards the Franco Alasia Vivai farm, located at Savigliano (44° 36 '8.82"N, 7° 38' 30.32"E; 320 m a.s.l.), in the North-West of Italy. In this area the climate is intermediate between Mediterranean sub-oceanic to sub-continental. The precipitation pattern is regular with two dry months (July and August) and temperatures ranging from 0 °C (winter) to 30 °C (summer). The soil is clay loam (according to the USDA soil classification scheme) typical of the Langhe, Monferrato and the 'hills of the Po river' an area with clay accumulation. In this site the harvesting was carried out in June 2010 on a 1-year-old plantation.

The study on systems 3 and 4 was conducted in Central Italy at Marciano della Chiana (43°18′21″N 11°47′15″E; 240 m a.s.l.) in a farm of the Regional Agency for Agriculture Development (ARSIA). The site has a temperate/mesothermal climate with warm summer. The average annual rainfall is around 750 mm distributed during the year. In winter, the minimum temperatures drop around 5 °C, while during summer the maximum can reach 25–28 °C. Following the USDA soil classification scheme, the soil is silty clay. The test was conducted on March 2009 with system 3 and on March 2010 with system 4. The plantation was 3- (2009) and 4- (2010) years-old.

Economic, Energy and Environmental Assessments

The economic, energy and environmental analysis was carried out considering boundaries strictly limited to the harvesting phase (Fig. 7.1). The functional unit used to quantify all inputs and outputs was set as 1 Mg of biomass produced during the four harvesting systems analysed. A 'hectare' functional unit has been used and reported as well for comparison.

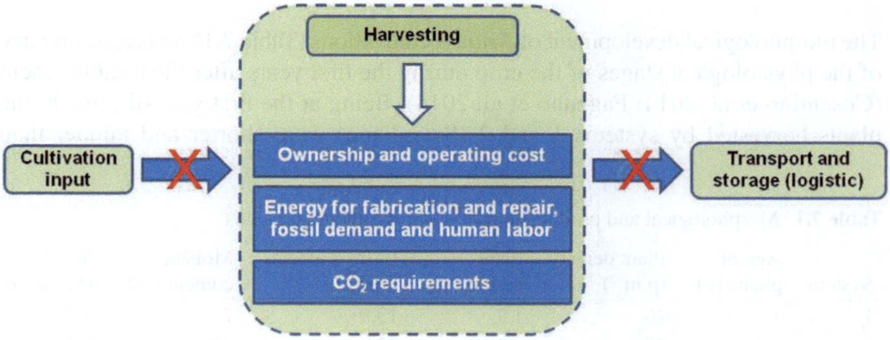

Fig. 7.1 Schematic representation of the boundaries of the studied systems

The evaluation of ownership (depreciation of the machine, interest on the investment, insurance and housing of the machine) and operating (labour, fuel, oil, repair and maintenance) costs was made following ASAE (2006) and Turhollow et al. (2009). For the interest, a rate of 5 % was applied; insurance and housing were calculated as 0.2 and 0.1 % of purchase price. Hourly cost of labour was assumed as 12 € h^{-1} including taxes and social security contributions. The price for fuel was set at 0.93 € L^{-1} whereas for lubricant was 3.6 € L^{-1}. For the average hourly consumption of engine fuel and oil as well as for the repair and maintenance costs (ASABE 2006; Turhollow et al. 2009), we considered the corresponding coefficients of engine load (Piccarolo 1989).

The use of the machines was assumed for an overall surface of 200 ha. For each implement/system the energy inputs were distinguished in: (a) energy costs due to the energy consumption for fabrication and repair; (b) fossil costs for fuel and lubricant; (c) energy cost for human labour. The coefficients for determining the energy requirements (MJ kg^{-1}) of the first two items were taken from Angelini et al. (2009). The energy required by the human labour was calculated as suggested by Patzek (2004).

The environmental impact analysis was conducted according to the LCA methodology based on the ISO 14040:2006 standard (ISO EN 2006) and by means of the software Simapro8.0.2 of PRé Consultants bv (3821 AD Amersfoort, The Netherlands). Ecoinvent database in Simapro was the source of secondary data of the life cycle impacts of the equipment production used. The emissions generated by tractors during harvesting (CO, NOx, PM, HC) were calculated according to OTAC (2010) and Van Belle (2006) for the CO_2. The Greenhouse Gas Protocol V1.01 method (WBCSD and WRI 2009) was applied to assess the kilograms of CO_2eq emitted.

Results and Discussion

Working Capacity

The morphological development of *Arundo* cultivations (Table 7.1) was representative of the physiological stages of the crop during the first years after the establishment (Cosentino et al. 2014; Fagnano et al. 2014). Being at the first year of growth, the plants harvested by systems 1 and 2 (Savigliano) were shorter and thinner than

Table 7.1 Morphological and productive data of the experimental fields

System	Age of plants (y)	Plant density (p m^{-2})	Plant height (m)	Basal diameter (mm)	Moisture content (%)	Yield (t_{dm} ha^{-1})[a]
1	1	20	1.7	13.9	76	8.5
2	1	20	1.7	13.9	45	8.5
3	3	18	4.2	19.2	51	22.9
4	4	21	3.5	15.5	49	13.4

[a]t_{dm}: tons on dry matter basis

Table 7.2 Main data concerning the performance of the machines used in each system

	System	Machine width (m)	Operative speed (km h^{-1})	Field capacity (ha h^{-1})	Operative time (h ha^{-1})
1	SPFH	4.50	3.90	1.34	0.75
2	Tractor+Mower	1.90	6.00	0.86	1.16
	SPFH+Pick-up	4.50	2.90	1.01	0.99
3	SPFH	4.50	1.33	0.39	2.56
4	Tractor+Shredder	1.80	4.28	0.69	1.45
	Tractor + Baler	2.48	1.94	0.44	2.27

those collected with the systems 3 and 4 at Marciano della Chiana. Considering the specific pattern of development, over the years the *Arundo* cultivation can assume a messy development and a layout less ordered than similar crops as maize. At Marciano della Chiana the plants were frequently arranged not orthogonally to the ground leading to a reduction in operative speed of the SPFH (system 3) when compared with the same SPFH in system 1 (Table 7.2). As a result, the field capacity of the SPFH in system 3 was one-third than the same SPFH used at Savigliano in system 1 (Table 7.2). The figures of the operative time highlighted the role of age, arrangement and yield on harvesting performances because with older and more productive crop the operative time increased.

The harvesting system 3 resulted with the lowest field capacity. After one year, in the same plantation, system 4 had better performance, even though, Bentini and Martelli (2013) got better results when they tested a similar shredding machine with a two-pass harvesting system on a 7-year-old cultivation. The authors reported a field capacity of 0.95 ha h^{-1} for the shredding and 0.90 ha h^{-1} for baling. In a previous work, Bentini and Zucchelli (2008) using the same system for shredding registered a field capacity of 0.4 ha h^{-1}, without providing information about the baling efficiency. In fact, the model used in system 4 had a smaller working width (1.8 *vs* 2.8 m) and a lower number of cutting elements (32 *vs* 64). This led to a lower field capacity of system 4 for shredding (0.69 ha h^{-1}) and for baling (0.44 ha h^{-1}).

Economic Analysis

The structure and the amount of the costs (Table 7.3) were influenced by the boundaries chosen for the study, the machines' performance (in particular, the field capacity) and the yield. As described in materials and methods, the boundaries limited the analysis at the machines present in the field during the harvesting for producing a feedstock ready to be delivered to the storage point. This means that, in the case of the one-pass, the values included the costs of the SPFH and the tractor-trailer unit as a whole, or, as in the case of the two-pass systems, the cost of all the machines. The single passage of the SPFH had the lowest cost per hour compared with the other two systems (2 and 3) involving the use of SPFH. Considering the cost per hectare,

Table 7.3 Ownership (depreciation, interest, insurance and housing of the machine), operating (labour, fuel, oil, repair and maintenance) and total cost for the harvesting systems analysed

System	Cost (€ h^{-1})			Cost (€ ha^{-1})		
	Ownership	Operating	Total	Ownership	Operating	Total
1	125.24	165.25	290.49	93.46	123.32	216.78
2	122.54	253.16	375.70	122.86	256.08	378.94
3	107.79	259.96	367.75	276.38	666.58	942.96
4	40.91	112.90	153.81	82.50	217.72	300.22

the differences increased as a result of the effect of the field capacity. A field capacity of 1.34 ha h^{-1} (system 1) led to a cost of 216.78 € ha^{-1}, whereas systems 2 and 3 required 378.93 and 942.95 € ha^{-1}. In the present work, the use of SPFH in one-pass had a cost of 25.5 € t_{dm}^{-1} (system 1) and 41.2 € t_{dm}^{-1} (system 3). The figure of the former is comparable with those reported for other energy crops, such as poplar, where the harvesting system is analogous (Spinelli et al. 2009). The use of a Claas Jaguar 880 equipped with a GBE-1 head on a 4-years-old poplar SRC required 448.46 € ha^{-1} (average yield 47.32 t ha^{-1}, moisture content 55.60 %) corresponding to 9.81 € t^{-1}, fresh matter basis (Verani et al. 2010). Working with the same forager mounting a GBE-2 head the cost calculated by Fiala and Bacenetti (2012) ranged from 19.07 to 19.78 € t_{dm}^{-1} depending on the extent of yield and field capacity. Equivalent figures (19.70 € t_{dm}^{-1}) were observed in the harvesting of 2–4 years-old willow or poplar with a New Holland FR 9060 and FR 9050 equipped with the cutting head 130 FB (Schweier and Becker 2012). Beyond the consistency (woody *vs* herbaceous), the main factor affecting the cost-effectiveness appeared the extent of yield. For the *Arundo*, the yield is expected to increase progressively in the years following the establishment until reaching a plateau (Angelini et al. 2009; Fagnano et al. 2014). In the same way, the harvesting cost should decrease until reaching a steady state in the presence of uniform crop condition.

The use of a shredder (system 4) determined the lowest hourly cost (153.81 € h^{-1}), but also in this case a low field capacity (in particular for baling phase) led to a doubling of cost per hectare (300.22 € ha^{-1}). The hourly cost of the system 4 was lower than that observed by Bentini and Martelli (2013) for shredding and baling (190.5 € h^{-1}, on the whole), but it was higher when calculated on hectare as a consequence of the better field capacity reported by the authors.

Energy and Environmental Analysis

Fuel and lubricant consumption (fossil cost) were the inputs more energy-demanding (Fig. 7.2). Systems 2 and 3 had the highest values of fossil cost owing to different causes. The need of three working units (tractor-mower, SPFH and tractor-trailer) entailed a greater energy expenditure of system 2 when compared with the system 1 tested in the same conditions. Instead, the lowest field capacity of System 3 required more time to complete the work and, hence, higher consumptions. On the

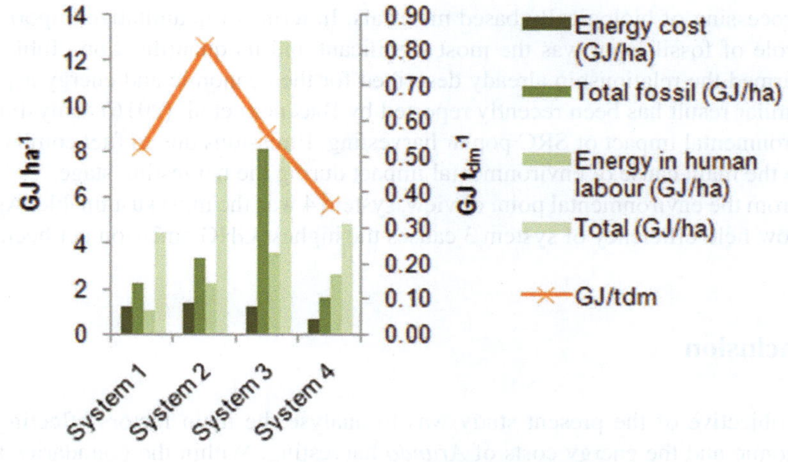

Fig. 7.2 Energy input required by the harvesting systems analysed

Table 7.4 CO_2eq (kg) produced by fossil and biogenic sources

	Fossil (CO_2eq kg)		Biogenic (CO_2eq kg)	
	ha	t_{dm}	ha	t_{dm}
System 1	179.26	21.09	0.37	0.04
System 2	265.29	31.21	0.56	0.07
System 3	623.45	27.23	1.27	0.06
System 4	121.97	9.10	0.37	0.03

other side, the use of machines with a lower engine power allowed the lowest fossil consumptions to the system 4. Considering the requirement of total energy per hectare, system 1 resulted the most favourable (4.46 GJ ha^{-1}).

Some works (Angelini et al. 2009; Mantineo et al. 2009) reported the energy requirements of *Arundo* harvesting (3.9 GJ ha^{-1} and 2.9 GJ ha^{-1}, respectively) in the Italian environments. The first (estimated in the Central Italy) is comparable with the system 1, whereas the second (South Italy) appears more energy-saving. However, the energy evaluations included the whole production chain where the harvesting was a single phase of the whole. Moreover, the technical specifications of this step were absent. As shown before, both the type of machine used and the crop condition are issues which can affect the efficiency of the harvesting, modifying the final result. A further matter of discussion involves the methodology adopted for the energy evaluation because different approaches (choice of the inputs, calculation of consumption, factors of conversion, unit of measure) can lead to ample differences for the same operation in the same crop.

The trend observed for the economic and energy evaluations was confirmed also for the greenhouse gases (GHG) emitted by the analysed systems (Table 7.4). A distinction was made between GHG emissions from fossil sources and biogenic carbon emissions. Biogenic CO_2 emissions are related to the natural carbon cycle, as well as to those resulting from combustion, digestion, fermentation, decomposition or processing of biologically based materials. In terms of quantitative importance, the role of fossil input was the most significant and its quantification (Table 7.4) confirmed the relationship already described for the economic and energy aspects. A similar result has been recently reported by Bacenetti et al. (2016) analysing the environmental impact of SRC poplar harvesting. Emissions due to fuel combustion were the main cause of environmental impact during the harvesting stage.

From the environmental point of view, system 4 was the most sustainable. Again, the low field efficiency of system 3 causes the highest GHG emission per hectare.

Conclusion

The objective of the present study was to analyse the main factors affecting the economic and the energy costs of *Arundo* harvesting. Within the boundaries fixed for the analisys (machines operating in the field during the harvesting) the extent of such costs appeared conditioned by crop condition, age of the plantation, yield, field capacity of the machines and complexity of the chain.

The use of a SPFH in one-pass is the straightforward solution and its use also for the harvesting of fodder crops or maize allows a full employment of the machine during the year with a further reduction in ownership costs. The analysis emphasised as the field efficiency of a SPFH is significantly influenced by the factors cited previously. The same model of SPFH showed a field efficiency clearly different by changing the age of plantation and the crop condition. Such an outcome has some implications about the economic and energy balance of the whole production chain. The harvesting cost (one of the main items of the balance) can be assumed constant with caution during the entire cycle because the field efficiency of the machine could change owing to the contingent status of the crop.

Moreover, other technical options can be available and have to be taken into account in designing a profitable supply chain. Beside their technical efficiency, the economic and energy evaluation appear also essential in order to identify the more sustainable. In our study, this is the case of system 2 (mowing plus chopping) which revealed less convenient when compared with the use of SPFH under the same conditions; or the system 4 (shredding plus baling) showing to be feasible at a reasonable cost. This last result implies also that harvesting systems involving two-passes can be more sustainable than harvesting in a single solution with SPFH. Limited to the present work, the discriminating factors were the ownership cost and the fossil consumption.

The different types of cost (economic, energy but also environmental) have proven strictly linked and varied following the same pattern. Hence, the choice of a correct mechanisation has substantial consequence on the environmental sustainability of harvesting and this is a key point when related to a renewable energy source. Technological improvements are needed in order to ameliorate and lower the cost, but the intensity and the convenience of this effort will rely on the importance that the crops will assume in the future as feedstock for bioenergy.

Acknowledgment This activity is supported by the European Community Framework Programme 7, Optimization of Perennial Grasses for Biomass Production (OPTIMA), contract no. 289642.

References

Acampora A, Assirelli A, Croce S, Gallucci F, Pari L (2014) Prove di raccolta di *Arundo* trinciato fresco e affienato. Supplement to L'Informatore Agrario 11:20–24

Angelini LG, Ceccarini L, Bonari E (2005) Biomass yield and energy balance of giant reed (*Arundo donax* L.) cropped in central Italy as related to different management practices. Eur J Agron 22:375–389

Angelini LG, Ceccarini L, Nassi o Di Nasso N, Bonari E (2009) Comparison of *Arundo donax* L. and *Miscanthus x giganteus* in a long-term field experiment in Central Italy: analysis of productive characteristics and energy balance. Biomass Bioenergy 33(4):635–643

ARSIA (2004) Le colture dedicate ad uso energetico: il progetto Bioenegy farm. Quaderno ARSIA, 6/2004, ARSIA Regione Toscana, Firenze

ASABE (2006) D497.5: Agricultural machinery management data. In: ASABE Standards. American Society of Agricultural Engineers, St. Joseph, MI, pp 360–367

Bacenetti J, Pessina D, Fiala M (2016) Environmental assessment of different harvesting solutions for Short Rotation Coppice plantations. Sci Total Environ 541:210–217

Bentini M, Martelli R (2013) Prototype for the harvesting of cultivated herbaceous energy crops, an economic and technical evaluation. Biomass Bioenergy 57:229–237

Bentini M, Zucchelli M (2008) Feeding furnaces: biomass harvest equipment. Mondo Macchina 17(1):32–39

Ceotto E, Di Candilo M (2010) Shoot cuttings propagation of giant reed (*Arundo donax* L.) in water and moist soil: The path forward? Biomass Bioenergy 34:1614–1623

Corno L, Pilu R, Adani F (2014) *Arundo donax* L.: a non-food crop for bioenergy and bio-compound production. Biotechnol Adv 32(8):1535–1549

Cosentino SL, Scordia D, Sanzone E, Testa G, Copani V (2014) Response of giant reed (*Arundodonax* L.) to nitrogen fertilization and soil water availability in semi-arid Mediterranean environment. Eur J Agron 60:22–32

Curt MD, Sanz, Mosquera F, Mauri PV, Plaza A, Aguado PL, Fernández J (2013) Harvest mechanisation of Arundo donax L. in Spain. In: Proc. 21st EU Biomass Conf. and Exhibition, Copenhagen, Denmark, 302–307

Dragoni F, Nassi o Di Nasso N, Tozzini C, Bonari E, Ragaglini G (2015) Aboveground yield and biomass quality of giant reed (*Arundo donax* L.) as affected by harvest time and frequency. BioEnergy Res 8(3): 1321–1331

Fagnano M, Impagliazzo A, Mori M, Fiorentino N (2014) Agronomic and environmental impacts of giant reed (*Arundodonax* L.): results from a long-term field experiment in hilly areas subject to soil erosion. Bioenergy Res 8:415–422

Fiala M, Bacenetti J (2012) Economic, energetic and environmental impact in short rotation coppice harvesting operations. Biomass Bioenergy 42:107–113

ISO EN (2006) 14040: 2006. Environmental management-Life cycle assessment-Principles and framework.European Committee for Standardization

Mantineo M, D'Agosta GM, Copani V, Patanè C, Cosentino SL (2009) Biomass yield and energy balance of three perennial crops for energy use in the semi-arid Mediterranean environment. Field Crop Res 114:204–213

McKendry P (2002) Energy production from biomass (part 1): overview of biomass. Bioresour Technol 83(1):37–46

Monti A, Zanetti F, Scordia D, Testa G, Cosentino SL (2015) What to harvest when? Autumn, winter, annual and biennial harvesting of giant reed, miscanthus and switchgrass in northern and southern Mediterranean area. Ind Crop Prod 75:129–134

OTAC EPA (2010) Exhaust and crankcase emission factors for non road engine modelling -Compression-Ignition. EPA-1,056-R-10-018, NR-009d, July 2010. http://www.epa. gov/otaq/models/nonrdmdl/nonrdmdl2010/1,056 r10018. Pdf [Report]

Pari L, Suardi A, Civitarese V, Giannini E (2011) Meccanizzazione della raccolta del prodotto trinciato di canna comune (*Arundo donax* L.), in:Luigi Pari (Ed.), Lo sviluppo delle colture energetiche in Italia. Contributo dei Progetti di Ricerca SUSCACE e FAESI. Ed. Nuova Cultura, Roma, pp. 723–738

Pari L, Assirelli A, Acampora A, Del Giudice A, Santangelo E (2015) A new prototype for increasing the particle size of chopped *Arundo donax* (L.). Biomass Bioenergy 74:288–295

Patzek TW (2004) Thermodynamics of the corn-ethanol biofuel cycle. Crit Rev Plant Sci 23: 519–567

Piccarolo P (1989) Criteri di scelta e di gestione delle macchine agricole. Macchine e Motori Agricoli 12:37–57

Ragaglini G, Dragoni F, Simone M, Bonari E (2014) Suitability of giant reed (*Arundo donax* L.) for anaerobic digestion: Effect of harvest time and frequency on the biomethane yield potential. Bioresour Technol 152:107–115

Schweier J, Becker G (2012) New Holland forage harvester's productivity inshort rotation coppice: evaluation of field studies from a German perspective. Int J For Eng 23:82–88

Spinelli R, Nati C, Magagnotti N (2009) Using modified foragers to harvest short-rotation poplar plantations. Biomass Bioenergy 33(5):817–821

Turhollow A, Wilkerson E, Sokhansanj S (2009) Cost methodology for biomass feedstocks: herbaceous crops and agricultural residues. Oak Ridge National Laboratory, Oak Ridge, TN, Technical Report No.ORNL/TM-2008/105

Van Belle J-F (2006) A model to estimate fossil CO_2 emissions during the harvesting of forest residues for energy with an application on the case of chipping. Biomass Bioenergy 30(12): 1067–1075

Verani S, Sperandio G, Di Matteo G (2010) Analisi del lavoro della Claas Jaguar 880 con testata GBE-1 nella raccolta di un pioppeto da biomassa. Forest@-Journal of Silviculture and Forest Ecology 7(1): 22

WBCSD, WRI (2009) Product life cycle accounting and reporting standard. Review draft for Stakeholder Advisory Group. The Greenhouse Gas Protocol Initiative. November 2009

Chapter 8
Biomass Potential of Fibre Nettle in Lithuania

Z. Jankauskienė and E. Gruzdevienė

Introduction

The stinging nettle (*Urtica dioica* L.) is a perennial herbaceous plant from the *Urticaceae* family. This plant is considered a weed in the gardens or in the fields of various agricultural crops, but potentially it can be grown as a valuable agricultural crop because all parts of the plant (stems, leaves, roots and seeds) could be successfully processed and used (Bacci et al. 2010; di Virgilio et al. 2015; Gruzdevienė and Jankauskienė 2012). Piers Warren described 101 uses of stinging nettles mentioning the benefit of nettles in the garden, kitchen, medicine, for animals, fibre applications, etc. (Warren 2006).

Fibre nettle, a cultivated improved form of the wild nettle, was developed in 1918–1959 by German researcher G. Bredemann (di Virgilio et al. 2008; Hartl and Vogl 2002). By the way of selection, fibre content was increased from about 5 % in wild nettle plants to up to 17–18 % of stem dry matter in the cultivated species (Hartl and Vogl 2002). Austria, Germany, Finland, United Kingdom, the Netherlands, Italy, Latvia and Lithuania are the countries which have shown scientific and practical interest in stinging nettle (Jankauskienė and Gruzdevienė 2015).

One of numerous possible uses of this plant could be for biomass production. In trials in Austria in 1998, in the second cultivation year, the stem dry matter yield of nettle clones ranged from 2.3 to 4.7 t ha^{-1}, and the dry matter yield in the third year ranged from 5.6 to 9.7 t ha^{-1} (Hartl and Vogl 2002). Austrian or German clones of stinging nettle in Germany produced yield of between 6 and 8 t ha^{-1} of dry matter of straw (Nebel et al. 2002). A German clone grown in central Italy, with higher temperatures and generally lower rainfall, gave a stalk yield of 15.4 t ha^{-1} dry matter (Bacci et al. 2009).

Z. Jankauskienė (✉) • E. Gruzdevienė
Upytė Experimental Station of the Lithuanian Research Centre for Agriculture and Forestry, Upytė, Lithuania
e-mail: soja@upyte.lzi.lt

S. Barth et al. (eds.), *Perennial Biomass Crops for a Resource-Constrained World*, DOI 10.1007/978-3-319-44530-4_8

Although the stinging nettle is a perennial plant its productivity may decline with age. Different economic life spans of stinging nettle crops have been reported varying from 3 to 4 years to 10–15 years and more (Tavano et al. 2011; Vogl and Hartl 2003).

The aim of our research was the evaluation of the biomass yield potential of stinging nettle crops established at different planting densities as well as at different ages.

Materials and Methods

Location. The investigation was carried out at the Upytė Experimental Station of the Lithuanian Research Centre for Agriculture and Forestry (55.642901, 24.227766 DMM). Research was started in 2008 and finished in 2014. The soil of the experimental site is an *Eutri-Endohypogleyic Cambisol, CMg-n-w-eu* (Buivydaitė et al. 2001), with a pH_{KCl} of 6.7–7.7 (potentiometrically), humus concentration of 1.89–2.33 % (ISO 10694:1995). The content of available phosphorus (P_2O_5) in the soil plough layer was between 137 and 245 mg kg^{-1}, and the content of available potassium (K_2O) was 122–152 mg kg^{-1} (determined in A-L extraction).

Trial design. The trial plots were established in 2008 and 2009. The plants of stinging nettle (clones of fibre nettle from the Netherlands) were cut in pieces and vegetatively multiplied (Jankauskienė and Gruzdevienė 2008, 2010). The 1st treatment (planting distance of 60 cm between the plants in the row × 60 cm inter-row spacing) was established in triplicate in May 2008. In 2009 (end of April), the 2nd and 3rd treatments were set up with the crop of planted at distances of 60 × 60 and 60 × 100 cm, respectively. As a result, crops of different planting density (60 × 60 and 60 × 100 cm) and age were established. The established trial plot were 20 m^2, but sampling was from plots of 4 m^2. When sampling in 2010, the 60 × 60 cm trial plots were in the 3rd cropping year (established in 2008) and the 60 × 60 cm and 60 × 100 cm crops were in the 2nd year (established in 2009). The stinging nettle crop was fertilised by N:P:K 16:16:16 at a rate of 200 kg ha^{-1} in the first half of May. In the autumn of 2008, the first yield of stinging nettle was determined. The beginning of seed ripening was selected as a time for nettle harvesting (end of August—middle of September).

Yield Assessments. The yield of dry biomass was assessed after sawing off and weighting all cut stems of each trial plot (area of 4 m^2) in three replications. One stem per plot was cut into small pieces and the sample for moisture content determination was taken directly in the field. Moisture content was evaluated by the oven-dry method (drying for 1.5 h at 105 °C).

Meteorological conditions. The meteorological conditions varied from year to year but during the growth seasons rainfall was abundant (Figs. 8.1 and 8.2).

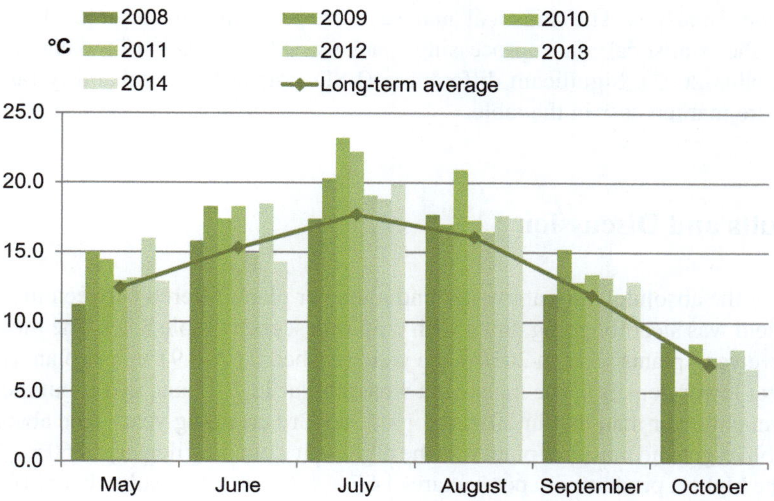

Fig. 8.1 Annual growing season temperature (°C) at Upytė, 2008–2014

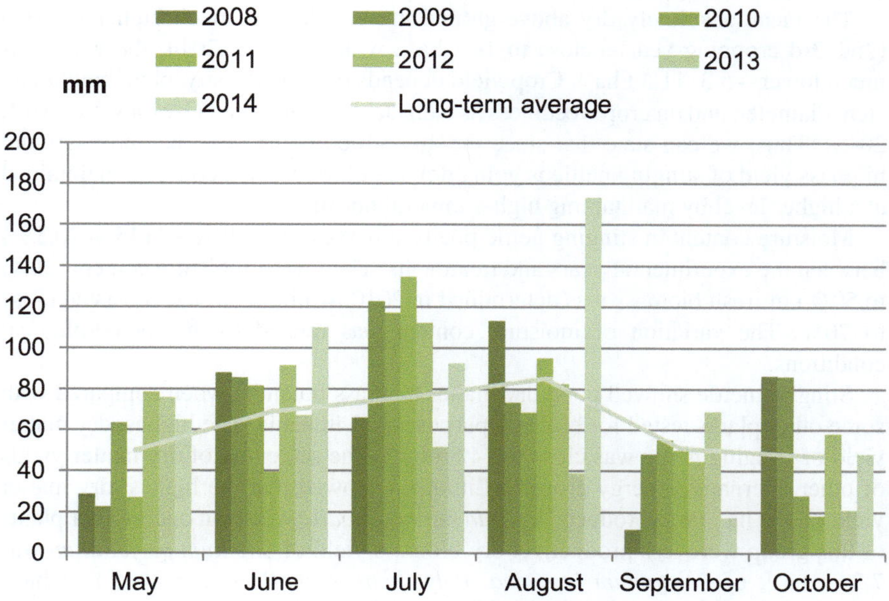

Fig. 8.2 Annual growing season precipitation (mm) at Upytė, 2008–2014

Statistical analysis. The statistical analyses were performed using *ANOVA* software from the statistical data processing package *SELEKCIJA* (Tarakanovas and Raudonius 2003). Significant differences (LSD test) at 0.05 probability (at least) level are marked as 'a' in the table.

Results and Discussion

Data of the absolutely dry above-ground mass per plant differed between the years; the yield was steadily going down with crop senescence (Table 8.1.). The most productive yang plants were in 2010—the yield reached 0.62–0.91 kg per plant (higher yield at lower density). The mean data were similar in 2011 and 2012, but the yield was much lower than that in 2010 (crop of 2nd–3rd cropping year). The absolutely dry above-ground mass per plant was the highest at a planting density of 60 × 100 cm.

The highest productivity per hectare (14.7–14.8 t ha^{-1}) of absolutely dry biomass was shown by the crop of 2nd year, 60 × 60. Since 3rd–4th cropping year dry matter yield started decreasing. Favourable growing season for nettle was in 2013 when again high yield of 8.8–11.6 tonnes of dry matter was obtained. The exceptional yield in 2013 was fixed despite of the fact that vegetation period was much dryer than others. Especially low yields (5.3–8.9 t ha^{-1}) of dry above-ground mass were obtained in the crop in 2014.

The mean absolutely dry above-ground mass yield was much higher in 2010 (2nd–3rd cropping year)—close to 14 t ha^{-1}, while in 2011–2014, the yield was much lower—5.3–11.5 t ha^{-1}. Crop yield depends on crop density, plant height and stem diameter, and on crop weediness as well (Jankauskienė and Gruzdevienė 2013, 2015). Thus, we can state that since the 4th–5th cropping year the above-ground biomass yield of stinging nettle is going down, but probably it could be maintained at a higher level by maintaining higher mineral nutrition.

Moisture content in stinging nettle plants at harvest time differed (45.0–70.5 %) between the experimental years and treatments. The lowest moisture content (close to 50 %) in fresh biomass was determined in 2010, while in 2011 it consisted close to 70 %. The variation of moisture content was related to the meteorological conditions.

Stinging nettle showed good dry matter biomass potential when compared with some other plants tested for biomass purposes in Lithuania. The highest dry matter yield of stinging nettle was close to 14.8 t ha^{-1}. The screening of dry matter yields of other alternative energy crops in Lithuania showed that the highest dry matter yield (9.58 t ha^{-1}) was produced by *Sida hermaphrodita* when cut once. Other plants of this group were less productive: dry biomass yield of *Silphium perfoliatum* was 7.29 t ha^{-1}, of *Polygonum japonica*, *Polygonum sachalinensis* 8.74–5.13 t ha^{-1}, respectively. Of the non-traditional plants, the highest biomass content was accumulated by *Artemisia dubia* (11.10 t ha^{-1}) and *Helianthus tuberosus* (8.56 t ha^{-1}). In the group of local plants, unfertilised *Galega orientalis* produced 10.98 t ha^{-1} of dry

Table 8.1 Dry above-ground mass and moisture content of fibre nettle at harvest 2008–2014

Treatments	Dry above-ground mass		Moisture content (%) in fresh above-ground mass
	Of 1 plant (kg plant^{-1})	Of all plants per 1 ha (kg ha^{-1})	
2008			
1th year, 60×60 cm	0.48	10,787	52.3
2009			
2nd year, 60×60 cm	0.65	14,716	61.5
1th year, 60×60 cm	0.49	10,927	61.3
1th year, 60×100 cm	0.53	7934	61.4
Mean	0.56	11,192	61.4
2010			
3rd year, 60×60 cm	0.62	13,891	45.0
2nd year, 60×60 cm	0.66	14,847	52.4
2nd year, 60×100 cm	0.91	13,615	47.1
Mean	0.73	14,118	48.2
2011			
4th year, 60×60 cm	0.30	6832	70.5
3rd year, 60×60 cm	0.25	5585	70.1
3rd year, 60×100 cm	0.60[a]	9036	70.2
Mean	0.38	7151	70.3
2012			
5th year, 60×60 cm	0.27	6112	68.7[a]
4th year, 60×60 cm	0.32	7150	61.4[a]
4th year, 60×100 cm	0.42	6235	65.3
Mean	0.34	6499	65.1
2013			
6th year, 60×60 cm	0.51	11,552	68.5
5th year, 60×60 cm	0.41	9172	67.3
5th year, 60×100 cm	0.59	8851	67.9
Mean	0.50	9858	67.9
2014			
7th year, 60×60 cm	0.40	8931	61.4
6th year, 60×60 cm	0.24	5312	59.8
6th year, 60×100 cm	0.39	5902	55.1
Mean	0.34	6715	58.8

[a]Significant at 0.05 probability level

biomass over three cuts. *Phalaris arundinacea*, fertilised with N 120 kg ha^{-1}, produced 9.89 t ha^{-1} (Slepetys et al. 2012).

The investigation and interest in biomass in Lithuania is growing. The investigation of the suitability of stinging nettle biomass preparation for burning—chopping, milling and pelleting was less complicated than that of hemp, and crushing and pressing of stinging nettle plants required less time and about 15–17 % less power (Jasinskas et al. 2013).

Conclusion

The perennial fibre stinging nettle (*Urtica dioica* L.) showed promising potential for biomass production between perennial plants. The highest annual dry matter yield was close to 14.8 t ha^{-1}. Over all experimental years, irrespective of the crop age, the per-plant productivity of fibre stinging nettle was higher at a planting density of 60×100 cm. The mean above-ground mass yield per plant decreased along with crop senescence. The biomass yield was generally higher in the treatments planted at 60×60 cm density. The increased crop weediness also had the negative impact on the decrease of nettle biomass yield. The nutrition requirement for stinging nettle needs to be investigated.

Acknowledgments The research was a part of the long-term LRCAF programme 'Biopotential and Quality of Plants for Multifunctional Use'.

References

Bacci L, Baronti S, Predieri S, di Virgilio N (2009) Fiber yield and quality of fiber nettle (*Urtica dioica* L.) cultivated in Italy. Ind Crops Prod 29(2–3):480–484

Bacci L, di Lonardo S, Albanese L, Mastriomei G, Perito B (2010) Effect of different extraction methods on fibre quality of nettle (*Urtica dioica* L.). Textile Res J 81(8):827–837

Buivydaitė VV, Vaičys M, Juodis J, Motuzas A (2001) The classification of Lithuania's soils. Vilnius, Lithuania, p 76 (in Lithuanian)

di Virgilio N, Papazoglou EG, Jankauskiene Z, di Lonardo S, Praczy M, Wielgusz K (2015) The potential of stinging nettle (*Urtica dioica*. L) as a crop with multiple uses. Ind Crops Prod 68:42–49

di Virgilio N, Predieri S, Gatti E, Bacci L, Baronti S, Romani A, Rossi F (2008) The stinging nettle (*Urtica dioica* L.): a neglected, multifunctional species for a Low-impact land use. Italian J Agron 3:443–444

Gruzdevienė E, Jankauskienė Z (2012) Fibre nettle (*Urtica dioica* L.)—a new fibre crop in the agricultural ecosystem in Lithuania. Human and nature safety 2012: proceedings of the International scientific conference, Part 2, Akademija (Kauno r.), 94–97 (in Lithuanian)

Hartl A, Vogl R (2002) Dry matter and fiber yields, and the fiber characteristics of five nettle clones (*Urtica dioica* L.) organically grown in Austria for potential textile use. Am J Altern Agric 17(4):195–200

Jankauskienė Z, Gruzdevienė E (2008) Stinging nettle (*Urtica dioica* L.)—an alternative fibre plant. Opportunities and challenges of National Economic Development: proceedings of the international conference. Rezekne, Latvia, 175–182

Jankauskienė Z, Gruzdevienė E (2010) Investigation of stinging nettle (*Urtica dioica* L.) in Lithuania. Latgale National Econ Res 1(2):176–186

Jankauskienė Z, Gruzdevienė E (2013) Investigation of changes of stinging nettle's (*Urtica dioica* L.) crop density. Environment technology. Resources: proceedings of the 9th international scientific and practical conference, vol. I. Rezekne, Latvia, 116–121

Jankauskienė Z, Gruzdevienė E (2015) Changes in the productivity of wild and cultivated stinging nettle (*Urtica dioica* L.) as influenced by the planting density and crop age. Zemdirbyste-Agriculture, 102(1):31–40

Jasinskas A, Simonavičiūtė R, Šarauskis E, Sakalauskas A, Čekanauskas S (2013) Assessment of unconventional bioenergy plant chopping, milling and pelleting quality indicators and physical-mechanical properties. Agron Res 11(2):307–318

Nebel K, Kokhan A, Selcuk R, Pichert T (2002) Textiles from fibre nettles. Production, processing and use of natural fibres. Potsdam-Bornim, 77–78

Slepetys J, Kadziuliene Z, Sarunaite L, Tilvikiene V, Kryzeviciene A (2012) Biomass potential of plants grown for bioenergy production. Renew Energy Energy Effic 2012:66–72

Tarakanovas P, Raudonius S (2003) Statistical analysis of agronomical research data with computer programs ANOVA, STAT, SPLIT-PLOT from packet SELEKCIJA and IRRISTAT. Lithuanian University of Agriculture (in Lithuanian)

Tavano D, Gallucci T, Amaggio G, Lagioia G (2011) Potential of dyeing and fiber plants in Apulia region. J Commodity Sci Technol Q 50(III):207–224

Vogl ChR, Hartl A (2003) Production and processing of organically grown fiber nettle (*Urtica dioica* L.) and its potential use in the natural textile industry: a review. Am J Altern Agric 18(3):119–128

Warren P (2006) 101 uses for stinging nettles. Wildeye, United Kingdom

Chapter 9
An Evaluation of Grass Species as Feedstocks for Combustion in Ireland

J. Finnan, J. Carroll, and B. Burke

Introduction

Growing evidence of the impact of greenhouse gas (GHG) emissions on climate change (Stocker et al. 2013) is forcing countries to consider renewable forms of energy, including bioenergy. Bioenergy feedstocks can be used to provide energy as heat, electricity or transport fuels. However, additional sources of biomass such as purpose grown energy crops will be needed to fulfil future global demand (Sims et al. 2006). Fast growing woody species such as willow and poplar together with perennial rhizomatous grasses such as switchgrass, reed canary grass and *Miscanthus x giganteus* have proved to be serious candidates as energy crops (Clifton-Brown et al. 2007). It has been suggested that grasses such as switchgrass and reed canary grass have an economic advantage over crops such as willow and Miscanthus as they are cheaper to establish (Riche 2005). Additionally, the general familiarity of the farming community with grass production and the widespread availability of harvesting machinery suitable for herbaceous biomass has also been cited as an advantage (McLaughlin and Kszos 2005). Interest in the use of grass as a feedstock for bioenergy has increased in Europe and North America and grass is currently used as a feedstock for biogas production as well as a solid biofuel for combustion (Prochnow et al. 2009a). Grass and herbaceous biomass can be fired as bales, as loose material, as briquettes or as pellets (Prochnow et al. 2009b). One of the most significant recent uses of grass biomass has been in Finland where reed canary grass has been grown as a feedstock for power plants (Pahkala et al. 2008). Additionally, there is significant interest in the use of grass for combustion in North America where switchgrass has been tested as a solid biofuel for heat and electricity production (CANMET 2008; Cherney 2010).

J. Finnan (✉) • J. Carroll • B. Burke
Crops Research Department, Crops, Environment and Land Use Programme, Teagasc Oak Park, Carlow, Republic of Ireland
e-mail: john.finnan@teagasc.ie

S. Barth et al. (eds.), *Perennial Biomass Crops for a Resource-Constrained World*, DOI 10.1007/978-3-319-44530-4_9

The United States Department of Energy identified switchgrass as a promising species for development as a herbaceous biomass fuel crop (Vogel 1996). High biomass yields from switchgrass have been reported in many studies conducted in the United States (Lemus et al. 2002; Vogel et al. 2002; Casler et al. 2004). Interest in the use of switchgrass as a feedstock for bioenergy has also developed in Europe and high yields of switchgrass (>10 t DM/ha) have been reported in Greece and Italy (Alexopoulou et al. 2008) and also in the United Kingdom (Christian et al. 2002). Reed canary grass is a widely adapted and distributed temperate grass which has attracted interest as a bioenergy crop in Sweden (Burvall 1997), the United Kingdom (Christian et al. 2006), Finland (Pahkala et al. 2008) as well as the United States (Casler et al. 2009).

Ireland is almost totally dependent on energy imports and a substantial area of energy crops will be required to meet bioenergy targets (Finnan 2010). Energy grasses which can be established relatively cheaply from seed and which can be harvested with conventional machinery may have a role to play in meeting Ireland's bioenergy targets but there is no information on how different grass species perform as feedstocks for combustion under Irish conditions. The objective of this study was to evaluate switchgrass ecotypes and reed canary grass varieties in Ireland and to compare their biomass yields to those of grass species which are traditionally grown for haymaking in Ireland.

Materials and Methods

Field trials were established in 2008 at two sites in Ireland; Oak Park, Carlow and Knockbeg, Co Laois. Both sites are located within 5 km of each other but are representative of two contrasting soil types, (Oak Park-Eutric cambisol, a light, free-draining soil) and (Knockbeg-luvisol, a heavy, moisture retentive soil). Both experimental sites were located on farms used for cereal crop production, the site in Oak Park was previously used for cereal crop production, potatoes had been grown on the site in the previous year. Winter wheat had been grown at the Knockbeg site in the previous year.

The field trials established on both sites were identical, the experimental design was a randomised complete block with four replications. Upland and lowland ecotypes of switchgrass, varieties of reed canary grass from Sweden (Bamse) and the United States (Chiefton) together with locally sourced varieties of cocksfoot and fall fescue were included as treatments. Details of the grasses included as treatments in the study are included in Table 9.1.

Grass varieties were sown at both sites on 20th May 2008 at a seeding rate of 20 kg/ha and rolled immediately afterwards. Plots measured 16.5 m in length and 5 m in width. Both trials were managed under a low input regime in order to minimise energy input and greenhouse gas emissions. Applications of 125 kg K/ha and 30 kg P/ha were spread on both trial areas on 27/3/2009. 70 kg N/ha was applied to both trials on 8th May 2009. Applications of 60 kg N/ha in 2010 (19th April),

Table 9.1 Grasses included in the study

Treatments (Species)	Treatments (common name)	Varietal name	Origin
Panicum virgatum	Switchgrass	Kanlow	United States (lowland ecotype) (Casler et al. 2004)
Panicum virgatum	Switchgrass	Shawnee	United States (upland ecotype) (Casler et al. 2004)
Phalaris arundinacea	Reed canary grass	Bamse	Sweden (Casler et al. 2009)
Phalaris arundinacea	Reed canary grass	Chiefton	United States (Casler et al. 2009)
Dactylis glomerata	Cocksfoot	Sparta	Seed source: Denmark (DLF Trifolium www.dlf.com)
Festuca arundinacea	Tall fescue	Olga	Seed source: Denmark (DLF Trifolium www.dlf.com)

Table 9.2 Harvest Dates, 2009 to 2012

	Tall fescue, Cocksfoot	Reed canary grass, switchgrass
2009	5–6/8/09	15/1/2010
2010	16/8/2010	20/10/2010
2011	1/9/2011	28/9/2011
2012	7–8/8/2012	22/10/2012

70 kg N/ha in 2011 (15th April) and 2012 (9th May) were spread on both trial areas. Both trials were sprayed with broad leaf herbicide on 24th June 2008 (Metsulfuron and Mecoprop-P), on 8th August 2008 (2-methyl-4-chlorophenoxyacetic acid (MCPA)) as well as on 2nd June 2011 (MCPA).

Harvesting

Strips of 16.5 m by 1.25 m were harvested from each plot after anthesis. Tall fescue and cocksfoot reached anthesis first and were typically harvested in August or early September, whereas reed canary grass and switchgrass reached maturity at a later date and were typically harvested in October (Table 9.2). The fresh weight of the harvested material was determined in the field using a Salter Brecknell WB6200 (1000 Armstrong Drive, Fairmont, MN, 56031, USA) weighing system before a sample was taken for dry matter determination. The biomass from all plots was harvested and removed from the experimental areas after the plots had been harvested.

Statistics

The results were analysed by analysis of variance using the Proc GLIMMIX procedure (SAS 2009). Pairwise differences between treatments were evaluated using Tukey's test.

Table 9.3 Meteorological data for each growing season (April to October)

	Mean Temperature (°C)	Rainfall (mm)	Solar Radiation (MJ/m2/day)
2009	12.8	659.7	12.4
2010	12.6	439.4	13.2
2011	12.5	408.1	12.4
2012	11.9	574.3	12.0

Results and Discussion

Meteorological Information

Mean temperature, rainfall and solar radiation for the growing season (April to October) of each year of the study are presented in Table 9.3. The mean temperature during the growing season was highest in 2009 (12.8 °C) and was lowest in 2012 (11.9 °C). Total rainfall was also highest during the 2009 growing season (659.7 mm) and was lowest in 2011 (408.1 mm). Solar radiation was highest in 2010 but lowest in 2012.

Biomass Yield

Both switchgrass varieties were slow to establish, the first leaves emerged above the ground 5 weeks after sowing. The switchgrass variety ‘Kanlow’ did not survive the winter following the first growing season. The lowest minimum air temperature during that winter was −7 °C while the lowest minimum soil (5 cm) temperature was −1.8 °C. Kanlow frequently suffers from substantial winter damage when grown North of 40 ^{0}N in the US Great Plains and Midwest (Vogel et al. 2014). In Europe, ‘Kanlow’ can experience winter survival problems particularly in the first year (Christian et al. 2001). ‘Kanlow’ originates from Oklahoma (Vogel et al. 2014) and its poor survival in Ireland most probably results from the fact that it was grown too far North of the area in which it was adapted. Survival decreases as lowland ecotypes are moved North of their zone of adaptation (Casler et al. 2004). However, Riche (2005) reported that ‘Kanlow’ established successfully at several sites in the south of England. In contrast, the switchgrass variety ‘Shawnee’ is an upland variety found in Illinois and consequently better adapted to ambient temperatures in Ireland. However, ‘Shawnee’ was slow to establish and only produced a harvestable yield after the third growing season. Yields from 2010 to 2012 were reasonably stable but only approximately half the magnitude of the better yielding grasses. Yields of the switchgrass variety ‘Shawnee’ (approximately 4 t DM/ha) were considerably lower than yields reported in the United States (Lemus et al. 2002; McLaughlin and

Table 9.4 Biomass yields (DM tonnes/ha) from 2009 to 2012. Figures followed by the same letter are not significantly different

	2009	2010	2011	2012	All years
Treatments					
Dactylis glomerata	8.1a	8.2a	6.4a	7.7a	7.4a
Festuca arundinacea	8.1a	7.3a	5.6a	8.1a	7.3a
Phalaris arundinacea Bamse	8.5a	8.4a	5.2ab	3.7b	6.5ab
Phalaris arundinacea Chiefton	9.0a	7.3a	5.1ab	2.3b	5.4b
Panicum virgatum Shawnee	–	4.7b	3.1b	4.1b	3.5c
Site					
Oak Park	8.0a	6.1b	4.3b	4.6b	5.3b
Knockbeg	8.6a	8.3a	5.8a	5.8a	6.7a
Statistics					
Site	ns	<0.0001	0.0048	0.0331	<0.0001
Treatment	ns	<0.0001	0.0044	<0.0001	<0.0001
Year					<0.0001
Site*Treatment	ns	ns	0.0146	ns	<0.05
Year*Site					ns
Year*Treatment					<0.0001
Year*Site*Treatment					ns

Kszos 2005; Vogel et al. 2002) and also lower than yields reported in the United Kingdom (Christian et al. 2002; Riche 2005) where establishment was also faster than that experienced at both Irish sites.

Dry matter yields during the 4 years of the study are presented in Table 9.4. Switchgrass was slow to develop and there was insufficient biomass to harvest in 2009. There were no significant effects of treatment or site on yield in 2009 when biomass yields ranged from 8.1 t DM/ha to 9 t DM/ha (tonnes of dry matter per hectare). There was a significant treatment effect in 2010 ($P<0.0001$) when the yield of the switchgrass variety 'Shawnee' was lower than that of all other grasses. There was also a significant treatment effect in 2011 ($P<0.01$) when switchgrass yields were significantly lower ($P<0.0001$) than those of cocksfoot and tall fescue and in 2012 ($P<0.05$) when the yields of both reed canary grass varieties and that of 'Shawnee' were significantly lower than those of cocksfoot and tall fescue. Site had a significant effect on yield from 2010 to 2012 when yields from the heavier soil at Knockbeg were greater than those from the lighter soil in Oak Park. There was a significant interaction between site and treatment in 2011 ($P<0.05$) as yields of 'Shawnee' were 7.6 t DM/ha in Knockbeg but only 2.5 t DM/ha in Oak Park. Over the last three years of the study, there was a significant treatment effect ($P<0.01$) as yields of Shawnee were significantly lower than all other grasses and yields of the reed canary grass variety 'Chiefton' were significantly lower than those of cocksfoot and tall fescue. There was a significant interaction between site and treatment ($P<0.01$) primarily as a result of very different switchgrass yields between and two sites. There was also a significant interaction between year and treatment ($P<0.0001$)

primarily because of the poor performance of switchgrass at the start of the study and because of the fact that yields of reed canary grass declined as the study progressed.

Reed canary grass yields did not differ significantly from yields of cocksfoot and tall fescue in 2009 and 2010. However, after 2010, the yields of both varieties of reed canary grass progressively declined whereas yields of cocksfoot and tall fescue were comparatively stable. During the first 2 years (2009–2010), reed canary grass yields were somewhat higher than those reported by Riche (2005) but similar to those reported by Christian et al. (2006) for reed canary grass grown in the United Kingdom. Similar yields have also been reported in studies conducted in the United States (Casler et al. 2009; Cherney et al. 2003). In general, reed canary grass showed poor persistence as yields declined rapidly after 2010. On the basis of these results, reed canary grass would need to be re-sown every 3–5 years.

Moisture Content

Biomass dry matter contents are shown in Table 9.5, dry matter contents were generally less than 50 %. There were significant treatment effects on dry matter content in all years of the study. In 2009, the reed canary grass variety 'Bamse' had a significantly lower dry matter content compared to the other grasses in the study. In 2010 and 2011, there was no significant difference in the dry matter content between the

Table 9.5 Biomass dry matter contents from 2009 to 2012. Figures followed by the same letter are not significantly different

	2009	2010	2011	2012	2010–2012
Treatments					
Dactylis glomerata	38.8 a	48.4 a	49.5 ab	34.1 b	44.0ab
Festuca arundinacea	38.5 a	46.3 a	56.3 a	37.9 ab	46.9a
Phalaris arundinacea Bamse	30.7 b	45.5 a	54.2 ab	44.2 a	48.0a
Phalaris arundinacea Chiefton	39.1 a	44.0 ab	53.8 ab	41.8 ab	46.6a
Panicum virgatum Shawnee	–	39.2 b	43.2 b	38.3 ab	40.3b
Site					
Oak Park	36.0 a	43.4 b	53.2 a	37.8 a	44.8a
Knockbeg	37.5 a	46.0 a	49.6 a	40.8 a	45.5a
Statistics					
Site	ns	<0.05	ns	ns	ns
Treatment	<0.01	<0.001	<0.05	<0.05	<0.01
Year					<0.0001
Site*Treatment	ns	ns	ns	ns	ns
Year*Site					ns
Year*Treatment					<0.05
Year*Site*Treatment					ns

reed canary grass varieties and the cocksfoot and tall fescue varieties but the dry matter content of the switchgrass variety 'Shawnee' was lower than all the other grasses. In 2012, the dry matter content of cocksfoot was lower than all of the other grasses and significantly lower than the reed canary grass variety 'Bamse'. Dry matter contents were generally lower than those reported by Riche (2005) and Christian et al. (2006) for reed canary grass harvested in the United Kingdom. Over the last three years of the study, there was a significant treatment effect ($P<0.01$). 'Shawnee' had significantly lower dry matter content compared to tall fescue and the reed canary grass varieties 'chiefton' and 'Bamse' although there was no significant difference in the dry matter contents of the two reed canary grass varieties, tall fescue and cocksfoot. There was a significant interaction between year and treatment ($P<0.001$) primarily due to differences in the dry matter content of the reed canary grass variety 'Bamse' between years.

Conclusions

Upland switchgrass ecotypes were slow to establish and produced poor biomass yields over the course of the study, whereas the biomass yields of reed canary grass varieties declined sharply after 3 years. In contrast, cocksfoot and tall fescue varieties produced good biomass yields over the course of the study. Additionally, such species mature earlier in the summer when better drying conditions generally prevail. Higher dry matter contents should consequently be achievable which will benefit both conservation and combustion. Earlier harvesting should also permit additional revenue generation from late season grazing. We conclude that cocksfoot and tall fescue varieties which are used in Ireland for hay-making offer a better alternative to the use of switchgrass ecotypes and reed canary grass varieties as feedstocks for combustion.

Acknowledgements The authors are grateful for the assistance and advice of Dr Andrew Riche from Rothamsted Research Centre, Harpenden, Hertfordshire.

References

Alexopoulou E, Sharma N, Papatheohari Y, Christou M, Piscioneri I, Panoutsou C, Pignatelli V (2008) Biomass yields for upland and lowland switchgrass varieties in the Mediterranean region. Biomass Bioenerg 32:926–933

Burvall J (1997) Influence of harvest time and soil type on fuel quality in reed canary grass (Phalaris arundinacea L.). Biomass Bioenerg 12:149–154

CANMET (2008) Combustion testing of switchgrass pellets on a 1MWth industrial grate furnace. CANMET energy technology centre, Ottawa. http://www.agrireseau.qc.ca/energie/documents/switchgrass%20(3).pdf. Accessed 8 Jan 2015

Casler MD, Vogel KP, Taliaferro CM, Wynia RL (2004) Latitudinal adaptation of switchgrass populations. Crop Sci 44:293–303

Casler MD, Cherney JH, Brummer CE (2009) Biomass yield of naturalized populations and cultivars of reed canary grass. Bioenerg Res 2:165–173
Cherney JH (2010) Grass combustion update. http://www.cce.cornell.edu/EnergyClimateChange/Documents/Cherney%20Grass%20for%20Combustion%202010%20update%20_3_%20_3_.pdf. Accessed 8 Jan 2015
Cherney JH, Cherney DJR, Casler MD (2003) Low intensity harvest management of reed canary grass. Agron J 95:627–634
Christian DG, Riche AB, Yates NE (2002) The yield and composition of switchgrass and coastal panic grass grown as a biofuel in Southern England. Bioresour Technol 83:115–124
Christian DG, Yates NE, Riche A (2006) The effect of harvest date on the yield and mineral content of *Phalaris arundinacea* L. (reed canary grass) genotypes screened for their potential as energy crops in southern England. J Sci Food Agric 86:1181–1188
Christian DG, Elbersen HW, El Bassam N, Sauerbeck G, Alexopoulou E, Sharma N, Piscioneri I, de Visser P, van der Berg D (2001) Management guide for planting and production of switchgrass as a biomass crop in Europe. In. Final report of FAIR 5-CT97-3701, Switchgrass (*Panicum virgatum* L.) as an alternative energy crop in Europe, Initiation of a productivity network. pp 70–73. http://www.switchgrass.nl/en/show/Switchgrass-as-an-Alternative-Energy-Crop-1.htm. Accessed 14 Dec 2015
Clifton-Brown JC, Bruer J, Jones MB (2007) Carbon mitigation by the energy crop Miscanthus. Glob Chang Biol 13:1–12
Finnan J (2010) The case for energy crops. In: Energy crops manual 2010, Teagasc, Carlow (ISSN -13 978-1-84170-556-9), pp 10–14
Lemus R, Brummer EC, Moore KJ, Molstad NE, Burras CL, Barker MF (2002) Biomass yield and quality of 20 switchgrass populations in southern Iowa, USA. Biomass Bioenerg 23:433–442
McLaughlin SB, Kszos LA (2005) Development of switchgrass (*Panicum virgatum*) as a bioenergy feedstock in the United States. Biomass Bioenerg 28:515–535
Pahkala K, Aalto M, Isolahti M, Poikola J, Jauhiainen L (2008) Large-scale energy grass farming for power plants—a case study from Ostrobothnia, Finland. Biomass Bioenerg 32:1009–1015
Prochnow A, Heiermann M, Plochl M, Linke B, Idler C, Amon T, Hobbs PJ (2009a) Bioenergy from permanent grassland—a review: 1. Biogas Bioresour Technol 100:4931–4944
Prochnow A, Heiermann M, Plochl M, Amon T, Hobbs PJ (2009b) Bioenergy from permanent grassland—a review: 2. Combust Bioresour Technol 100:4945–5954
Riche A (2005) A trial of the suitability of switchgrass and reed canary grass as biofuel crops under UK conditions. ETSU B/CR/00655/00/00
SAS (2009) Version 9.4, SAS Institute Inc., Cary, NC
Sims REH, Hastings A, Schlamadinger B (2006) Energy crops: current status and future prospects. Glob Chang Biol 12:2054–2076
Stocker TF, Qin D, Plattner G-K, Alexander LV, Allen SK, Bindoff NL, Bréon F-M, Church JA, Cubasch U, Emori S, Forster P, Friedlingstein P, Gillett N, Gregory JM, Hartmann DL, Jansen E, Kirtman B, Knutti R, Krishna Kumar K, Lemke P, Marotzke J, Masson-Delmotte V, Meehl GA, Mokhov II, Piao S, Ramaswamy V, Randall D, Rhein M, Rojas M, Sabine C, Shindell D, Talley LD, Vaughan DG, Xie S-P (2013) Technical Summary. In: Climate Change 2013: The Physical Science Basis. Contribution of Working Group I to the Fifth Assessment Report of the Intergovernmental Panel on Climate Change [Stocker, T.F., D. Qin, G.-K. Plattner, M. Tignor, S.K. Allen, J. Boschung, A. Roy, M.M., Corscadden, K.W. (2012) An experimental study of combustion and emissions of biomass briquettes in a domestic wood stove. Applied Energy, 99, 206–212
Vogel KP (1996) Energy production from forages (or American agriculture—back to the future). J Soil Water Conserv 51:137–139
Vogel KP, Brejda JJ, Walters DT, Buxton DR (2002) Switchgrass biomass production in the Midwest USA: Harvest and nitrogen management. Agron J 94:413–420
Vogel KP, Mitchell RB, Casler MD, Sarath G (2014) Registration of 'liberty' switchgrass. J Plant Reg 8:242–247

Part III
Genotypic Selection of Perennial Biomass Crops for Crop Improvement

Chapter 10
New Breeding Collections of *Miscanthus sinensis*, *M. sacchariflorus* and Hybrids from Primorsky Krai, Far Eastern Russia

T.R. Hodkinson, E. Petrunenko, M. Klaas, C. Münnich, S. Barth, S.V. Shekhovtsov, and S.E. Peltek

Introduction

Miscanthus is one of the most promising sources of biomass and is being bred for growth in a wide range of environments including both productive agricultural land and areas considered marginal for conventional farming (Jørgensen 2011; Nijsen et al. 2012). Marginal land is hard to define but biologically it includes areas that are suboptimal, unpredictable, or poor for crop production due to factors such as climate, drought, nutrient deficiency, pollution, salinity, or waterlogging (Cai et al. 2011; Donnelly et al. 2011; Gopalakrishnan et al. 2013). The potential of these areas is enormous for biomass production. For example, China possesses 100 million hectares of marginal and degraded land especially in northern and western regions that have the potential to produce approximately one billion tonnes per annum of *Miscanthus* biomass feedstock (Sang and Zhu 2011). To breed *Miscanthus* suitable for agricultural and marginal land it will be essential to

T.R. Hodkinson (✉)
Department of Botany, School of Natural Sciences, Trinity College Dublin, The University of Dublin, Dublin, Ireland
e-mail: Trevor.Hodkinson@tcd.ie

E. Petrunenko
Botanical Garden-Institute FEB RAS, Vladivostok, Russia

M. Klaas • S. Barth
Teagasc, Crops, Environment and Land Use Programme, Oak Park Crops Research Centre, Carlow, Ireland

C. Münnich
Biotechnik und Pflanzenvermehrung GmbH, OT Klein Wanzleben, Magdeburger Landstr. 33, Wanzleben-Börde 9164, Germany
e-mail: cora.muennich@aol.de

S.V. Shekhovtsov • S.E. Peltek
Institute of Cytology & Genetics SB RAS, Novosibirsk 630090, Russia

S. Barth et al. (eds.), *Perennial Biomass Crops for a Resource-Constrained World*, DOI 10.1007/978-3-319-44530-4_10

incorporate and utilize a diverse range of germplasm in breeding programs to manipulate the traits of interest (reviewed in Jones et al. 2014). The EU FP7 project GrassMargins aims to utilize such diversity and develop genotypes suitable for growth on European marginal land (www.grassmargins.com).

The genus *Miscanthus* is diverse, and there is clearly potential to utilize the natural adaptive variation of the species (Hodkinson et al. 2015). Taxonomically, *Miscanthus* is a genus of about ten species. Some authors include additional species from Africa and the Himalayas (Clayton et al. 2006), but it has been shown by DNA sequencing that the genus is not monophyletic if these are included (Hodkinson et al. 1997, 2002a; Swaminathan et al. 2010). *Miscanthus sensu stricto* (*s.s.*, in the strict sense) therefore contains the following species (Hodkinson et al. 2015):

Miscanthus floridulus (Labil.) Warb. ex K. Schum. and Lauterb. [type]
M. intermedius (Honda) Honda
M. longiberbis Nakai
M. lutarioriparius L.Liu ex Renvoize and S.L. Chen*
M. oligostachyus Stapf.
M. paniculatus (B.S. Sun) Renvoize and S.L. Chen
M. sacchariflorus (Maxim.) Hack.
M. sinensis Anderss.
M. tinctorius (Steud.) Hack.
M. transmorrisonensis Hayata
M. ×*giganteus* Greef et Deuter ex Hodkinson and Renvoize
*Note *M. lutarioriparius* is likely to be an infraspecific taxon of *M. sacchariflorus*

Only those species with basic chromosome number of 19 should be included in *Miscanthus s.s.* (Hodkinson et al. 2002a, c; 2014). Polyploidy is common in the genus and species form a polyploid complex with diploids, triploids, tetraploids, pentaploids, and hexaploids (Adati and Shiotani 1962; Hodkinson et al. 2001; Hodkinson et al. 2002b). The type species of the genus is *M. floridulus* (Labil.) Warb. (= *M. japonicus* Anderss; basionym *Saccharum floridulum* Labillardière described in 1824). *Miscanthus s.s.* is native to eastern Asia, south-eastern Asia and the South Pacific with the highest species diversity recorded in eastern Asia especially China and Japan (Chen and Renvoize 2006; Sun et al. 2010). Its native distribution includes a wide range of climatic zones running latitudinally from temperate south east Russia at 50°N to tropical Polynesia at 22°S. Longitudinally it extends from Burma, Andaman and Nicobar Islands at 92°E to Fiji at 179°W (Hodkinson et al. 2015). Some species such as *M. floridulus* generally grow at sea level or in warm tropical climates, but others such as *M. paniculatus* can tolerate high altitudes of up to 3100 m on mountains in Guizhou, Sichuan, and Yunnan of China (Chen and Renvoize 2006). *Miscanthus* has also been become naturalized following human introduction in many regions of the world including Eurasia, New Zealand, and the Americas (Meyer et al. 2010; Quinn et al. 2010; Barney et al. 2012).

Miscanthus species are perennial and rhizomatous (Fig. 10.1) with erect cane-like stems growing up to 7 m tall (in *M. lutarioriparius*=*M. sacchariflorus*; Chen and Renvoize 2006). They are sometimes tufted with short rhizomes and can be

Fig. 10.1 Line drawings of (**a**) *Miscanthus sacchariflorus* and (**b**) *M. sinensis* (from Sun et al. (2010), with permission, license number 3715490204490). (**a**) A, rhizome and culm; B, panicle and leaf; C, paired spikelets; D, back and ventral sides of a lower glume; E, ventral side of an upper glume; F, back of a lower lemma; G, back of an upper lemma without awn; and H, lodicule. GR, seed (**b**) A, panicle and leaf; B, paired spikelets; C, back of a lower glume; D, ventral side of an upper glume; E, ventral side of an upper lemma with awn; F, stamens and gynoecium; GR, seed

spreading with long racemes (Fig. 10.1). The inflorescence is terminal and contains plumose racemes with spikelets that are pedicellate and paired, one with long pedicel and one with short pedicel. The inflorescence axis may be long and have relatively short racemes as in *M. floridulus* or may be short with long racemes. When the axis is short it is known as a subdigitate inflorescence, as in most *M. sacchariflorus* and *M. sinensis* (Koyama 1987; Osada 1993; Chen and Renvoize 2006; Sun et al. 2010; Hodkinson et al. 2015).

Miscanthus × *giganteus* is most commonly grown for biomass and is a hybrid between *M. sacchariflorus* and *M. sinensis*. Its taxonomy and systematics are discussed in Hodkinson and Renvoize (2001) and Hodkinson et al. (2002a, b, 2014). The most widely grown genotype of *M.* ×*giganteus* is triploid (Linde-Laursen 1993) and is morphologically most similar to *M. sacchariflorus* in its inflorescence characters but lacks its spreading rhizomes and is more like *M. sinensis* in habit.

Three taxa, *M. sacchariflorus*, *M. sinensis*, and *Miscanthus* × *giganteus* are under the greatest scrutiny for biomass production (Jones and Walsh 2001; Clifton-Brown et al. 2008) and germplasm from a diversity of habitats in their natural range is most likely to provide the natural genetic resources required in breeding programs for various desired traits (Jones et al. 2014). In north-west Europe for example, temperature and length of growing season are particularly limiting (Clifton-Brown et al. 2008). Cold tolerance can be expected from populations at high latitude and altitudes in the native range. We therefore targeted the Russian Far East for the collection of new germplasm as it has a diverse range of habitats and is near the northern limit of *Miscanthus* distribution. Both *M. sacchariflorus* and *M. sinensis* are native to this area and are particularly abundant in Primorsky Krai and Sakhalin (Komarov et al. 1963). The collecting trip to Primorsky Krai was led by Trevor Hodkinson and assisted by Ekaterina Petrunenko, Manfred Klaas, Cora Münnich, and Sergei Shekhovtsov as part of the EU GrassMargins project in September 2014 with the aim of collecting new *Miscanthus* genotypes adapted to a wide range of environments. The results of the expedition are outlined in this paper.

Materials and Methods

Herbarium specimens from the Institute of Biology and Soil Sciences, FEB RAS, Vladivostok, and the Royal Botanic Gardens, Kew UK were screened for geolocality information. On the basis of the search, and also additional records from GBIF (gbif.org), we planned and undertook collecting of *Miscanthus* along a route through Primorsky Krai designed to maximize diversity of samples by covering a wide range of habitat types and altitudes including south west of the region (Khasansky district), Khanka lowland (Khankaysky, Khorolsky districts), and the Sikhote Alin Ridge (Fig. 10.2). These included coastal saline and sandy regions, disturbed open ground, forest margins, wet floodplains, and ditches. Associated passport information was recorded at each site and GPS coordinates and altitude recorded using a Garmin GPSMAP 62 or an Apple iPhone5.

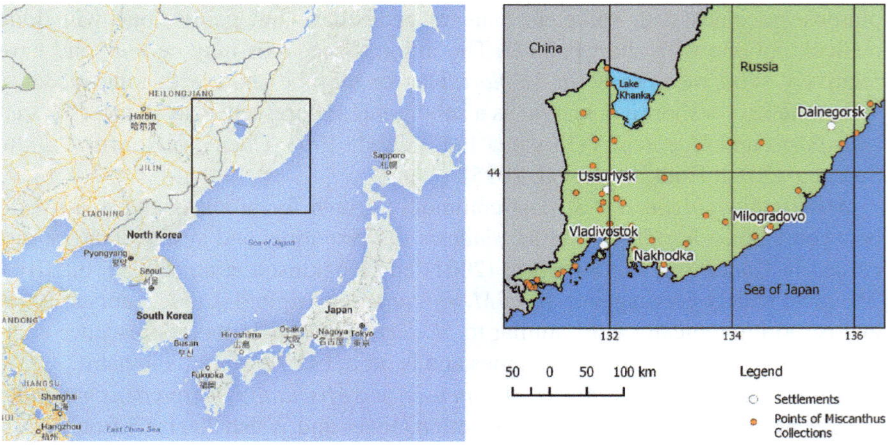

Fig. 10.2 Sampling sites in Primorsky Krai, Russia for *Miscanthus sacchariflorus*, *M. sinensis* and hybrids

Intense sampling was undertaken at ten populations by collecting ca. 30 genotypes along transects, sampling plants every 5 m. Additional collections were also made on route to maximize sampling coverage (spot sampling). For population sampling (Table 10.1a), approximately 30 individuals were sampled for seed by placing whole inflorescences in paper envelopes, at least two individuals were sampled for rhizomes by taking about 10 cm of rhizome and placing it in a small ziplock bag of potting compost, and at least two samples per population were taken and pressed in a standard plant press for herbarium specimens (to confirm identity of the material). For spot sampling (Table 10.1b); seed, rhizome, and herbarium samples were taken for each individual plant. Rhizome material was grown into mature plants that are now maintained in the field at SB RAS Novosibirsk, Russia and Teagasc, Ireland. Identity of plant material was confirmed by checking gross morphology with a stereomicroscope (Leica MZ-12), with graticule, using mainly spikelet characters as described in detail in Hodkinson et al. (2014). Particular emphasis was given to spikelet characters such as the absence/presence of awns and length of the callus hairs as these are particularly useful for species differentiation especially when combined with culm bud and rhizome characters (Lee 1964a, b, c; Scally et al. 2001a, b).

Results and Discussion

A total of 361 plants were sampled in Primorsky Krai (Table 10.1) along a latitudinal range from 42.61 to 45.24 and longitude range from 130.65 to 136.04. Altitude of sampling site varied from sea level to 786 m, including the range of altitudes between (Fig. 10.3). However, *Miscanthus* became rare at the higher altitudes and was not found above 786 m despite efforts to find such populations. At higher altitudes *Miscanthus* populations did not form thick dense stands as is found at low altitudes. This could be explained directly by climatic conditions or indirectly due to the decrease in disturbed open grounds (such as agriculture lands) at higher altitudes and the prevalence of forest vegetation, creating conditions unsuitable for *Miscanthus* growth. Some populations were coastal, found either directly on sand spit or beach ecosystems or on land adjacent to the coast within salt spray distance or subjected to occasional sea flooding

Rhizome material was sent to RAS Novosibirsk, Russia and then divided and sent to Teagasc, Ireland for growing on. These plants are now maintained in the living collection at both these institutes. Inflorescences with seed and herbarium specimens are stored in Trinity College Dublin, Ireland Herbarium (TCD). Although the material has yet to undergo physiological evaluation, it is assumed that the genotypes are locally adapted to the variety of environments they were sampled from. We attempted to incorporate as much adaptive and genetic diversity as possible in our sampling regime. Field trials and laboratory based controlled experiments on *Miscanthus* collections housed elsewhere, using a range of genotypes, have revealed variation in agronomic traits such as yield (Jeżowski et al. 2011; Gauder et al. 2012),

Table 10.1 *Miscanthus* collections made in September 2014 in Primorsky Krai, Far Eastern Russia

	Number	Name	Date	Location	Lat	Long	Alt (m)
Population samples (a)							
Population 1	1–25	*M. sinensis* and *sacchariflorus*	08/09/14	Sadgorod Station Vladivostok	43.262158	132.041311	26.62
Population 2	37–68	*M. sinensis* and *sacchariflorus*	09/09/14	East of Mayachnoye	42.609235	130.776914	0
Population 3	72–103	*M. sinensis* and *sacchariflorus*	09/09/14	East of Slavyanka	42.869752	131.437481	47.12
Population 4	110–151	*M. sacchariflorus*	10/09/14	Near Monakino (Ussuriysk district)	43.76133	131.453222	273.1
Population 5	154–185	*M.sacchariflorus*	11/09/14	North of Galyonki	44.085711	131.727682	85.85
Population 6	190–222	*M. sacchariflorus*	11/09/14	Close to Chinese border	45.240108	131.961044	103.04
Population 7	227–258	*M. sacchariflorus*	12/09/14	Near Dostoyevka	44.315067	133.462891	152.27
Population 8	261–290	*M. sacchariflorus*	13/09/14	Coast SE of Dalnegorsk	44.349268	135.804589	0
Population 9	293–322	*M. sacchariflorus*	14/09/14	NW of Vetka	43.783829	135.091991	0
Population 10	331–360	*M. sacchariflorus*	15/09/14	North of Lukyanovka	43.182406	132.695666	142.95
Spot samples (b)							
	26–28	*M. sacchariflorus*	08/09/14	Btw Sanatornaya and Okeanskaya stations	43.234092	131.987626	0.47
	29	*M. sacchariflorus*	08/09/14	Near Volno-Nadezhdinskoye	43.383384	132.001122	81.51
	30–32	*M. sinensis* and *sacchariflorus*	08/09/14	South of Olenevod	43.557578	131.847387	23.12
	33	Hybrid	08/09/14	Barabash	43.1775	131.529044	25.03
	34	*M. sacchariflorus*	09/09/14	North of Mayachnoye	42.669587	130.652358	0
	35	*M. sinensis*	09/09/14	South of Mayachnoye	42.61284	130.718234	24.72
	36	Hybrid	09/09/14	South of Mayachnoye	42.612414	130.718421	17.3
	69	*M. sacchariflorus*	09/09/14	North of Mayachnoye	42.641054	130.691634	4.52

	71	Hybrid	09/09/14	East of Slavyanka	42.868665	131.428785	16.6
	104	Hybrid	09/09/14	East of Slavyanka	42.870287	131.437256	49.48
	105	*M. sinensis*	09/09/14	Near Ryazanovka	42.801152	131.245324	59.51
	106	*M. sacchariflorus*	09/09/14	SW of Ryazonvka	42.772059	131.157649	132.82
	107	*M. sacchariflorus*	10/09/14	Near Baranovskiy	43.642138	131.895028	24.72
	108	*M. sacchariflorus*	10/09/14	East of Linevichi	43.748473	131.873335	21.05
	109	*M. sacchariflorus*	10/09/14	Near Monakino (Ussuriysk district)	43.76026	131.454471	275.71
	152	*M. sacchariflorus*	10/09/14	Near Ussuriysk	43.690882	132.111905	28.63
	153	*M. sacchariflorus*	10/09/14	Near Kamenushka	43.637874	132.219489	56.1
	186–187	*M. sacchariflorus*	11/09/14	South of Nesterovka	44.401283	131.769104	121
	188	*M. sacchariflorus*	11/09/14	West of Rubinovka	44.711266	131.568692	454.36
	189	*M. sacchariflorus*	11/09/14	West of Rubinovka	45.055682	131.992568	68.8
	223	*M. sacchariflorus*	11/09/14	Kamen-Rybolov	44.72708	132.0403	118
	224	*M. sacchariflorus*	11/09/14	South of Khorol	44.3889	132.07529	131
	225	*M. sacchariflorus*	12/09/14	Near Gorbatka	43.951211	132.316014	177.95
	226	*M. sacchariflorus*	12/09/14	East of Orlovka	43.939567	132.898836	257.69
	259	*M. sacchariflorus*	12/09/14	North of Kamenka	44.363061	133.981584	221.12
	260	*M. sacchariflorus*	12/09/14	Close to Kavalerovo	44.36362	134.481845	256.1
	291	*M. sacchariflorus*	13/09/14	Between Dalnegorsk and Plastun	44.473862	136.036245	0
	292	*M. sacchariflorus*	13/09/14	North of Plastun	44.812747	136.266257	59.4
	323	*M. sacchariflorus*	14/09/14	Shcherbakovka	43.559628	134.633803	126.95

(continued)

Table 10.1 (continued)

	Number	Name	Date	Location	Lat	Long	Alt (m)
Spot samples (b)							
	324	*M. sacchariflorus*	14/09/14	Milogradovo	43.34627	134.634599	32.78
	325	*M. sacchariflorus*	14/09/14	Danilchenkovo	43.229673	134.378008	8.27
	326	*M. sacchariflorus*	14/09/14	North of Lazo	43.400797	133.893328	226.07
	327	*M. sacchariflorus*	14/09/14	NE of Monakino (Partizansk district)	43.482352	133.577514	786.05
	328	*M. sacchariflorus*	14/09/14	Between Orel and Vodopadnoye	43.141042	133.25324	158.51
	329	*M. sacchariflorus*	15/09/14	NW of Nakhodka	42.885545	132.889295	215.15
	330	*M. sacchariflorus*	15/09/14	South of Bolshoy Kamen	43.070949	132.413365	70.44
	360	*M. sacchariflorus*	15/09/14	North of Lukyanovka	43.182406	132.695666	142.95
	361	*M. sacchariflorus*	15/09/14	North of Shkotovo	43.354726	132.367605	19.68

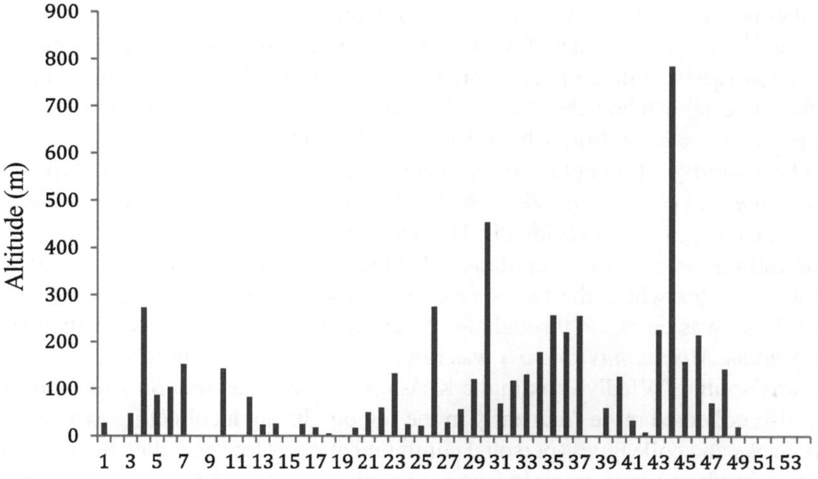

Fig. 10.3 Altitudinal distribution of collecting sites of *Miscanthus* in Primorsky Krai, Russia (as coded in Table 10.1)

drought tolerance (Clifton-Brown and Lewandowski 2002), temperature control of leaf growth (Farrell et al. 2006), frost and cold tolerance (Clifton-Brown and Jones 1997; Głowacka et al. 2014a; Zub et al. 2012), flowering time (Jensen et al. 2011), senescence (Robson et al. 2011), chemical composition, morphology (Jørgensen 1997; Hodgson et al. 2010; Allison et al. 2011; Zhao et al. 2013, 2014; Zhang et al. 2012), and seed germination (Dwiyanti et al. 2014). These studies have demonstrated huge phenotypic variation in and among *Miscanthus* species (Zub and Brancourt-Hulmel 2010; Jones et al. 2014) that can be utilized in breeding. Our collections therefore represent important new resources for breeding institutes and genetic resource scientists. Although many genetic resource collections of *Miscanthus* exist in Europe and the Americas outside of Asia, there is neither a directory of *Miscanthus* collections nor a coordinated program for the conservation of its genetic resources. There is a therefore a need for a formalized network of genetic resource collections and better seed/field bank coordination at an international level.

Little is known about population genetic structuring and biogeography in *Miscanthus* (Hodkinson et al. 2013). However, a recent study by Clark et al. (2014) sampled of over 600 *M. sinensis* accessions covering China, South Korea, and Japan using a high-density set of SNP markers and ten plastid microsatellites. The markers detected six genetic clusters from geographically distinct regions. Four clusters were from mainland Asia (Southeast China, Yangtze-Qinling, Sichuan Basin, Korea/North China) and two were from Japan (Southern and Northern). Another study by Slavov et al. (2014) used SNP and SSR markers to study putatively neutral genetic diversity in a large breeding collection of *Miscanthus* from China, Japan, and Korea. They also included phenotypic traits relating to biomass, cell wall composition, morphology, and phenology. Their data revealed considerable population genetic differentiation over geographical space from Korea to Japan with a longitudinal cline (from 124° to 142° E) accounting for a high proportion of the molecular variation. In contrast, they found that latitudinal and altitudinal variation best explained variation in the phenotypic traits. It will therefore be important, in future studies, to establish how the Russian *Miscanthus* collections made in our expedition relates to the material from China, Korea, and Japan.

The majority of samples we collected out of a total of 361 genotypes were *Miscanthus sacchariflorus* (287), with a lower frequency of *M. sinensis* (70) and genotypes assigned as hybrids (4). However, the frequency of hybrids might be an underestimation because morphological evidence of introgression was detected at a number of sites where the two species were found in sympatry. *Miscanthus sacchariflorus* was common throughout the expedition range and locally abundant in many areas. *Miscanthus sinensis* was rarer but locally abundant in some areas, particularly south of Vladivostok in the Khasansky district. *Miscanthus sacchariflorus* was also collected more frequently in the region, during a collecting expedition by Erik Sacks and colleagues (personal communication). However, they did not sample the area south of Vladivostok in Khasansky where we found *M. sinensis*, and potential interspecific hybrids, at a higher frequency.

The collections provide new material that can be used to improve *M. sinensis* or *M. sacchariflorus* through infra-specific crossing. In addition, the two species are the parents of the interspecific hybrid *M.* ×*giganteus* so they can be used as parental material in controlled crosses aiming to generate new *M.* ×*giganteus* genotypes. Furthermore, the potential wild hybrids discovered during the expedition could, in themselves, represent important new *M.* ×*giganteus* genotypes and these are currently being evaluated by the GrassMargins team. Diversity in *M.* ×*giganteus* collections is of particular concern and there is an extremely narrow genetic base to the crops. Głowacka et al. (2014b) used nuclear and chloroplast SSRs in combination with restriction-site associated DNA sequencing (RAD-Seq), to estimate genetic similarity in over 30 *M.* ×*giganteus* accessions of unknown provenance (legacy cultivars) from collections in North America and Europe, alongside a number of newly bred *M.* ×*giganteus* genotypes and found that genetic variation in the legacy cultivars was extremely low. A total of 27 of these legacy cultivars were inferred as clones matching the *M.* ×*giganteus* type specimen of Hodkinson and Renvoize (2001). Therefore, the material collected here will help address the alarmingly low levels of genetic diversity in *M.* ×*giganteus* germplasm as well as being useful for the wider biological community (Salamin et al. 2001).

Miscanthus × *giganteus* is believed to have been first collected in Yokohama Japan by a Danish plant collector (Nielsen 1990) and subsequently introduced around the world. Japan is therefore considered a likely source of new, natural allotriploid *M.* ×*giganteus*. Recently, Nishiwaki et al. (2011) investigated natural occurrences of triploidy in sympatric Japanese populations of tetraploid *M. sacchariflorus* and diploid *M. sinensis*. They measured seed set of *M. sinensis* and *M. sacchariflorus* growing in sympatry and determined their ploidy with flow cytometry. Triploid seeds were found on inflorescences of *M. sacchariflorus*. These plants have potential as new sources of variation in breeding programs. However, they originate from the warm, moist region of Southern Japan. The authors speculate that more cold tolerant *M.* ×*giganteus* would be expected from more northerly, cooler regions of Japan (Nishiwaki et al. 2011). The area we sampled in Primorsky Krai, Russia, is approximately 10° further north than the Japanese populations so might also be expected to have evolved a greater degree of cold tolerance than the Japanese hybrids due to the colder climate of the region.

Our collection also offers the potential to study the process and direction of interspecific hybridization between the two parents of *M.* ×*giganteus*. Early investigations using a limited sample size by Hodkinson et al. (2002b) studied plastid genome variation in *Miscanthus* using gene sequencing and showed that *M. sacchariflorus* was the plastid genome donor (maternal parent) of the commonly grown *M.* ×*giganteus* genotype. Generally, plastid DNA is maternally inherited in grasses (McGrath et al. 2007; Diekmann et al. 2008) and *M.* ×*giganteus* was shown to have the plastid type of *M. sacchariflorus* in all included samples. Therefore, the allotriploid *M.* ×*giganteus* inherited its plastid (and by extrapolation mitochondrial DNA) from an *M. sacchariflorus* lineage. In more recent studies (De Cesare et al. 2010, 2011; Clark et al. 2014), the hybrid has the plastid genome of *M. sinensis* showing that hybridization is possible in both directions (with either species acting as maternal parent). There is no reason to assume a unidirectional bias in the female parent of wild interspecific hybrids between *M. sacchariflorus* and *M. sinensis*, but it is premature to make assumptions about the process at the moment. Unidirectional hybridization can be caused by several factors including nuclear cytoplasmic DNA incompatibility effects (Anderson and Maan 1995) or by population factors. For example if *M. sinensis* was rare and *M. sacchariflorus* common (or if phenological differences created such a pattern), the vast number of seeds set would be from *M. sacchariflorus* ovule donors (Hodkinson et al. 2015).

Conclusion

We have collected major new germplasm resources of *Miscanthus* from Primorsky Krai in the Russian Far East, growing at, or near, the northern limit of their natural distribution and in a diverse range of habitat types. A total of 361 genotypes were collected and ten populations sampled at high intensity. 70 individuals were *M. sinensis*, and 287 *M. sacchariflorus*. Potential hybrids have been identified using morphological traits. We are currently undertaking molecular analysis to determine the diversity and phylogeography of these taxa and also to study the incidence and process of interspecific hybridization leading to new *M.* ×*giganteus* genotypes. We expect that the collections collected here, as part of the EU GrassMargins project, will become significant germplasm for *Miscanthus* breeding of several traits and particularly those requiring optimization for biomass production on marginal land.

Acknowledgements We thank Dr Maria Vorontsova at the Royal Botanic Gardens, Kew, UK, for help with setting up the research collaboration between Ireland and Vladivostok Botanic Garden, Russia. Thanks also to Ruby Prickett for proving some collection locality data. This work was supported by European Framework Programme 7, grant number 289461 (GrassMargins).

References

Adati S, Shiotani I (1962) The cytotaxonomy of the genus *Miscanthus* and its phylogenetic status. Bulletin of the Faculty of Agriculture Mie University 25:1–24

Allison GG, Morris C, Clifton-Brown JC, Lister SJ, Donnison IS (2011) Genotypic variation in cell wall composition in a diverse set of 244 accessions of *Miscanthus*. Biomass and Bioenergy 35:4740–4747

Anderson JA, Maan SS (1995) Interspecific nuclear-cytoplasmic compatibility controlled by genes on group 1 chromosomes in durum wheat. Genome 38:803–808

Barney JN, Mann JJ, Kyser GB, DiTomaso JM (2012) Assessing habitat susceptibility and resistance to invasion by the bioenergy crops switchgrass and *Miscanthus × giganteus* in California. Biomass and Bioenergy 40:143–154

Cai X, Zhang X, Wang D (2011) Land availability for biofuel production. Environ Sci Technol 45:334–9

Chen SL, Renvoize SA (2006) Miscanthus. Flora of China 22:581–583

Clark LV, Brummer JE, Głowacka K, Hall M, Heo K, Long SP, Peng J, Yamada T, Yoo JH, Yu CY, Zhao H, Sacks EJ (2014) A footprint of past global climate change on the population genetic structure of *Miscanthus sinensis*. Ann Bot 114:97–107

Clayton, WD, Vorontsova MS, Harman, KT, Williamson H (2006) GrassBase—The Online World Grass Flora. http://www.kew.org/data/grasses-db.html

Clifton-Brown JC, Jones MB (1997) The thermal response of leaf extension rate in genotypes of the C4-grass *Miscanthus*: an important factor in determining the potential productivity of different genotypes. J Exp Bot 48:1573–1581

Clifton-Brown JC, Lewandowski I (2002) Screening *Miscanthus* genotypes in field trials to optimise biomass yield and quality in Southern Germany. Eur J Agron 16:97–110

Clifton-Brown JC, Chiang YC, Hodkinson TR (2008) *Miscanthus* genetic resources and breeding potential. In: Vermerris W (ed) Genetic improvement of bioenergy crops. Springer Science, New York, pp 273–290

De Cesare M (2011). PhD Thesis. University of Dublin, Trinity College Dublin, Ireland

De Cesare M, Hodkinson TR, Barth S (2010) Chloroplast DNA markers (cpSSRs, SNPs) for *Miscanthus*, *Saccharum* and related grasses (Panicoideae, Poaceae). Mol Breeding 26:539–544

Diekmann K, Hodkinson TR, Fricke E, Barth S (2008) An optimized chloroplast DNA extraction protocol for grasses (Poaceae) proves suitable for whole plastid genome sequencing and SNP detection. PLoS One 3(7), e2813

Donnelly A, Styles D, Fitzgerald J, Finnan J (2011) A proposed framework for determining the environmental impact of replacing agricultural grassland with *Miscanthus* in Ireland. GCB Bioenergy 3:247–263

Dwiyanti MS, Stewart JR, Nishiwaki A, Yamada T (2014) Natural variation in *Miscanthus sinensis* seed germination under low temperatures. Grassland Sci 60:194–198

Farrell AD, Clifton-Brown JC, Lewandowski I, Jones MB (2006) Frost tolerance and thermal response of leaf growth in four *Miscanthus* genotypes. The impact of genotypic variation on potential yield. Ann Appl Biol 149:337–345

Gauder M, Graeff-Hönninger S, Lewandowski I, Claupein W (2012) Long-term yield and performance of 15 different *Miscanthus* genotypes in southwest Germany. Ann Appl Biol 160: 126–136

Głowacka K, Adhikari S, Peng J, Gifford J, Juvik JA, Long SP, Sacks EJ (2014a) Variation in chilling tolerance for photosynthesis and leaf extension growth among genotypes related to the C4 grass *Miscanthus × giganteus*. J Exp Bot 65:5267–5278

Głowacka K, Clark LV, Adhikari S, Peng J, Stewart JR, Nishiwaki A, Yamada T, Jørgensen U, Hodkinson TR, Gifford J, Juvik JA, Sacks EJ (2014b) Genetic variation in *Miscanthus × giganteus* and the importance of estimating genetic distance thresholds for differentiating clones. GCB Bioenergy 7:386–404

Gopalakrishnan G, Cristina Negri M, Snyder SW (2013) A novel framework to classify marginal land for sustainable biomass feedstock production. J Environ Q 40:1593–600

Hodgson EM, Lister SJ, Bridgwater AV, Clifton-Brown JC, Donnison IS (2010) Genotypic and environmentally derived variation in the cell wall composition of *Miscanthus* in relation to its use as a biomass feedstock. Biomass and Bioenergy 34:652–660

Hodkinson TR, Renvoize SA (2001) Nomenclature of *Miscanthus × giganteus* (Poaceae). Kew Bulletin 56:759–760

Hodkinson TR, Chase MW, Renvoize SA (1997) Systematics of *Miscanthus*. Asp Appl Biol 49:189–198

Hodkinson TR, Chase MW, Renvoize SA (2001) Genetic resources of *Miscanthus*. Asp Appl Biol 65:239–248

Hodkinson TR, Chase MW, Lledó D, Salamin N, Renvoize SA (2002a) Phylogenetics of *Miscanthus*, *Saccharum* and related genera (Saccharinae, Andropogoneae, Poaceae) based on DNA sequences from ITS nuclear ribosomal DNA. J Plant Res 115:381–392

Hodkinson TR, Chase MW, Takahashi C, Leitch I, Bennett MD, Renvoize SA (2002b) The use of DNA sequencing (ITS and *trnL-F*), AFLP, and fluorescent in situ hybridization to study allopolyploid *Miscanthus* (Poaceae). Am J Bot 89:279–286

Hodkinson TR, Renvoize SA, Chase MW (2002c) Characterization of a genetic resource collection for *Miscanthus* (Saccharinae, Andropogoneae, Poaceae) using AFLP and ISSR PCR. Ann Bot 89:627–636

Hodkinson TR, De Cesare M, Barth S (2013) Nuclear SSR markers for *Miscanthus*, *Saccharum*, and related grasses (Saccharinae, Poaceae). Appl Plant Sci 1:1300042

Hodkinson TR, Klaas M, Jones MB, Prickett R, Barth S (2015). *Miscanthus*: a case study for the utilization of natural genetic variation. Plant Genet Resour. 13:219–237

Jensen E, Farrar K, Thomas-Jones S, Hastings A, Donnison I, Clifton-Brown J (2011) Characterization of flowering time diversity in *Miscanthus* species. GCB Bioenergy 3:387–400

Jeżowski S, Głowacka K, Kaczmarek Z (2011) Variation on biomass yield and morphological traits of energy grasses from the genus *Miscanthus* during the first years of crop establishment. Biomass and Bioenergy 35:814–821

Jones MB, Walsh M (2001) *Miscanthus* for energy and fibre. London: James and James Ltd, The Cromwell Press. pp 192

Jones MB, Finnan J, Hodkinson TR (2014) Morphological and physiological traits for higher biomass production in perennial rhizomatous grasses grown on marginal land. GCB Bioenergy 7:375–385

Jørgensen U (1997) Genotypic variation in dry matter accumulation and content of N, K and Cl in *Miscanthus* in Denmark. Biomass and Bioenergy 12:155–169

Jørgensen U (2011) Benefits versus risks of growing biofuel crops: the case of *Miscanthus*. Curr Opin Environ Sustain 3:24–30

Komarov VL, Rozhevits RY, Shishkin BK (1963) Flora of the USSR. The Botanical Institute of the Academy of Sciences of the USSR, Leningrad, USSR

Koyama T (1987) Grasses of Japan and its neighboring regions: an identification manual. Kodansha, Ltd., Tokyo, p 370

Lee YN (1964a) Taxonomic studies on the genus *Miscanthus*: relationships among the section, subsection and species, part 1. J Jpn Bot 39:196–205

Lee YN (1964b) Taxonomic studies on the genus *Miscanthus*: relationships among the section, subsection and species, part 2. J Jpn Bot 39:257–265

Lee YN (1964c) Taxonomic studies on the genus *Miscanthus*: relationships among the section, subsection and species, part 3. J Jpn Bot 39:289–298

Linde-Laursen IB (1993) Cytogenetic analysis of *Miscanthus* '*Giganteus*', an interspecific hybrid. Hereditas 119:297–300

McGrath S, Hodkinson TR, Barth S (2007) Extremely high cytoplasmic diversity in natural and breeding populations of *Lolium* (Poaceae). Heredity 99:531–544

Meyer MH, Paul J, Anderson NO (2010) Competitive ability of invasive *Miscanthus* biotypes with aggressive switchgrass. Biol Invas 12:3809–3816

Nielsen PN (1990) Elefantengrassanbau in Danemark—Praktikerbericht. Pflug und Spaten 3:1–4

Nijsen M, Smeets E, Stehfest E, Vuuren DP (2012) An evaluation of the global potential of bioenergy production on degraded lands. GCB Bioenergy 4:130–147

Nishiwaki A, Mizuguti A, Kuwabara S, Toma Y, Ishigaki G, Miyashita T, Yamada T, Matuura H, Yamaguchi S, Rayburn AL, Akashi R, Stewart JR (2011) Discovery of natural *Miscanthus* (Poaceae) triploid plants in sympatric populations of *Miscanthus sacchariflorus* and *Miscanthus sinensis* in southern Japan. Am J Bot 98:154–9

Osada T (1993) Illustrated grasses of Japan. Heibonsia Ltd., Tokyo, pp 158–205

Quinn LD, Allen DJ, Stewart JR (2010) Invasiveness potential of *Miscanthus sinensis*: implications for bioenergy production in the United States. GCB Bioenergy 2:310–320

Robson P, Mos M, Clifton-Brown JC, Donnison IS (2011) Phenotypic variation in senescence in *Miscanthus*: towards optimising biomass quality and quantity. Bioenergy Res 5:95–105

Salamin N, Hodkinson TR, Savolainen V (2001) Towards building the tree of life: a simulation study for all angiosperm genera. Syst Biol 54:183–196

Sang T, Zhu W (2011) China's bioenergy potential. GCB Bioenergy 3:79–90

Scally L, Hodkinson TR, Jones MB (2001a) Origins and taxonomy of *Miscanthus*. In: Jones MB, Walsh M (eds) *Miscanthus* for energy and fibre. The Cromwell Press, London, UK, James and James Ltd, pp 1–9

Scally L, Waldren S, Hodkinson TR, Jones MB (2001b) Morphological and molecular systematics of the genus *Miscanthus*. Asp Appl Biol 65:231–237

Slavov GT, Nipper R, Robson P, Farrar K, Allison GG, Bosch M, Clifton-Brown JC, Donnison IS, Jensen E (2014) Genome-wide association studies and prediction of 17 traits related to phenology, biomass and cell wall composition in the energy grass *Miscanthus sinensis*. New Phytologist 201:1227–39

Sun Q, Lin, Q, Yi Z-L, Yang Z- R, Zhou F-S (2010) A taxonomic revision of *Miscanthus s.l.* (Poaceae) from China. Bot J Linnean Soc 164:178–220

Swaminathan K, Alabady MS, Varala K, De Paoli E, Ho I, Rokhsar DS, Arumuganathan AK, Ming R, Green PJ, Meyers BC, Moose SP, Hudson ME (2010) Genomic and small RNA sequencing of *Miscanthus x giganteus* shows the utility of sorghum as a reference genome sequence for Andropogoneae grasses. Genome Biology 11: R12

Zhang T, Wyman CE, Jakob K, Yang B (2012) Rapid selection and identification of *Miscanthus* genotypes with enhanced glucan and xylan yields from hydrothermal pretreatment followed by enzymatic hydrolysis. Biotechnol Biofuels 5:56

Zhao H, Li Q, He J, Yu J, Yang J, Liu C, Peng J (2013) Genotypic variation of cell wall composition and its conversion efficiency in *Miscanthus sinensis*, a potential biomass feedstock crop in China. GCB Bioenergy 6:768–776

Zhao H, Huai Z, Xiao Y, Wang X, Yu J, Ding G, Peng J (2014) Natural variation and genetic analysis of tiller angle gene MsTAC1 in *Miscanthus sinensis*. Planta 240:161–175

Zub HW, Brancourt-Hulmel M (2010) Agronomic and physiological performances of different species of *Miscanthus*, a major energy crop. A review. Agron Sustain Dev 30:201–214

Zub HW, Arnoult S, Younous J, Lejeune-Hénaut I, Brancourt-Hulmel M (2012) The frost tolerance of *Miscanthus* at the juvenile stage: differences between clones are influenced by leaf-stage and acclimation. Eur J Agron 36:32–40

Chapter 11
Creation of Novel Tetraploid *Miscanthus sinensis* Genotypes

C. Münnich, M. Klaas, V. Bartels, and C. Gebhardt

Introduction

Miscanthus is emerging as one of the most promising lignocellulosic biomass crops for temperate climates due to its efficient C_4 photosynthesis, resulting in high biomass accumulation, and its relative cold hardiness. In Europe, one genotype (*M.* × *giganteus*) dominates cultivation. It is a spontaneous sterile (3n) hybrid from the parents *M. sinensis* and *M. sacchariflorus* (Hodkinson et al. 2002) which was collected in the 1930s in Japan. New hybrids would introduce greater genetic diversity to the crop, allowing for better adaptation to different soil and climatic conditions, but also increasing resilience against the pests and diseases (Jørgensen 2011). Different diseases have been observed on *Miscanthus* (Hosoya et al. 2013; Gilley et al. 2012a, b; Scauflaire et al. 2013; Heaton 2010; Bradshaw et al. 2010) but with no significant effect so far on yield. Sterility as found in *M.* × *giganteus* (Słomka et al. 2012) is desirable for new crops to reduce the risk of spread into natural habitats (Jørgensen 2011). Newly developed polyploids derived from different accessions of *M. sinensis* and *M. sacchariflorus* could be used to broaden the genetic base of natural *M.* × *giganteus* (Sacks et al. 2013), creating more sterile genotypes from their fertile parent species. *Miscanthus sinensis* on its own is also a species with high biomass potential. While most of the genotypes are diploid, *M. sinensis* 'Goliath' is triploid and can yield biomass comparable to *M.* × *giganteus* (Jørgensen and Muhs 2001). Rounsaville et al. (2011) found that some triploid clones provide

C. Münnich (✉) • V. Bartels • C. Gebhardt
Tinplant Biotechnik und Pflanzenvermehrung GmbH,
Magdeburger Landstraße 33, Klein Wanzleben D-39164, Germany
e-mail: cora.muennich@aol.de

M. Klaas
Teagasc, Crops, Environment and Land Use Programme, Oak Park Crops Research Centre,
Carlow, Ireland

S. Barth et al. (eds.), *Perennial Biomass Crops for a Resource-Constrained World*, DOI 10.1007/978-3-319-44530-4_11

highly infertile alternatives to existing diploid cultivars. Triploid genotypes of *M. sinensis* can result from crossing diploid with tetraploid genotypes. Tetraploid *M. sinensis* can be obtained by chromosome doubling as shown by Petersen et al. (2002, 2003) and Głowacka et al. (2010b). The aim of this study was to produce tetraploid *Miscanthus* genotypes as a basis for further crossings to produce sterile triploid clones and thus avoid seed spreading, and for further experiments to produce novel high biomass yielding hybrids.

Material and Methods

Plant Material

The treatment was applied to 19 late-flowering genotypes of *M. sinensis* and one genotype of *M. sacchariflorus*. The genotypes of *M. sinensis* included two ornamentals (Malepartus and Silberturm), six genotypes obtained from European collections (supplied by Teagasc, Ireland and AU Foulum, Denmark), and one provided by the Polish Academy of Science, and ten genotypes collected at various locations in Japan and grown at AU Foulum since 1997. The *M. sacchariflorus* (MSA) genotype was collected in Far East Russia and was provided by the Siberian branch of Russian Academy of Science. All genotypes were grown in 5 L pots in the greenhouse until April 2014. The pots were then transferred outside to a wind protected place, watered once per day and fertilized once per week.

Callus Induction and Colchicine Treatment

186 immature inflorescences were harvested as 100–150 mm long stem segments when the flag leaf was visible. The stem segments contained the complete immature inflorescence. The outer layer of leaves was removed, leaving the last enclosing leaf in place. These stem segments were surface sterilized for 7 min with 1.5 % HgCl and rinsed three times in sterile deionized water. Afterwards the stem cuttings were opened and the immature inflorescences were removed and cut into 5 mm long segments. The segments were cultured on callus induction medium. The callus induction medium was made from modified Murashige and Skoog (MS) medium (Murashige and Skoog 1962) containing 20 g/L sucrose and supplemented with 5 mg/L 2,4-dichlorophenoxyacetic acid (2,4-D), adjusted to pH 5.8 and then autoclaved at 121 °C for 7 min. The tissue culture tubes (diameter 2 cm, height 10 cm) contained crumpled paper to avoid the direct placement of the segments into the liquid medium. The callus was cultivated and multiplied at 26 °C in the dark. Every 2 or 3 weeks the calli were divided and subcultured onto fresh medium until approximately 100 callus tubes were assembled per genotype. Then the callus was placed

on solid colchicine medium, made of modified MS medium containing 20 g/L sucrose, 2 g/L Gelrite and 5 mg/L 2,4-D, adjusted to pH 5.8, autoclaved at 121 °C for 7 min, and then supplemented with colchicine from filter sterilized stock to 313 μM. The calli stayed on colchicine medium for 4 days in the dark.

Shoot Regeneration from Calli

After colchicine treatment the calli were transferred to regeneration medium under light conditions. The regeneration medium consists of modified MS, 20 g/L sucrose, supplemented by either 3 or 6 mg/L 6-benzylaminopurine (BAP), or 3 mg/L kinetin, or 100 mg/L adenine. Afterwards the regeneration medium was adjusted to pH 5.8 and autoclaved at 121 °C for 7 min. The tissue culture tubes contained crumpled paper to avoid the direct placement of the calli into the liquid medium. The calli were transferred every 2–3 weeks to fresh medium. After 16 weeks of regeneration, the occurrence of shoots was recorded and shoots were randomly chosen from each replication and transferred to rooting medium. The rooting medium contained 0.5× modified MS, 20 g/L sucrose, 0.5 mg/L indole-3-butyric acid (IBA). For the first culture cycle the shoot was placed on crumpled paper in culture tubes, later submerged directly in liquid medium. After 4–7 weeks of sub culturing, the clusters start to break up into single rooted shoots. These plantlets were transferred to soil pots of 10 cm diameter and cultivated in the greenhouse until 3–5 leaves had emerged. The plants were watered each day and fertilized once per week.

Ploidy Analysis

The plants were analyzed after full adaptation to greenhouse conditions. Leaf samples of approximately 2 cm length were taken from three different leaves of the same plant. The analysis was done by the company Plant Cytometry Services, Netherlands.

Results

Callus Induction

13 genotypes from a total of 20 were able to form callus (Table 11.1) under our conditions. These included ornamental genotypes as well as accessions held at European collections, and accessions collected in the wild in Japan and Russian Far East. The callus from 12 genotypes was able to multiply (Table 11.1).

Table 11.1 Success of callus induction and reaction of calli after treatment with colchicine and regeneration (3 mg/L BAP)

				After colchicine and regeneration		
Genotype	Origin/latitude	Spotted callus	Able to multiply	No reaction	Root development	Shoot development
MSA	Far East	X	X	X		
Silberturm	Breeder Pagels	X	No			
Malepartus	Breeder Pagels	X	X	X		
D-60	Japan/36° N	X	X			X
D-62	Japan/36° N	X	X			X
D-81	Japan/36° N	X	X		X	
D-83	Japan/36° N	X	X			X
D-85	Japan/36° N			X		
D-88	Japan/36° N	X	X	X		
D-89	Japan/36° N	X	X		X	X
D-90	Japan/36° N					
D-109	Japan/37° N					
D-110	Japan/37° N					
GoFal	European c.	X	X		X	
M81 RH	European c.	X	X		X	
15	European c.	X	X		X	
18	European c.	X	X		X	X
P75	European c.					
S43	European c.					
Pl17	PAS					

Note: *European c.* European collections, *PAS* Polish Academy of Science

Shoot Regeneration

After colchicine treatment and a shoot regeneration phase at 3 mg/L BAP, the callus was scored for occurrence of shoots, roots or no reaction (Table 11.1). For three genotypes only shoots were found. Two genotypes developed shoots and roots simultaneously, and four genotypes developed only roots. Four additional genotypes remained at the callus stage, and showed no further differentiation on the shoot regeneration medium. To stimulate shoot development in the last eight genotypes, the callus was treated with different cytokines after colchicine exposure (see Material and Methods section). These treatments did not improve the ability of callus to regenerate, and no further shoots were found (data not shown). The ability to form shoots was depended on the genotypes and independent from the origin.

Table 11.2 Success of obtaining tetraploid plants after colchicine treatment and regeneration and the untreated control 18 and D-62

Genotype	Analyzed plants	Tetraploid plants	Tetraploid (%)	Chimaeras
D-60	170	**31**	*18*	4
D-62	100	**38**	*38*	
D-83	4	**4**	*100*	
D-89	112	**42**	*38*	3
18	71	**34**	*48*	1
Control-18	*20*	*0*	–	
Control-D-62	*20*	*0*	–	

Ploidy Analysis

Altogether 457 plants from five shoot producing genotypes were analyzed. For each of these genotypes tetraploid plants were found. The percentage of tetraploid plants ranged from 18 % to 100 % (Table 11.2). Chimaera plants were rare. As control, regenerated plants derived from callus without colchicine treatment were analyzed. Only diploid plants were found (Table 11.2).

Characterization of Plants

In autumn 2015, the tetraploid plants were at the 5–7 leaf stage. So far no obvious differences between diploid and tetraploid plants were observed. For further experiments the plants have been transferred to greenhouses at Teagasc (Ireland), AU Foulum (Denmark), PAS (Poland), and Tinplant (Germany). The plants will be planted on fields and evaluated in 2016.

Discussion

Callus can be embryogenic, shoot-forming, rooty, and non-morphologic (Dalton 2013). In our experiments these different types of callus were found: shoot-forming, rooty, and non-morphologic types. Only 5 of 12 callus-forming genotypes were able to form shoots. These genotypes showed no obvious morphological similarities. Also no relationship with origin was observed. The latitudes from the genotypes collected by AU Foloum, Denmark are known (Table 11.1), however they cannot be distinguished in terms of callus formation from other genotypes which were introduced to Europe a long time ago. The rate of shoot regeneration in our experiments was lower than reported by Dalton (2013). According to Wang et al. (2011; cited in Dalton (2013)), the propensity of a genotype to produce embryogenic callus is correlated to the mean 5 year temperature at the collection site. Accessions from Southern latitudes with mean temperatures of 17–25 °C performed best. We used plants grown outdoors as starting material for our experiments. It is possible that the low mean temperatures in 2014 with 11 °C for spring and 18 °C for summer did not reach the optimal temperature range for *Miscanthus*. Possibly these temperatures were too low to achieve a high percentage of shoot regeneration, and plants grown in the greenhouse at higher mean temperatures should be preferred as starting material. Similar to our experiments, Perera et al. (2015) also used field maintained plants as starting material, at Mississippi State University, and they obtained a high percentage of up to 77 % regenerable calli.

Dalton (2013) surveyed the literature on the influence of the explants size of immature inflorescences on their ability to regenerate, ranging from 0.05 to 200 mm in length. It was shown that the smaller explants are more effective in their ability to regenerate shoots. In contrast, in our experiments we found that callus developed better from larger inflorescence explants (data not shown). Dalton (2013) found that most callus induction media recorded in the literature contain a low concentration of cytokinin. Perhaps inclusion of cytokinin would increase the success rate in the development of shoot-forming callus for those genotypes that were recalcitrant in our experiments. Another reason for the variability in callus formation could be genotype specific differences in the development of different types of callus. This is supported by our findings that the use of different cytokinins had no influence on the ability to regenerate after colchicine treatment and suggests an influence of the genotype on development. Similarly, Petersen et al. (2003) concluded that the efficiency of tetraploid formation was genotype dependent, primarily due to differences in the ability to form regenerable callus and to regenerate intact plants from callus.

As described by Petersen et al. (2002, 2003), we applied 313 μM colchicine for 4 days to induce chromosome doubling. The successful chromosome doubling in our experiment was genotype dependent relating to its ability to regenerate as tetraploids were obtained for all genotypes which formed shoots under our conditions. Similarly to Petersen et al. (2002), only few chimaeras were found.

We were unable to regenerate shoots from *M. sacchariflorus* callus. This should be addressed in further experiments by fine tuning the conditions for callus formation and shoot regeneration, as chromosome doubling in *M. sacchariflorus* has already been demonstrated by Chae et al. (2013).

Chromosome doubling can lead to gigantism, a desirable trait in its own right for a biomass crop, and can also be the basis for the construction of new genomic combinations (Sacks et al. 2013). Głowacka et al. (2010a) found in field experiments for two genotypes that polyploid *M. sinensis* had significantly wider leaves, and that the beneficial impact of increased ploidy was genotype dependent (Głowacka et al. 2010b). They and Głowacka et al. (2009) also demonstrated that induced tetraploids are fertile and are a useful tool for the production of new hybrids.

The plants from our experiments will be planted in field trials in the next growing season to assess their potential for higher biomass. In parallel, newly developed tetraploid plants are cultivated for crossings with diploid *M. sinensis* to create promising triploid for future agronomic evaluations.

Acknowledgments This work was supported by FP7 KBBE.2011.3.1-02 Grant number 289461 (GrassMargins).

References

Bradshaw JD, Prasifka JR, Steffey KL, Gray ME (2010) First report of field populations of two potential aphid pests of the bioenergy crop *Miscanthus* × *giganteus*. Fla Entomol 93(1): 135–137

Chae WB, Hong SJ, Gifford JM, Rayburn AL, Widholm JM, Juvik JA (2013) Synthetic polyploidy production of *Miscanthus sacchariflorus*, *Miscanthus sinensis*, and *Miscanthus* × *giganteus*. GCB Bioenergy 5:338–350

Dalton SJ (2013) Biotechnology of *Miscanthus*. In: Jain SM, Dutta Gupta S (eds) Biotechnology of neglected and underutilized crops. Springer, Dordrecht, pp 243–294

Gilley MD, Tomaso-Peterson M, Allen TW, Baldwin BS (2012a) Frequency of fungi associated with giant *Miscanthus* in 2011. In: APS annual meeting, Providence, RI, 4–8 August 2012, 182–O

Gilley MD, Tomaso-Peterson M, Allen TW, Baldwin BS (2012b) Diseases of giant *Miscanthus* and switchgrass in Mississippi. In: APS annual meeting, Providence, RI, 4–8 August 2012, 403-P

Głowacka K, Jeżowski S, Kaczmarek Z (2009) Polyploidization of *Miscanthus sinensis* and *Miscanthus* × *giganteus* by plant colchicine treatment. Ind Crops Prod 30:444–446

Głowacka K, Jeżowski S, Kaczmarek Z (2010a) Impact of colchicine application during callus induction and shoot regeneration on micropropagation and polyploidisation rates in two *Miscanthus* species. In Vitro Cell Dev Biol Plant 46:161–171

Głowacka K, Jeżowski S, Kaczmarek Z (2010b) In vitro induction of polyploidy by colchicine treatment of shoots and preliminary characterization of induced polyploids in two *Miscanthus* species. Ind Crops Prod 32:88–96

Heaton E (2010) Implications of plant diseases in biofuel production. Phytopathology 100(6):S181

Hodkinson TR, Chase MW, Takahashi C, Leitch I, Bennett MD, Renvoize SA (2002) The use of DNA sequencing (ITS and *trnL-F*), AFLP, and fluorescent *in situ* hybridization to study allopolyploid *Miscanthus* (*Poaceae*). Am J Bot 89:279–286

Hosoya T, Hosaka K, Saito Y, Degawa Y, Suzuki R (2013) *Naemacyclus culmingenus*, a newly reported potential pathogen to *Miscanthus sinensis*, new to Japan. Mycoscience 54(6): 433–437

Jørgensen U (2011) Benefits versus risks of growing biofuel crops: the case of *Miscanthus*. Curr Opin Environ Sustain 3:24–30

Jørgensen U, Muhs HJ (2001) *Miscanthus* breeding and improvement. In: Jones BM, Walsh M (eds) Miscanthus for energy and fibre. James & James, London, pp 67–85

Murashige T, Skoog F (1962) A revised medium for rapid growth and bio assays with tobacco tissue cultures. Physiol Plant 15:473–497

Perera D, Barnes DJ, Baldwin BS, Reichert NA (2015) Direct and indirect in vitro regeneration of *Miscanthus* × *giganteus* cultivar freedom: effects of explants type and medium on regeneration efficiency. In Vitro Cell Dev Biol Plant 51:294–302

Petersen KK, Hagberg P, Kristiansen K, Forkmann G (2002) *In vitro* chromosome doubling of *Miscanthus sinensis*. Plant Breed 121:445–450

Petersen KK, Hagberg P, Kristiansen K (2003) Colchicine and oryzalin mediated chromosome doubling in different genotypes of *Miscanthus sinensis*. Plant Cell Tiss Org Cult 73:137–146

Rounsaville TJ, Touchell DH, Ranney TG (2011) Fertility and reproductive pathways in diploid and triploid *Miscanthus sinensis*. HortScience 46(10):1353–1357

Sacks E, Juvik JA, Lin Q, Stewart R, Yamada T (2013) The gene pool of *Miscanthus* species and its improvement. In: Paterson AH (ed) Genomics, plant genetics and genomics: crops and models 11. Springer, New York, pp 73–101

Scauflaire J, Gourgue M, Foucart G, Renard F, Vandeputte F, Munaut F (2013) *Fusarium miscanthi* and other *Fusarium* species as causal agents of *Miscanthus* × *giganteus* rhizome rot. Eur J Plant Pathol 137:1–3

Słomka A, Kuta E, Płażek A, Dubert F, Żur I, Dubas E, Kopeć P, Żurek G (2012) Sterility of *Miscanthus* × *giganteus* results from hybrid incompatibility. Acta Biol Cracov Ser Bot 54(1):113–120

Wang X, Yamada T, Kong FJ, Abe Y, Hoshino Y, Sato H, Takamizo T, Kanazawa A, Yamada T (2011) Establishment of an efficient in vitro culture and particle bombardment-mediated transformation systems in *Miscanthus sinensis* Anderss., a potential bioenergy crop. GCB Bioenergy 3:322–332

Chapter 12
A Review of Frost and Chilling Stress in *Miscanthus* and Its Importance to Biomass Yield

S. Fonteyne, I. Roldán-Ruiz, H. Muylle, T. De Swaef, D. Reheul, and P. Lootens

Introduction

Miscanthus, a perennial rhizomatous C4 grass originating in East Asia, has become an important biomass crop in the last few decades. It can achieve high biomass yields with only limited input of fertilizers or pesticides and is suitable for growth on marginal land (Lewandowski et al. 2000; Heaton et al. 2010; Jones et al. 2015). Currently, most miscanthus production fields use *Miscanthus* × *giganteus*, a sterile hybrid of *M. sinensis* and *M. sacchariflorus*. Although several *M.* × *giganteus* clones are available, most commercial plantings descend from a single Japanese clone first introduced to Denmark (Głowacka et al. 2014b). There is therefore an urgent need to breed new miscanthus genotypes, as using only a single clone over broad geographical ranges implies a risk of vulnerability to diseases and pests and does not allow growers to plant varieties adapted to local environmental conditions. These genotypes should ideally combine a high yield potential with high resource use efficiency, the traits that have made *M.* × *giganteus* a successful biomass crop.

The high yield potential of *M.* × *giganteus* in temperate regions has been, at least partly, ascribed to its cold tolerance: it has higher CO_2 assimilation rates at cool temperatures than other C4 crops of the temperate zone (Long and Spence 2013; Sage et al. 2015). This is certainly true of the *M.* × *giganteus* clones currently used in commercial plantations, but research has shown the possibility of identifying

S. Fonteyne
Plant Sciences Unit, Institute for Agricultural and Fisheries Research (ILVO), Melle, Belgium

Department of Plant Production, Ghent University (UGent), Ghent, Belgium

I. Roldán-Ruiz • H. Muylle • T. De Swaef • P. Lootens (✉)
Plant Sciences Unit, Institute for Agricultural and Fisheries Research (ILVO), Melle, Belgium
e-mail: Peter.Lootens@ilvo.vlaanderen.be

D. Reheul
Department of Plant Production, Ghent University (UGent), Ghent, Belgium

S. Barth et al. (eds.), *Perennial Biomass Crops for a Resource-Constrained World*, DOI 10.1007/978-3-319-44530-4_12

genotypes in the miscanthus germplasm pool with an even higher level of cold tolerance (Clifton-Brown and Jones 1997; Głowacka et al. 2015a). This would allow expansion of the potential miscanthus growing area and would limit the risk of winter mortality in the first year after planting (Clifton-Brown and Lewandowski 2000; Hastings et al. 2009b; Peixoto et al. 2015). Increased growth at low temperatures could also prolong the growing season, allowing plants to capture more light energy, thereby increasing yield (Dohleman and Long 2009; Robson et al. 2013). Available knowledge on cold tolerance in miscanthus is fragmentary, however, with limited understanding of its physiological basis. In particular, variation available in the miscanthus germplasm pool has not been sufficiently explored. Consensus on the relationship between cold tolerance and biomass yield in miscanthus is also lacking. Here we systematically review the state of knowledge of the effects of low temperature on miscanthus survival and growth, and we discuss the expected and proven relationships between cold tolerance and biomass yield. Knowledge gaps are identified and directions for future research are suggested.

Cold Stress

Ruelland et al. (2009) divide cold stress in chilling stress and frost stress (Fig. 12.1). Chilling stress occurs when plants are exposed to temperatures that are too low for growth but still are above 0 °C. At chilling temperatures biochemical processes are disrupted: the speed of metabolic reactions decreases, enzymes become less active and less stable, and the rigidity of the cell membranes increases (Ruelland et al. 2009;

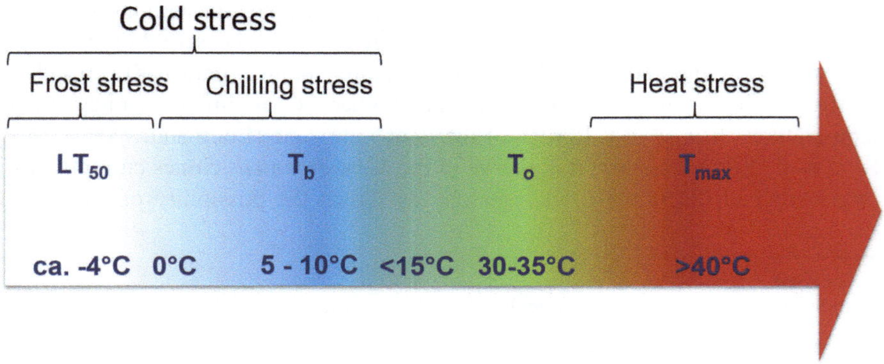

Fig. 12.1 Graphical representation of the temperature effect on plant growth. At the optimal temperature (T_o) plant growth is maximal. When temperature increases above T_o, heat stress occurs. Above the maximum temperature (T_{max}) growth ceases. If temperature is lower than T_o, first chilling stress will occur. Below the base temperature (T_b) growth ceases. Below 0 °C frost stress starts occurring. Below the LT_{50} 50 % of the plants dies. Approximate temperature stress ranges in miscanthus are given

Yadav 2010). This all disrupts the metabolic equilibria in the cell. For example, the capture of light energy during the light reactions of photosynthesis is less temperature-dependent than the use of this energy in the dark reactions. As a consequence, the chloroplast electron transfer chain can become overreduced under high light intensities at low temperature. This leads to the production of reactive oxygen species (ROS), which can damage cells and cause photoinhibition, especially if the activity of ROS scavenging enzymes and the synthesis of antioxidants is also reduced. Frost stress occurs at below zero temperatures and is mainly related to cell dehydration. At below zero temperatures, extracellular ice crystals are formed and attract water from the cells. When these crystals grow larger they can cause mechanical damage by penetrating the symplast (Ruelland et al. 2009). The temperatures at which a plant or plant organ experiences these different kinds of stresses depend on the species, and even on the genotype. For example, chilling stress severely reduces growth in sugarcane below 20 °C, while some miscanthus genotypes can still grow at 5 °C (Clifton-Brown and Jones 1997) and *M.* × *giganteus* rhizomes die around −3.4 °C while some other perennial grasses can survive temperatures below −20 °C (Clifton-Brown and Lewandowski 2000; Belintani et al. 2012; Friesen et al. 2015). In general, miscanthus is more tolerant to cold stress than other C4 grasses such as maize, sorghum, or sugarcane (Long and Spence 2013; Sage et al. 2015), but variability for this trait has also been described in the miscanthus germplasm pool (Clifton-Brown and Jones 1997; Clifton-Brown and Lewandowski 2000; Farrell et al. 2006; Friesen et al. 2014; Głowacka et al. 2014a, 2015b), as discussed below.

How Tolerant Is Miscanthus to Cold Stress?

The genus *Miscanthus,* which includes approximately 20 species, has a broad geographic distribution ranging between 50°N to 22°S (Hodkinson et al. 2002). The most studied miscanthus species are *M. sinensis* and *M. sacchariflorus,* since these are the parent species of *M.* × *giganteus*. The natural geographic distribution of these species ranges between eastern Russia in the north to Papua New Guinea in the south. They can grow in a wide range of climatic conditions, so it is expected that considerable genetic variation would exist for climatic adaptation within the genus. This view is supported by results of different studies. For example, seeds of Japanese *M. sinensis* accessions from higher latitudes germinate earlier than those from southern accessions, under both high and low temperature (Dwiyanti et al. 2014), indicating adaption to local climatic conditions. Yan et al. (2011) have evaluated seedlings of *M. sinensis, M. sacchariflorus* and *M. lutarioriparius* from populations originating across China at three locations representing the northern plains, the loess plateau and the warmer regions of central China. They found that accessions from northern locations showed greater winter survival. Clear interspecific variability was detected, as the majority of the *M. sacchariflorus* genotypes was able to survive the first winter in the northern plains, while most of the *M. sinensis* and *M. lutarioriparius* genotypes did not. In contrast, Anzoua et al. (2015)

detected no significant differences regarding winter survival among 43 *M. sinensis* accessions collected across Japan, where the species is found in climate conditions ranging from subarctic to subtropic (An et al. 2008; Stewart et al. 2009). When grown in Hokkaido, a location characterized by long winters and cool summers, the survival rate of accessions from subarctic and subtropic regions did not differ significantly. These apparently contradictory conclusions are probably due to the particular set of accessions compared and the location chosen for field evaluation.

Adaptation to local conditions has also been demonstrated along altitudinal gradients. For example, in Taiwan miscanthus grows from the coastal lowlands up to altitudes of 3200 m. Below an elevation of 2200 m, *M. floridulus* is the most common species, while above 2200 m *M. transmorrisonensis* is the most common (Chou and Chang 1988). *M. floridulus* genotypes collected at 1200 m above sea level do not survive the colder winter when transplanted to a location at 2600 m (Chou et al. 1991). Similarly, *M. transmorrisonensis* has a higher photosynthesis and biomass accumulation at 10 and 15 °C, while at 25 and 30 °C it is higher for *M. floridulus* (Kao et al. 1998). Miscanthus genotypes from coastal Taiwan reach maximum photosynthetic rates at higher temperatures than genotypes from higher altitudes; the latter reach higher values of photosynthesis at lower temperatures (Weng and Ueng 1997; Weng and Hsu 2001). This adaptation to different climatological conditions is also reflected in specific adaptations such as thicker leaves, fewer stomata and a thicker wax layer in accessions from higher altitudes (Kao and Chang 2001; Weng and Hsu 2001).

All this evidence demonstrates the existence of climatic adaptation within the genus *Miscanthus* which could be exploited in breeding. However genotypes adapted to colder conditions seem to be characterized by lower yields when grown in milder climates. For example, in a collection of 23 genotypes collected across China and grown in Wuhan (30°33′ N, 114°25′ E), biomass yield was negatively correlated with the latitude of the locality where the accession had been collected (Yu et al. 2013b). A study of 459 *M. sinensis* accessions collected across China yielded similar results (Zhao et al. 2013). It is therefore important to consider possible trade-offs between cold tolerance and yield (see below).

Frost Tolerance in Miscanthus

In temperate and continental climates, frost can affect miscanthus plants in different ways (Fig. 12.2). Shoots can be damaged by frost at the beginning or at the end of the growing season, while rhizomes can be killed by severe frost during winter. Frost damage to rhizomes is thought to be the main cause of winter mortality (Clifton-Brown and Lewandowski 2000; Peixoto et al. 2015) and it is one of the biggest problems in miscanthus production in northern, colder areas, especially in the first year after planting (Lewandowski et al. 2000).

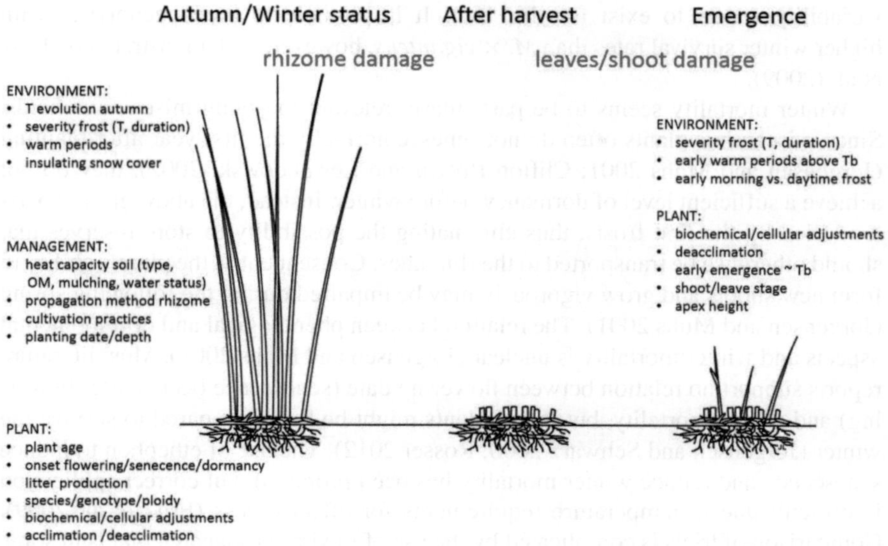

Fig. 12.2 Schematic overview of the factors contributing to frost stress and frost tolerance in miscanthus. *T* temperature, T_b base temperature

Frost Tolerance at the Rhizome Level and Winter Mortality

Winter mortality has been investigated in several field trials established in multiple locations with different levels of winter severity. Within the "European Miscanthus Improvement" (EMI) project, field trials including four *M.* × *giganteus*, one *M. sacchariflorus*, five *M sinensis* × *M. sacchariflorus* hybrids, and five *M. sinensis* genotypes were established in Portugal, England, Germany, Denmark, and Sweden. In Portugal, England and Germany, winter losses of the *M.* × *giganteus* genotypes did not surpass 1 %, while in Sweden and Denmark mortalities up to 100 % were observed. Furthermore, in the 15 field trials of the European Miscanthus Network, *M.* × *giganteus* showed good winter survival in southern Europe, but unreliable survival in the trials in northern Europe (Christian and Haase 2001; Clifton-Brown et al. 2001b). In the USA, Maughan et al. (2012) reported winter survival rates of 99 and 100 % for *M.* × *giganteus* in the warmer locations in New Jersey and Kentucky and 79 % and 25 % in the colder Nebraska and Illinois locations. Some studies have demonstrated a higher winter survival for *M. sinensis* (Clifton-Brown et al. 2001a), but this was not supported by Rosser (2012), who reported higher winter mortality for the genotypes of this species in a trial in Ontario, Canada. In this trial 20 miscanthus genotypes were planted at three locations of varying winter severity. The diploid *M. sinensis* × *M. sacchariflorus* hybrids investigated displayed the highest survival rates, followed by *M.* × *giganteus* and *M. sinensis* (Rosser 2012). Generalizations at the species level are therefore not possible, and large intra-species variability seems to exist for this trait. It is possible to breed genotypes with higher winter survival rates than *M.* × *giganteus*, however, as demonstrated by Fritz et al. (2009).

Winter mortality seems to be particularly relevant in young miscanthus fields. Since miscanthus plants often do not senesce normally the first year after planting (Jørgensen and Muhs 2001; Clifton-Brown and Lewandowski 2002), they do not achieve a sufficient level of dormancy before winter. Instead, the aboveground parts are killed by the first frosts, thus eliminating the possibility to store reserves that should otherwise be transported to the rhizomes. Consequently, the plant's ability to form new shoots and grow vigorously may be impaired during the following spring (Jørgensen and Muhs 2001). The relation between phenological and developmental aspects and winter mortality is unclear (Jørgensen and Muhs 2001). Most literature reports support no relation between flowering date (senescence occurs after flowering) and winter mortality, but larger plants might be better prepared to survive the winter (Jørgensen and Schwarz 2000; Rosser 2012). The use of ethephon to induce senescence and reduce winter mortality has been proposed, but correct application is difficult due to temperature requirements for effectiveness (Fritz et al. 2009). Comparison of trials is complicated by the use of in vitro propagated plants in some of the field trials because the effect of the phytohormones used in in vitro culture can be long lasting. In some trials plants propagated through rhizomes have shown greater winter survival rates than in vitro propagated plants (Fritz et al. 2009), but in other trials rhizome-propagated plants had higher winter mortality (Christian and Haase 2001). Hormonal treatments during in vitro propagation and hormonal status at planting can affect the survival capacity of the plants (Christian and Haase 2001).

Miscanthus rhizomes are able to survive subzero temperatures to a certain extent, but frost tolerance of miscanthus seems to be low compared to other rhizomatous grasses, such as switchgrass or prairie cordgrass (*Spartina pectinata*) (Hope and McElroy 1990; Peixoto et al. 2015). Within the genus, LT_{50}, the temperature at which 50 % of the plants are killed, ranged from −3.4 °C for *M.* × *giganteus* to −6.3 °C for certain *M. sinensis* genotypes (Clifton-Brown and Lewandowski 2000). The frost tolerance of the rhizomes as investigated in the laboratory was correlated with winter survival in the EMI trial described above, with the two *M. sinensis* hybrids surviving winter at all sites whereas *M.* × *giganteus* and *M. sacchariflorus* only survived on sites where soil temperatures at 5 cm depth did not fall below −2.8 °C (Clifton-Brown et al. 2001a). This indicates the possibility of predicting the frost stress tolerance of miscanthus rhizomes under field conditions using controlled freezing experiments.

Peixoto et al. (2015) tested five hybrid miscanthus genotypes and reported LT_{50} values between −1.5 and −6.7 °C depending on genotype and harvesting date. For some genotypes no difference in LT_{50} between harvesting dates was observed, but for some others, LT_{50} values were 2–3 °C higher when rhizomes were harvested in summer, suggesting that at least in some genotypes deacclimatization happens. In addition, it has been shown that the speed at which temperature decreases influences mortality rates. Indeed, when Peixoto et al. (2015) applied a staged cooling protocol, in which the temperature was decreased by 2.5 °C every 24 h, observed LT_{50}

values between −6.3 and −14.4 °C. Furthermore, hardening *M.* × *giganteus* plants at 12 or 5 °C increases the tolerance of rhizomes to freezing stress up to −3 °C (Płażek et al. 2011) and rhizomes of plants that have survived one winter can survive severer frost events than rhizomes that have not yet overwintered.

Several physiological mechanisms have been suggested to mediate frost tolerance in miscanthus. For example, it has been shown that hardening increases abscisic acid (ABA) contents in the leaves and the rhizomes and decreases moisture content in the rhizomes. Hardening also increases the amount of low molecular antioxidants and phenolic compounds in leaves and rhizomes, and decreases catalase activity (Płażek et al. 2011). Withers (2015) has demonstrated the accumulation of raffinose, linoleic acid (C18:2n6) and alpha-linolenic acid (C18:3n3) during cold acclimation. Linoleic acid and alpha-linolenic acid are known to stabilize cell membranes at low temperatures and increase their fluidity.

Frost Tolerance of New Shoots

Late frosts in spring affect shoot growth. When shoots emerge during early warm periods, they can be killed by subsequent frost events. This is particularly relevant if resprouting is compromised by a shortage of reserves in the rhizome, as discussed above during the first year after planting. If shoots are killed completely by a late frost, any yield advantage of early emergence will be lost. It is therefore important to understand how plants react to freezing temperatures at the start of the growing season. Farrell et al. (2006) exposed leaves of young, hardened plants of one *M.* × *giganteus*, one *M. sacchariflorus*, and two *M. sinensis* hybrids to controlled freezing temperatures up to −10 °C. Considerable variation was found with LT_{50} values ranging from −6 °C to −9 °C. Frost tolerance was different for plants at different developmental stages, with plants in the third or fifth leaf stage being more tolerant than those with six or seven leaves. In plants with three to five leaves, the shoot apex is most probably still underground, where it is protected from frost damage (Zub et al. 2012a). These four genotypes were part of the 15 genotypes planted at five locations by the EMI project. The *M.* × *giganteus* (LT_{50} −8 °C) and the *M. sacchariflorus* (LT_{50} −7 °C) did not survive the first winter in Sweden and Denmark, while the two *M. sinensis* hybrids (LT_{50} of −6 °C and −9 °C) did survive (Clifton-Brown et al. 2001a). Interestingly, leaf frost tolerance was not associated with greater winter survival, but was rather more related to rhizome frost tolerance as discussed above.

Chilling Stress in Miscanthus

C4 photosynthesis evolved from C3 photosynthesis to reduce photorespiration, the process in which RuBisCo uses O_2 as a substrate instead of CO_2. The affinity of RuBisCo for O_2 increases with temperature and decreases with increasing intracellular CO_2 concentration. As C4 species are common in warmer climates than in temperate and cold climates, it has long been assumed C4 photosynthesis is inherently chilling sensitive. However, the chilling sensitivity of C4 species such as maize or sorghum seems to be a consequence of their tropical or subtropical origins rather than an inherent characteristic of C4 photosynthesis (Long and Spence 2013; Sage et al. 2015). It is the growing season temperature, rather than the winter temperature, that determines survival and production of C4 species. In this section, an overview is provided of the effects of chilling temperature on the photosynthetic apparatus, metabolism and plant development, and yield in miscanthus.

Effect of Chilling Stress on Plant Growth and Development

Plant growth is strongly dependent on the temperature of the environment (T_e) (Fig. 12.1). If this temperature is below the base temperature (T_b) or above the maximum temperature (T_{max}), growth will cease. Above T_b growth increases up to an optimum temperature (T_o), and when temperature gets higher than T_o growth decreases. The effect of temperature on the growth and development of crops is generally expressed in thermal time, or the summation of the number of degrees the mean daily temperature is above T_b over a certain period. Shoot production of four miscanthus genotypes was reduced strongly at low temperatures, with only one genotype producing shoots on more than 50 % of its rhizomes at 7 °C (Farrell et al. 2006). The degree days needed for emergence was different for different genotypes and varied between 60 and 180° days, with T_b between 6 and 8.6 °C. This corresponds with the T_b of 6.8 °C that (Zub et al. 2012b) calculated using the formula developed by Yan and Hunt (1999). Clifton-Brown et al. (2000) calculated 10 °C as a base temperature for *M.* × *giganteus*, as this temperature gave the highest correlation between the increase in leaf area index and accumulated degree days. As T_b seems to vary according to genotype, it is necessary to use the appropriate T_b when determining growing season duration of a particular genotype, expressed in thermal time. However, this is not common practice and a general T_b value of 10 °C (Clifton-Brown et al. 2004; Angelini et al. 2009; Jensen et al. 2013; Arundale et al. 2014) or 0 °C (Miguez et al. 2008; Hastings et al. 2009a; Maughan et al. 2012) is used in most studies. Inter-genotype differences have also been reported with regards to plant elongation rates at chilling temperatures (Clifton-Brown and Jones 1997; Głowacka et al. 2014a), with 43-fold differences at 10/5 °C between the most chilling tolerant genotype and the least chilling tolerant genotype among a set of 50 (Głowacka et al. 2014a). In this latter study *M.* × *giganteus* was amongst the genotypes showing the least reduction in growth rate when transferred from 25 °C to 10/5 °C.

Effects of Low Temperature on the Photosynthetic Apparatus

M. × *giganteus* can achieve high rates of photosynthesis, with assimilation rates up to 35 μmol $m^{-2}s^{-1}$ under field conditions (Beale et al. 1996). The temperature optimum for photosynthesis lies around 30–35 °C for both *M.* × *giganteus* and maize (Naidu et al. 2003), but compared to other C4 crops such as maize or sugarcane, *M.* × *giganteus* can still achieve high photosynthetic rates at low temperatures (Beale et al. 1996; Naidu and Long 2004; Friesen et al. 2014). While maize displays an 80 % reduction in assimilated CO_2 when grown at 14/11 °C with respect to growth at 25 °C, Naidu et al. (2003) and Wang et al. (2008a) report nearly the same light-saturated rates of CO_2 (A_{sat}) for *M.* × *giganteus* grown at 14 °C or at 25 °C. *M.* × *giganteus* grown at 10 °C however, does show a marked decrease in the quantum efficiency of CO_2 fixation (Φ_{CO2}) and assimilation (Farage et al. 2006). The ratio between the quantum efficiency of electron transport in photosystem II (Φ_{PSII}) and Φ_{CO2}(Φ_{PSII}/ Φ_{CO2}) is similar in *M.* × *giganteus* grown at 25 °C, 14 °C, or 12 °C but it increases when grown at 10 °C. This indicates an increase in linear electron transport at lower temperature, i.e., more electrons are transported through PSII than are used for the assimilation of CO_2. These electrons can be directed to alternative electron sinks, for instance O_2 via the Mehler reaction, which generates superoxide and hydrogen peroxide (Naidu and Long 2004; Farage et al. 2006; Fonteyne et al. 2015). So *M.* × *giganteus* suffers chilling stress and has a risk of oxidative damage at temperatures below 12 °C. There are however large differences among miscanthus species and genotypes in chilling tolerance of photosynthesis. Several studies have shown that carbon assimilation rate declines relatively little in *M.* × *giganteus* after chilling shock compared to *M. sinensis* and *M. sacchariflorus* (Purdy et al. 2013; Friesen et al. 2014; Głowacka et al. 2014a; Fonteyne et al. 2015). Large differences in chilling tolerance of photosynthesis are found among miscanthus species and genotypes. Several studies have shown that carbon assimilation rate declines relatively little in *M.* × *giganteus* after chilling shock compared to *M. sinensis* and *M. sacchariflorus* (Purdy et al. 2013; Friesen et al. 2014; Głowacka et al. 2014a, b; Fonteyne et al. 2015).

Stomatal conductance limitations do not appear to lie at the basis of the decrease in assimilation rate at low temperatures in miscanthus (Naidu et al. 2003). This is supported by the conclusions of Głowacka et al. (2015b), in a study of 11 miscanthus genotypes and of Głowacka et al. (2014a) in a comparison of 13 miscanthus genotypes. Stomatal conductance decreased in all genotypes, but this was mainly a consequence of lower assimilation rates, rather than a cause. The main reason for low assimilation rates was light induced chilling damage of photosystem II. Głowacka et al. (2015a) and Friesen et al. (2014) used chlorophyll fluorescence measurements in the field and found that genotypes that show a higher chilling tolerance measured by chlorophyll fluorescence under controlled conditions also tend to show less cold stress under field conditions. Measurements of photosynthesis under controlled conditions are thus representative of field performance.

Biochemical Adaptations to Chilling Temperatures

The chilling tolerance of *M.* × *giganteus* is a result of its ability to maintain high rates of photosynthesis at low temperatures, while at the same time effectively dissipating excess light energy. At low temperatures, the activity and stability of enzymes such as ribulose-1,5-bifosfaat carboxylase oxygenase (RuBisCo), phosphoenolpyruvate carboxylase (PEPc) and pyruvate phosphate dikinase (PPDK) becomes limiting for C4 photosynthesis (Matsuoka et al. 2001; Sage and Kubien 2007). At low CO_2 concentration PEPc is the rate limiting enzyme, but in miscanthus under chilling stress the intracellular CO_2 concentration is generally not limiting (Głowacka et al. 2015b). This implies that enzymatic activity and stability of RuBisCo and PPDK are probably more important.

PPDK contents in leaves of both maize and *M.* × *giganteus* decline when transferred from 25 °C to 14 °C. While PPDK in maize leaves remains at a low level, PPDK contents in *M.* × *giganteus* leaves increase again after the initial decline, reaching significantly higher levels compared to 25 °C after 7 days of cold treatment (Wang et al. 2008b). This is accompanied by higher PPDK mRNA abundances and higher photosynthetic rate. Higher concentrations of PPDK increase the stability of the protein and thus its activity at low temperatures (Wang et al. 2008b). This agrees with the observation that the amount of RuBisCo, PEPc and PPDK is significantly lower in maize grown at 14 °C compared to 25 °C, while in *M.* × *giganteus* no difference in protein concentration between these growing temperatures could be observed. Similarly, in chilling tolerant sugarcane genotypes (Naidu et al. 2003) PPDK mRNA expression and enzyme activity increase under chilling stress, while they decline in sensitive genotypes (Du et al. 1999; Nogueira et al. 2003).

Although two PPDK genes have been described in *M.* × *giganteus*, their protein products are highly similar, and they even display a high level of sequence similarity when compared to orthologous genes of miscanthus, maize, and sugarcane (Du et al. 1999; Naidu et al. 2003; Wang et al. 2008b). Furthermore, there seems to be no functional difference between recombinant PPDK from both miscanthus and maize expressed in *E. coli* (Wang et al. 2008b). Likewise, no differences were found with regards to catalytic properties, activity and leaf concentrations of RuBisCo from maize and *M.* × *giganteus* grown at 14 °C or 25 °C (Wang et al. 2008a). However, higher RuBisCo contents were detected under chilling stress in *M. sinensis* (Spence 2012) and *M.* × *giganteus* (Spence et al. 2014). When exposed to 14 °C, *M. sinensis* accessions from colder climates displayed higher PPDK and RuBisCo contents (38 and 50 % higher, respectively), while PPDK content declined by 28 % and RuBisCo content did not change in comparison to a genotype from a warmer climate. Furthermore, in a microarray experiment, Spence et al. (2014) showed that in *M.* × *giganteus* the expression of all genes coding for photosynthetic proteins or proteins protecting PSII tested was higher under chilling stress. *M.* × *giganteus* thus counteracts the lower activity and stability of these proteins at low temperature by increasing the mRNA levels for their synthesis (Spence et al. 2014). In conclusion, increasing enzyme content under chilling stress is probably a general strategy of miscanthus in response to cold, but this should be confirmed in other species and genotypes.

Tolerance to chilling stress not only involves maintaining high levels of photosynthesis at low temperatures, but it is also necessary to avoid chilling-induced damage of the photosynthetic apparatus. High light intensities at low temperature can cause photoinhibition of photosystem II. Correspondingly, dissipation of excess light energy to heat through reversible photoprotective processes such as conversion of violaxanthin to zeaxanthin is increased in *M.* × *giganteus* under chilling stress (Farage et al. 2006). Compared to more chilling sensitive miscanthus genotypes, *M.* × *giganteus* shows a more pronounced response under chilling stress (Friesen et al. 2014). The role of reactive oxygen species (ROS) in chilling stress in miscanthus has not received much attention in the past, but ROS are also likely to play a role in miscanthus. This is supported by the relatively high tolerance of *M. sinensis* genotypes to oxidative stress caused by high heavy metal concentrations (Scebba et al. 2006; Ezaki et al. 2008).

Chilling stress has a marked influence on the carbohydrate concentrations in miscanthus leaves. Purdy et al. (2013) report an increase of soluble sugars and starch after a chilling shock in four miscanthus genotypes. These changes were ascribed to a decrease in growth, resulting in a decreased demand for carbohydrates (Purdy et al. 2013). The soluble sugar concentration remains high as long as the plants are under chilling stress (Mortaignie 2014), supporting the view that sugars might play a protective role. Indeed, in *M.* × *giganteus* and *M. sinensis* Goliath grown at 12 °C, raffinose is present in the leaf, while it is absent at 20 °C (Fonteyne et al. 2015). Raffinose is known to protect cells against the effects of chilling stress by stabilizing cell membranes (Valluru and Van den Ende 2008; Janská et al. 2010) and against ROS damage (Nishizawa et al. 2008). In our view, the role of soluble sugar changes in chilling tolerance of miscanthus deserves further investigation. The role of raffinose as protective agent is of particular interest.

Will Higher Cold Stress Tolerance Lead to Higher Yields?

Investigating the cold tolerance of miscanthus is of great interest in itself because of its close phylogenetic relation with cold sensitive crops such as maize or sugarcane (Friesen et al. 2014; Głowacka et al. 2015b). However, from the application point of view the main question remains whether increasing cold tolerance will lead to higher biomass yields in miscanthus. From a theoretical point of view, the capacity to form shoots and grow at low temperature, results in an increase of the canopy duration, allowing to intercept a higher amount of radiation and thus higher biomass yield. For example, while miscanthus has a lower photosynthetic capacity than maize, its larger leaf area, combined with a much longer growing season, allows it to accumulate more biomass (Dohleman and Long 2009). The positive effect of low temperature tolerance can however be counteracted if it results in a yield penalty that reflects the costs of improved low temperature tolerance (Trudgill et al. 2005). Although there is evidence that improved abiotic resistance does not always have a yield penalty, at least in Arabidopsis (Raineri et al. 2015; Yu et al. 2013a), currently available literature does not allow to determine unequivocally whether this is the case for miscanthus.

The miscanthus growing season is essentially delimited by the last spring frost and the first autumn/winter frost (Jørgensen and Muhs 2001). Within these limits, genotypes that emerge early and grow fast at low temperature in early spring have an advantage because early canopy closure leads to more interception of solar radiation (Clifton-Brown and Jones 1997; Sage et al. 2015). Farrell et al. (2006) modeled miscanthus biomass production and concluded that breeding new cultivars with an improved growth rate at low temperature could potentially increase yields. They showed that a hypothetical genotype with a similar growth rate as *M.* × *giganteus* at optimal temperature but lower base temperature for growth, a lower thermal time requirement for emergence, and a better leaf frost tolerance could yield up to 25 % more than *M.* × *giganteus* in southern Germany. Similarly, in a simulation by Davey et al. (2015), a lower base temperature or an earlier leaf emergence predicted a significantly higher yield. Nevertheless, these simulation studies did not take into account a possible trade-off between growth rate and base temperature, as reported by Clifton-Brown and Jones (1997), in a study of the growth rate of 32 miscanthus genotypes at different temperatures, and of Farrell et al. (2006) in four genotypes. According to the conclusions of these two studies, genotypes with higher growth at low temperatures were unlikely to be higher yielding compared to the other genotypes due to relatively lower growth rates at higher temperatures. Whether this relation is of general application in miscanthus, and whether there is a genetic linkage between these two aspects or just a correlation specifically for the set of genotypes investigated should be the topic of future research. Furthermore, it is currently unknown if breeding for increased cold tolerance will have impact, either positive or negative, on tolerance to other abiotic stresses, such as drought or nutrient deficiency. To our knowledge there are currently no reports available about possible interactions between different abiotic stress tolerances in miscanthus. Some information on this topic is expected soon, as one of the objectives of the EU-FP7 research project OPTIMISC (Lewandowski et al. 2015) is to identify miscanthus genotypes that are tolerant to multiple types of stress.

A study of the results of field experiments reveals strong genotype × environment effects. Indeed, Zub et al. (2011) and Zub et al. (2012b) found the highest yields among the 21 genotypes they examined in late emerging genotypes with high maximum growth rates in summer and negatively related to growth duration (Zub et al. 2012b). This is contradicted by Robson et al. (2013), who reported that canopy duration has a positive effect on yield and concluded that both early emergence and late senescence lead to higher yields, after examining 244 genotypes. These different conclusions might be related to differences in the locations at which these two experiments were carried out. In the French location where Zub et al. (2012b) carried out the experiment, it is possible that early emergence and growth are less relevant than in Wales where Robson et al. (2013) established their field trials. Weather data was not reported for either study, however. The differences in the methodologies and genotypes used in the trials might also explain the different conclusions (Robson et al. 2013). Furthermore, in the EMI trial described above (Clifton-Brown et al. 2001a), the two genotypes with the highest T_b among a set of 15 did not survive the first winter at the trials in the two coldest locations but yielded almost 10 tons/ha more than the two genotypes with lower T_b in the trial at the warmest location (Farrell et al. 2006). The trials thus demonstrated the need for varieties adapted to local environmental conditions to maximize yield potential.

Rhizome frost tolerance will not increase yield on a plant basis, but improved winter survival can increase yield in areas where frost kill is of concern. The real relevance of frost tolerance at the rhizome level for miscanthus plantings is difficult to estimate, because in the investigated areas, temperatures in the soil (at the level of the rhizomes) seldom reach the lethal temperatures given by Clifton-Brown et al. (2000). For example, Eitzinger et al. (2002) reported that while air temperatures dropped to −10 °C in a miscanthus field in Austria, soil temperatures never dropped below zero (Eitzinger and Kössler 2002). Probably in most field trials investigated to date, lethal soil temperatures were not reached, which explains at least partially the large variation in winter mortality observed among locations and years. Hastings et al. (2009a) identified a once-in-10-year occurrence of lethal frost as the upper limit for commercial exploitation of miscanthus. Using the MISCANFOR miscanthus yield model, they showed that a hypothetical improved variety with a LT_{50} of −6 °C would have a substantially larger growing area compared to *M.* × *giganteus* with at LT_{50} of −3.4 °C (Hastings et al. 2009b). Kucharik et al. (2013) simulated of minimum soil temperature in the Midwest region of the USA and showed that in the northern parts of this region, these minimum soil temperatures are often reached. However, soil cover by miscanthus straw or snow can have an insulating effect of 2–6 °C, protecting the rhizomes from lethal frost. A thick straw cover is not present the first winter after planting, however, which may explain higher mortalities. Furthermore, in older plantings, the large density of rhizomes may have an insulating effect, protecting the inner rhizomes from frost damage (Peixoto et al. 2015). Earlier emergence increases the risk of exposure of young buds and leaves to frost, and the risk that the apex is already above ground when a late frost strikes. To date it has not been quantified what the effect hereof could be on final yield. It appears however necessary to take shoot frost tolerance into account when breeding for earlier emergence.

Conclusion

Cold stress is one of the main constraints to miscanthus productivity. Frost stress can lead to winter mortality and can kill shoots early in the growing season. Chilling temperatures limit emergence, growth and photosynthesis in spring. Miscanthus is relatively tolerant to chilling stress, unlike other C4 crops. *M.* × *giganteus* seems to be one of the most chilling tolerant genotypes compared to the others reported in literature. Its high rates of photosynthesis at low temperature are not an effect of special protective mechanisms but rather of increased production of photosynthetic enzymes. The photosynthesis of miscanthus under chilling stress has mainly been studied in *M.* × *giganteus*, but recently genotypes with even higher photosynthetic rates at low temperature have been reported. Considerable variation for frost tolerance and for growth and photosynthesis under chilling stress has been reported, including genotypes more tolerant to cold stress than *M.* × *giganteus*. It should thus be possible to develop varieties with higher cold stress tolerance. Improved cold stress tolerance can potentially lead to higher biomass yields, through earlier emergence and canopy formation. It is however unknown at the moment whether this is associated to a yield penalty. More research is needed to determine the impact of increased cold tolerance on yield potential.

Acknowledgments The authors of this review were supported by the European Union's seventh framework project OPTIMISC (grant agreement no° 289159).

References

An G-H, Miyakawa S, Kawahara A, Osaki M, Ezawa T (2008) Community structure of arbuscular mycorrhizal fungi associated with pioneer grass species *Miscanthus sinensis* in acid sulfate soils: habitat segregation along pH gradients. Soil Sci Plant Nutr 54:517–528

Angelini LG, Ceccarini L, Nassi o Di Nasso N, Bonari E (2009) Comparison of *Arundo donax* L. and *Miscanthus* × *giganteus* in a long-term field experiment in central Italy: analysis of productive characteristics and energy balance. Biomass Bioenergy 33:635–643

Anzoua KG, Suzuki K, Fujita S, Toma Y, Yamada T (2015) Evaluation of morphological traits, winter survival and biomass potential in wild Japanese *Miscanthus sinensis* Anderss. Populations in northern Japan. Grassland Sci 61:83–91

Arundale RA, Dohleman FG, Heaton EA, Mcgrath JM, Voigt TB, Long SP (2014) Yields of *Miscanthus* × *giganteus* and *Panicum virgatum* decline with stand age in the Midwestern USA. GCB Bioenergy 6:1–13

Beale CV, Bint DA, Long SP (1996) Leaf photosynthesis in the C4-grass *Miscanthus* × *giganteus*, growing in the cool temperate climate of southern England. J Exp Bot 47:267–273

Belintani NG, Guerzoni JTS, Moreira RMP, Vieira LGE (2012) Improving low-temperature tolerance in sugarcane by expressing the ipt gene under a cold inducible promoter. Biol Plant 56:71–77

Chou C, Chang F (1988) Population study of *Miscanthus floridulus* II. Ecotypic variation of *M. floridulus* and *M. transmorrisonensis* as affected by altitude in Nantou, Taiwan. Bot Bull Acad Sin 29:301–314

Chou C-H, Lee Y, Chiu CY, Wang YC, Hsu FH (1991) Growth performance of *M. floridulus* and *M. transmorrisonensis* and their acclimation to temperatures and water stresses. Bot Bull Acad Sin 32:87–96

Christian DG, Haase E (2001) Agronomy of *Miscanthus*. In: Jones MB, Walsh M (eds) Miscanthus for energy and fibre. Earthscan, London, pp 21–45

Clifton-Brown JC, Jones MB (1997) The thermal response of leaf extension rate in genotypes of the C4-grass *Miscanthus*: an important factor in determining the potential productivity of different genotypes. J Exp Bot 48:1573–1581

Clifton-Brown JC, Lewandowski I (2000) Overwintering problems of newly established *Miscanthus* plantations can be overcome by identifying genotypes with improved rhizome cold tolerance. New Phytol 148:287–294

Clifton-Brown JC, Lewandowski I (2002) Screening *Miscanthus* genotypes in field trials to optimise biomass yield and quality in Southern Germany. Eur J Agron 16:97–110

Clifton-Brown JC, Neilson B, Lewandowski I, Jones MB (2000) The modelled productivity of *Miscanthus* × *giganteus* (GREEF et DEU) in Ireland. Ind Crops Prod 12:97–109

Clifton-Brown JC, Lewandowski I, Andersson B, Basch G, Christian DG, Kjeldsen JB, Jørgensen U, Mortensen JV, Riche AB, Schwarz K-U, Tayebi K, Teixeira F (2001a) Performance of 15 *Miscanthus* genotypes at five sites in Europe. Agron J 93:1013–1019

Clifton-Brown JC, Long SP, Jørgensen U (2001b) *Miscanthus* productivity. In: Jones MB, Walsh M (eds) Miscanthus for energy and fibre. Earthscan, London, pp 46–67

Clifton-Brown JC, Stampfl PF, Jones MB (2004) Miscanthus biomass production for energy in Europe and its potential contribution to decreasing fossil fuel carbon emissions. Glob Chang Biol 10:509–518

Davey CL, Jones LE, Squance M, Purdy SJ, Maddison AL, Cunniff J, Donnison I, Clifton-Brown JC (2015) Radiation capture and conversion efficiencies of *Miscanthus sacchariflorus, M. sinensis* and their naturally occurring hybrid *M.* × *giganteus*. GCB Bioenergy. doi:10.1111/gcbb.12331

Dohleman FG, Long SP (2009) More productive than maize in the Midwest: How does Miscanthus do it? Plant Physiol 150:2104–2115

Du Y, Nose A, Wasano K (1999) Effects of chilling temperature on photosynthetic rates, photosynthetic enzyme activities and metabolite levels in leaves of three sugarcane species. Plant Cell Environ 22:317–324

Dwiyanti MS, Stewart JR, Nishiwaki A, Yamada T (2014) Natural variation in *Miscanthus sinensis* seed germination under low temperatures. Grassland Sci 60:194–198

Eitzinger J, Kössler C (2002) Microclimatological characteristics of a miscanthus (*Miscanthus cv. giganteus*) stand during stable conditionsat night in the nonvegetative winter period. Theor Appl Climatol 72:245–257

Ezaki B, Nagao E, Yamamoto Y, Nakashima S, Enomoto T (2008) Wild plants, *Andropogon virginicus* L. and *Miscanthus sinensis* Anders, are tolerant to multiple stresses including aluminum, heavy metals and oxidative stresses. Plant Cell Rep 27:951–961

Farage PK, Blowers D, Long SP, Baker NR (2006) Low growth temperatures modify the efficiency of light use by photosystem II for CO_2 assimilation in leaves of two chilling-tolerant C4 species, *Cyperus longus* L. and *Miscanthus* × *giganteus*. Plant Cell Environ 29:720–728

Farrell AD, Clifton-Brown JC, Lewandowski I, Jones MB (2006) Genotypic variation in cold tolerance influences the yield of *Miscanthus*. Ann Appl Biol 149:337–345

Fonteyne S, Lootens P, Muylle H, Van den Ende W, De Swaef T, Reheul D, Roldán-Ruiz I (2015) Chilling tolerance and early vigor related characteristics evaluated in two Miscanthus genotypes. Photosynthetica. doi:10.1007/s11099-016-0193-y

Friesen PC, Peixoto MM, Busch FA, Johnson DC, Sage RF (2014) Chilling and frost tolerance in *Miscanthus* and *Saccharum* genotypes bred for cool temperate climates. J Exp Bot 63: 3749–3758

Friesen PC, Peixoto MM, Lee DK, Sage RF (2015) Sub-zero cold tolerance of *Spartina pectinata* (prairie cordgrass) and *Miscanthus* × *giganteus*: candidate bioenergy crops for cool temperate climates. J Exp Bot 66:4403–4413

Fritz M, Formowitz B, Jodl S, Eppel-Hotz A, Kuhn W (2009) Miscanthus als nachwachsender rohstoff—ergebnisse aus bayerischen forschungsarbeiten. Technologie- und Förderzentrum im Kompetenzzentrum für Nachwachsende Rohstoffe (TFZ), Straubing

Głowacka K, Adhikari S, Peng J, Gifford J, Juvik JA, Long SP, Sacks EJ (2014a) Variation in chilling tolerance for photosynthesis and leaf extension growth among genotypes related to the C4 grass *Miscanthus* × *giganteus*. J Exp Bot 65:5267–5278

Głowacka K, Clark LV, Adhikari S, Peng J, Ryan Stewart J, Nishiwaki A, Yamada T, Jørgensen U, Hodkinson TR, Gifford J, Juvik JA, Sacks EJ (2014b) Genetic variation in *Miscanthus* × *giganteus* and the importance of estimating genetic distance thresholds for differentiating clones. GCB Bioenergy 7:386–404

Głowacka K, Jørgensen U, Kjeldsen JB, Korup K, Spitz I, Sacks EJ, Long SP (2015a) Can the exceptional chilling tolerance of C4 photosynthesis found in *Miscanthus* × *giganteus* be exceeded? Screening of a novel *Miscanthus* Japanese germplasm collection. Ann Bot 115:981–990

Głowacka K, Ahmed A, Sharma S, Abbott T, Comstock JC, Comstock JC, Long SP, Sacks EJ (2015b). Can chilling tolerance of C4 photosynthesis in Miscanthus be transferred to sugarcane? GCB Bioenergy. doi:10.1111/gcbb.12283

Hastings A, Clifton-Brown J, Wattenbach M, Mitchell CP, Smith P (2009a) The development of MISCANFOR, a new *Miscanthus* crop growth model: towards more robust yield predictions under different climatic and soil conditions. GCB Bioenergy 1:154–170

Hastings A, Clifton-Brown J, Wattenbach M, Mitchell CP, Stampfl P, Smith P (2009b) Future energy potential of *Miscanthus* in Europe. GCB Bioenergy 1:180–196

Heaton EA, Dohleman FG, Miguez AF, Juvik JA, Lozovaya V, Widholm J, Zabotina OA, Mcisaac GF, David MB, Voigt TB, Boersma NN, Long SP (2010) *Miscanthus*: a promising biomass crop. Adv Bot Res 56:76–124

Hodkinson TR, Chase MW, Renvoize SA (2002) Characterization of a genetic resource collection for *Miscanthus* (*Saccharinae*, *Andropogoneae*, *Poaceae*) using AFLP and ISSR PCR. Ann Bot 89:627–636

Hope H, McElroy A (1990) Low-temperature tolerance of switchgrass (*Panicum virgatum* L.). Can J Plant Sci 70:1091–1096

Janská A, Marsík P, Zelenková S, Ovesná J (2010) Cold stress and acclimation—what is important for metabolic adjustment? Plant Biol 12:395–405

Jensen EF, Robson PRH, Norris J, Cookson A, Farrar K, Donnison IS, Clifton-Brown JC (2013) Flowering induction in the bioenergy grass *Miscanthus sacchariflorus* is a quantitative short-day response, whilst delayed flowering under long days increases biomass accumulation. J Exp Bot 64:541–552

Jones MB, Finnan J, Hodkinson TR (2015) Morphological and physiological traits for higher biomass production in perennial rhizomatous grasses grown on marginal land. GCB Bioenergy 7:375–385

Jørgensen U, Muhs H-J (2001) *Miscanthus* breeding and improvement. In: Jones MB, Walsh M (eds) Miscanthus for energy and fibre. Earthscan, London, pp 68–85

Jørgensen U, Schwarz K-U (2000) Why do basic research? A lesson from commercial exploitation of *miscanthus*. New Phytol 148:190–193

Kao W, Chang K-W (2001) Altitudinal trends in photosynthetic rate and leaf characteristics of *Miscanthus* populations from central Taiwan. Aust J Bot 49:509–514

Kao W, Tsai T, Chen W (1998) A comparative study of *Miscanthus floridulus* (Labill) Warb and *M. transmorrisonensis* Hayata: photosynthetic gas exchange, leaf characteristics and growth in controlled environments. Ann Bot 81:295–299

Kucharik CJ, VanLoocke A, Lenters JD, Motew MM (2013) *Miscanthus* establishment and overwintering in the Midwest USA: a regional modeling study of crop residue management on critical minimum soil temperatures. PLoS One 8, e68847

Lewandowski I, Clifton-Brown JCJ, Scurlock JMO, Huisman W (2000) *Miscanthus*: European experience with a novel energy crop. Biomass Bioenergy 19:209–227

Lewandowski I, Kalinina O, Kiesel A, Clifton-Brown J, Farrar K, Donnison I, Huxley L, Nunn C, Trindade L, Dolstra O, van der Weijde T, van der Linden G, Chen C-L, Roland-Ruiz I, Muylle H, Lootens P, Fonteyne S, Harding G, Mos M, Schwarz KU, Meyer H, Muller-Samann K, Xi Q, Özgüven M, Khokhlov N, Tarakanov I, Schüle H (2015) OPTIMISC—developing miscanthus production systems for marginal lands. In: Proceedings of the 23rd European biomass conference and exhibition, Vienna, Austria, pp 6–8

Long SP, Spence AK (2013) Toward cool C(4) crops. Annu Rev Plant Biol 64:701–722

Matsuoka M, Furbank RT, Fukayama H, Miyao M (2001) Molecular engineering of C4 photosynthesis. Annu Rev Plant Physiol Plant Mol Biol 52:297–314

Maughan M, Bollero G, Lee DK, Darmody R, Bonos S, Cortese L, Murphy J, Gaussoin R, Sousek M, Williams D, Williams L, Miguez F, Voigt T (2012) *Miscanthus* × *giganteus* productivity: the effects of management in different environments. GCB Bioenergy 4:253–265

Miguez FE, Villamil MB, Long SP, Bollero GA (2008) Meta-analysis of the effects of management factors on *Miscanthus* × *giganteus* growth and biomass production. Agric For Meteorol 148:1280–1292

Mortaignie E (2014) Evaluatie van het effect van koudestress op de jeugdgroei van Miscanthus. Master thesis, Ghent University

Naidu SL, Long SP (2004) Potential mechanisms of low-temperature tolerance of C4 photosynthesis in *Miscanthus* × *giganteus*: an in vivo analysis. Planta 220:145–155

Naidu SL, Moose SP, Al-shoaibi AK, Raines CA, Long SP (2003) Cold Tolerance of C4 photosynthesis in *Miscanthus* × *giganteus*: adaptation in amounts and sequence of C4 photosynthetic enzymes. Plant Physiol 132:1688–1697

Nishizawa A, Yabuta Y, Shigeoka S (2008) Galactinol and raffinose constitute a novel function to protect plants from oxidative damage. Plant Physiol 147:1251–1263

Nogueira TS, De Rosa VE, Menossi M, Ulian C, Arruda P (2003) RNA expression profiles and data mining of sugarcane response to low temperature. Plant Physiol 132:1811–1824

Peixoto MDM, Friesen PC, Sage RF (2015) Winter cold-tolerance thresholds in field-grown Miscanthus hybrid rhizomes. J Exp Bot 66:4415–4425

Płażek A, Dubert F, Janowiak F, Krępski T, Tatrzańska M (2011) Plant age and in vitro or in vivo propagation considerably affect cold tolerance of *Miscanthus* × *giganteus*. Eur J Agron 34:163–171

Purdy SJ, Maddison AL, Jones LE, Webster RJ, Andralojc J, Donnison I, Clifton-Brown J (2013) Characterization of chilling-shock responses in four genotypes of Miscanthus reveals the superior tolerance of *M.* × *giganteus* compared with *M. sinensis* and *M. sacchariflorus*. Ann Bot 111:999–1013

Raineri J, Ribichich KF, Chan RL (2015) The sunflower transcription factor HaWRKY76 confers drought and flood tolerance to Arabidopsis thaliana plants without yield penalty. Plant Cell Rep 34:2065–2080

Robson PRH, Farrar K, Gay AP, Jensen EF, Clifton-Brown JC, Donnison IS (2013) Variation in canopy duration in the perennial biofuel crop *Miscanthus* reveals complex associations with yield. J Exp Bot 64:2373–2383

Rosser B (2012) Evaluation of miscanthus winter hardiness and yield potential in Ontario. Master thesis, University of Guelph

Ruelland E, Vaultier MN, Zachowski A, Hurry V (2009) Cold signalling and cold acclimation in plants. Adv Bot Res 49:36–126

Sage RF, Kubien DS (2007) The temperature response of C3 and C4 photosynthesis. Plant Cell Environ 30:1086–1106

Sage RF, de Melo Peixoto M, Friesen P, Deen B (2015) C4 bioenergy crops for cool climates, with special emphasis on perennial C4 grasses. J Exp Bot 66:4195–4212

Scebba F, Arduini I, Ercoli L, Sebastiani L (2006) Cadmium effects on growth and antioxidant enzymes activities in *Miscanthus sinensis*. Biol Plant 50:688–692

Spence AK (2012). Understanding the physiological and molecular basis of chilling tolerance accros species of the C4 genera Miscanthus and Spartina. Ph.D. dissertation, University of Illinois

Spence AK, Boddu J, Wang D, James B, Swaminathan K, Moose SP, Long SP (2014) Transcriptional responses indicate maintenance of photosynthetic proteins as key to the exceptional chilling tolerance of C4 photosynthesis in *Miscanthus* × *giganteus*. J Exp Bot 65:3737–3747

Stewart JR, Toma Y, Fernández FG, Nishiwaki A, Yamada T, Bollero G (2009) The ecology and agronomy of *Miscanthus sinensis*, a species important to bioenergy crop development, in its native range in Japan: a review. GCB Bioenergy 1:126–153

Trudgill DL, Honek A, Li D, Straalen NM, Van Straalen NM (2005) Thermal time—concepts and utility. Ann Appl Biol 146:1–14

Valluru R, Van den Ende W (2008) Plant fructans in stress environments: emerging concepts and future prospects. J Exp Bot 59:2905–2916

Wang D, Naidu SL, Portis AR, Moose SP, Long SP (2008a) Can the cold tolerance of C4 photosynthesis in *Miscanthus* × *giganteus* relative to *Zea mays* be explained by differences in activities and thermal properties of Rubisco? J Exp Bot 59:1779–1787

Wang D, Portis AR, Moose SP, Long SP (2008b) Cool C4 photosynthesis: pyruvate Pi dikinase expression and activity corresponds to the exceptional cold tolerance of carbon assimilation in *Miscanthus* × *giganteus*. Plant Physiol 148:557–567

Weng JH, Hsu FHF (2001) Gas exchange and epidermal characteristics of *Miscanthus* populations in Taiwan varying with habitats and nitrogen application. Photosynthetica 39:35–41

Weng JH, Ueng RG (1997) Effect of temperature on photosynthesis of *Miscanthus* clones collected from different elevations. Photosynthetica 34:307–311

Withers KK (2015) Morphological adaptations and membrane stabilizing mechanisms of overwintering miscanthus (*Poaceae*). Ph.D. dissertation, University of Guelph

Yadav SK (2010) Cold stress tolerance mechanisms in plants. A review. Agron Sustain Dev 30:515–527

Yan W, Hunt L (1999) An equation for modelling the temperature response of plants using only the cardinal temperatures. Ann Bot 84:607–614

Yan J, Chen W, Luo F, Ma H, Meng A, Li X, Zhu M, Li S, Zhou H, Zhu W, Han B, Ge S, Li J, Sang T (2011) Variability and adaptability of *Miscanthus* species evaluated for energy crop domestication. GCB Bioenergy 4:49–60

Yu L, Chen X, Wang Z, Wang S, Wang Y, Zhu Q, Li S, Xiang C (2013a) Arabidopsis *Enhanced Drought Tolerance1*/HOMEODOMAIN GLABROUS11 confers drought tolerance in transgenic rice without yield penalty. Plant Physiol 162:1378–1391

Yu L, Ding G, Huai Z, Zhao H (2013b) Natural variation of biomass yield and nutrient dynamics in *Miscanthus*. Field Crops Res 151:1–8

Zhao H, Wang B, He J, Yang J, Pan L, Sun D, Peng J (2013) Genetic diversity and population structure of *Miscanthus sinensis* germplasm in China. PLoS One 8, e75672

Zub HW, Arnoult S, Brancourt-Hulmel M (2011) Key traits for biomass production identified in different *Miscanthus* species at two harvest dates. Biomass Bioenergy 35:637–651

Zub HW, Arnoult S, Younous J, Lejeune-Hénaut I, Brancourt-Hulmel M (2012a) The frost tolerance of *Miscanthus* at the juvenile stage: differences between clones are influenced by leaf-stage and acclimation. Eur J Agron 36:32–40

Zub H, Rambaud C, Bethencourt L, Brancourt-Hulmel M (2012b) Late emergence and rapid growth maximize the plant development of *Miscanthus* clones. Bioenergy Res 5:841–854

Chapter 13
Creation and Evaluation of Novel Cold Tolerant *Miscanthus* Hybrids

C. Münnich, K. Kørup, M. Klaas, S. Barth, J.B. Kjeldsen, J. Finnan, S. Fonteyne, M. Jankowska, and U. Jørgensen

Introduction

Among potential bioenergy crops in Europe, miscanthus has previously been found to have high biomass production (Heaton et al. 2004). As a C_4 plant, miscanthus has high water- and nitrogen-use efficiencies (Lewandowski et al. 2000) securing high productivity from low inputs. However, C_4 photosynthesis also leads to a limitation of growth caused by low temperatures in spring (Naidu et al. 2003). Still, C_4 photosynthesis is not inherently limited to warm climates (Long and Spence 2013) as experiments have shown that *Miscanthus* × *giganteus* was very productive in countries with a cooler climate such as Britain at 52°N (Beale and Long 1995). Furthermore, experiments in China demonstrated that miscanthus has a high potential to grow at sites that are colder than their native habitats (Yan et al. 2012).

The original version of this chapter was revised. An erratum to this chapter can be found at DOI 10.1007/978-3-319-44530-4_25

C. Münnich (✉)
Tinplant Biotechnik und Pflanzenvermehrung GmbH,
Magdeburger Landstraße 3, D-39164 Klein Wanzleben, Germany
e-mail: cora.muennich@aol.de

K. Kørup • J.B. Kjeldsen • U. Jørgensen
Department of Agroecology, Aarhus University, Blichers Allé 20, Postboks 50, DK-8830 Tjele, Denmark

M. Klaas • S. Barth • J. Finnan • M. Jankowska
Teagasc, Crops, Environment and Land Use Programme, Oak Park Crops Research Centre, Carlow, Ireland

S. Fonteyne
Plant Sciences Unit, Institute for Agricultural and Fisheries Research (ILVO), Melle, Belgium

Department of Plant Production, Ghent University (UGent), Ghent, Belgium

S. Barth et al. (eds.), *Perennial Biomass Crops for a Resource-Constrained World*, DOI 10.1007/978-3-319-44530-4_13

Depending on the regions, various genome combinations are conceivable for Northern Europe (*M. sinensis* with *M. sinensis* or *M. sacchariflorus* day-neutral), Middle Europe (*M. sinensis* with *M. sacchariflorus* large-stemmed), or Southern Europe (*M. sacchariflorus* large-stemmed with *M. sinensis*, *M. sacchariflorus*, or *M. floridulus*) (Deuter 2000). Inclusion of *M. lutarioriparius* in crossing programs is considered promising because of its biomass accumulation (Yan et al. 2012), frost survival and flooding tolerance (Schalitz and Behrendt 2012). *Miscanthus lutarioriparius* is poorly studied and could be a synonym of *M. sacchariflorus* (Głowacka et al. 2015a). However, it is kept separate here following Hodkinson et al. (2015).

Selection of new hybrids with improved characteristics can be guided by phenotyping offspring in the field. Additionally, differences in cold tolerance between offspring can also be evaluated using chlorophyll (Chl) *a* fluorescence imaging. Chlorophyll fluorescence is a measure of re-emitted light from photosystem II (PSII). After dark adaptation of the plants, the ratio F_v/F_m is measured, and this represents the maximum quantum yield of PSII primary photochemistry. For non-stressed leaves this value is between 0.79 and 0.83 (Maxwell and Johnson 2000). It has previously been found that leaves of miscanthus exposed to colder temperature had decreased quantum efficiency (Kao et al. 1998; Friesen et al. 2014; Głowacka et al. 2015b). The breeding strategy is to increase early vigor and to extend the cultivation period and thereby possibly broaden the growth area to include cooler regions (Fracheboud et al. 1999).

The aim of the present study was to cross different miscanthus species and to screen the breeding material for cold tolerance in the field and under controlled conditions. Additionally we wanted to assess the suitability of both the CF-Imager and the Mini-PAM system for these tests in a large scale application to miscanthus plants. This work was undertaken as part of the EU GrassMargins project (www.grassmargins.eu).

Material and Methods

Plant Material

Three different groups of crossings were performed in greenhouses at Tinplant, Germany (*M. sinensis* × *M. sinensis, M. sinensis* × *M. sacchariflorus* (day-neutral), and *M. sinensis* × *M. lutarioriparius*), and survival in the field of the resulting progeny was evaluated in Ireland and Denmark. In Denmark, the plants were additionally measured for shoot length and evaluated for onset of flowering.

The *M. sinensis* × *M. sinensis* crossings were primarily made between genotypes from two different gene pools. Genotypes from a number of European collections were supplied by Teagasc (Ireland) while Aarhus University (AU), Denmark provided genotypes collected from various locations in Japan that had been grown at AU Foulum since 1997.

The *M. sinensis* × *M. sacchariflorus* (day-neutral) crossings used a genotype of *M. sacchariflorus* (day-neutral, provided by Siberian branch of Russian Academy of Science) which was taller than the in Europe often used *M. sacchariflorus* 'Robustus', but the vegetation period was shorter than from *M. sinensis*.

The *M. lutarioriparius* genotype for the *M. sinensis* × *M. lutarioriparius* crossings was provided by Kai-Uwe Schwarz (JKI Braunschweig, Germany) and had been collected at 29° 19′ N 112° 57′ E in China. *M. lutarioriparius* was crossed with a group of *M. sinensis* genotypes from the gene pool supplied from Ireland and Denmark. Only the seeds from *M. lutarioriparius* were harvested.

Plant Cultivation for Chl a Fluorescence Measurements

A total of 903 plants were grown from seeds from crosses of *M. sinensis* × *M. sacchariflorus* (day-neutral) in spring 2014 in small plastic boxes size 12 × 17 cm with six plants in each box. They were transported to AU Foulum, Denmark, on February 27th 2015 in a dormant growth stage and stored at 6 °C in a dark room until March 30th. Then they were moved to the greenhouse, where the maximum and minimum temperature for the first two weeks was 15 °C and 6 °C, respectively. Thereafter the temperature was increased to 20/12 °C day/night.

Chlorophyll a Fluorescence Measurements

On April 27th the plants were moved into two controlled-environment rooms with the dimensions 2.40 m × 1.97 m × 2.05 m (MB-Teknik, Brøndby, Denmark). They were kept at the following environmental conditions in a 13/11 h day/night cycle: T = 12/12 °C, photon flux = 850 μmol m^{-2} s^{-1}, relative humidity = 75/85 % and CO_2-conc. = 400 ppm.

The plants were grown at these conditions for six days and thereafter they were measured for Chl *a* fluorescence with a CF-Imager (Technologica Ltd, Essex, UK). When possible, the youngest fully developed leaf (as judged by ligule emergence) of each plant was measured on the adaxial side. The measurements were performed when the plants were in the dark-adapted state. The minimal fluorescence (F_0) was measured in the absence of actinic light (0 μmol m^{-2} s^{-1}) where the PSII reaction centers are in the 'open' state. The maximal fluorescence (F_m) was measured with a saturation pulse of 6164 μmol m^{-2} s^{-1}, which drives the reaction center into closure. F_v/F_m was calculated from F_m and the variable fluorescence in the dark-adapted state (F_m-F_0) as $F_v/F_m = (F_m-F_0)/F_m$.

Thereafter the temperature was decreased to 10/8 °C in a 13/11 h day/night cycle. The other environmental conditions were as described above. Two days later the Chl *a* fluorescence was measured using a Mini-PAM fluorometer (Walz GmbH, Effeltrich, Germany). Two leaves of each of the seed plants were measured in the dark-adapted state at 8 °C. The measurement was performed on the middle part of the leaf where it was broadest. F_0 was determined at very low photosynthetic active radiation and F_m was measured by applying a 0.8 s pulse at a high light level of approximately 4000 μmol m^{-2} s^{-1}. F_v/F_m was calculated as an average of the measurements on two leaves.

Results

Field Evaluation for Winter Survival and Growth

A total of 826 seedlings of the *M. sinensis* (Msi) × *M. sinensis* and *M. sinensis* × *M. sacchariflorus* (Msa) crossings were planted in Denmark on June 13th 2014 and evaluated for survival in the field in Denmark on August 21st 2015, while 186 seedlings from the *M. sinensis* × *M. lutarioriparius* cross were planted on August 14th 2015.

A total of 751 plants from 44 crossings planted in 2014 survived the winter, which is approximately 91 % survival. The offspring were planted according to the original groups of parents from which the crossings were made. The survival rate within each group varied from 33 to 100 %.

In Ireland, 816 seedlings were planted in a field early in May 2014 and evaluated for survival on August 26th 2014 and June 16th 2015. The survival rate for Ireland was approximately 79 %, with 648 out of the 816 plants surviving. The survival rate within each group varied from 14 to 100 %. Differences in survival rates can be attributed to the uneven soil quality of the field which is fairly stony in patches. For seven *M. sinensis* × *M. sinensis* crossings all offspring survived the winter in both Ireland and Denmark. One of these offspring (EU026) had also the longest shoot with 203 cm at evaluation on August 24th 2015. Offspring of these crosses may be most interesting for further investigations.

Chlorophyll a Fluorescence Measurements

Chlorophyll *a* fluorescence of 881 miscanthus seedlings was measured at 12 °C using the CF-Imager, and at 8 °C using the Mini-PAM. The F_v/F_m values from both types of measurements were well correlated even though they were measured at different temperatures (Fig. 13.1). This correlation was also observed, to a lesser extent, for the groups of genotypes with high and low F_v/F_m values (Fig. 13.5). Both measurements resulted in a broad range of F_v/F_m for plants under mild (at 12 °C: 0.825–0.439) and severe (at 8 °C: 0.741–0.315) stress, indicating genetic variability for cold stress tolerance in this plant material.

The offspring can be differentiated depending on the seed mother. Hybrids with a *M. sacchariflorus* mother (Msa, one genotype) tended to have higher F_v/F_m values compared to hybrids with a *M. sinensis* mother (Msi, seven genotypes) (Fig. 13.2). A similar relation was observed for the mean F_v/F_m values (Fig. 13.3, CF-Imager at 12 °C; similar results for the Mini-PAM at 8 °C not shown). However, the offspring

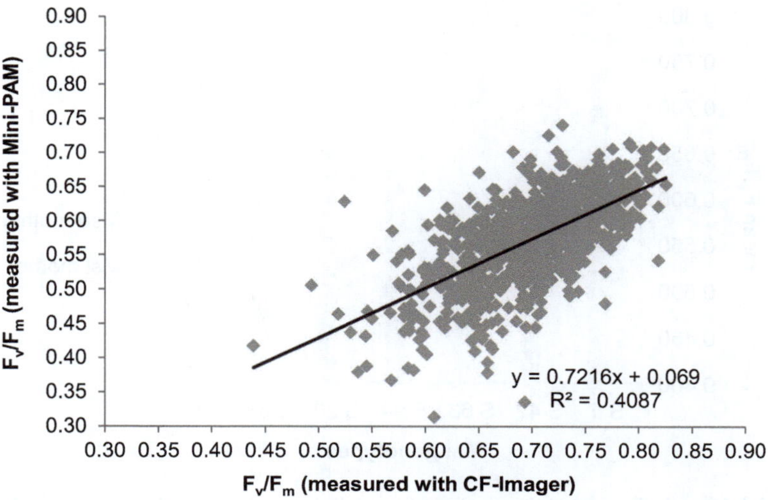

Fig. 13.1 Correlation between F_v/F_m for 881 miscanthus seedlings measured with a CF-Imager at 12 °C and a Mini-PAM at 8 °C

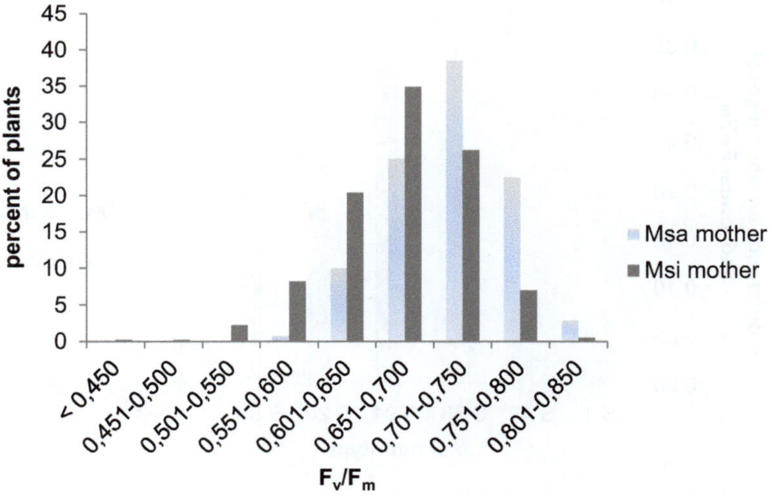

Fig. 13.2 Percent of plants ($n=881$) measured for F_v/F_m with the CF-Imager at 12 °C, and separated according to the mother from whom the seeds were harvested (Msa—*M. sacchariflorus*, Msi—*M. sinensis*)

from the seven Msi genotypes showed clear differences: plants from seeds from genotype S1 had the highest, and from S94 the lowest mean F_v/F_m (Fig. 13.3, CF-Imager results). The mean F_v/F_m values from the offspring harvested from Msa mother had smaller differences. Still there is considerable variation in F_v/F_m values between and within populations at 12 °C (Fig. 13.4, CF-Imager at 12 °C, similar results for the Mini-PAM at 8 °C not shown) which is correlated with the Msi genotype.

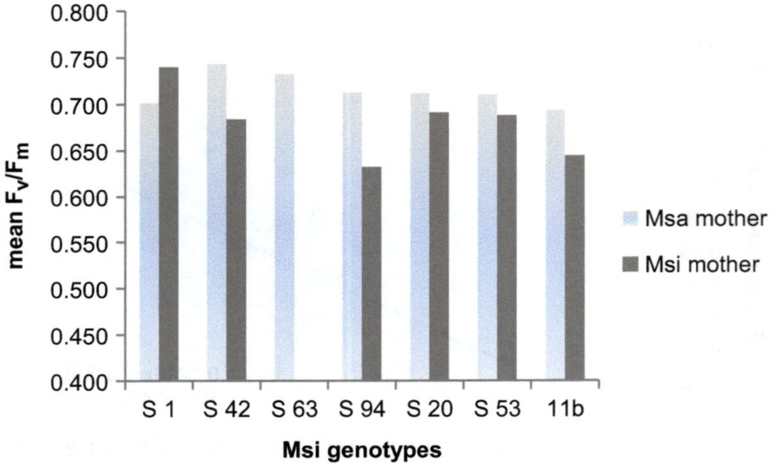

Fig. 13.3 Mean F_v/F_m measured with CF-Imager at 12 °C, and separated according to Msi genotypes and to the mother from whom the seeds were harvested, $n=881$ (Msa—*M. sacchariflorus*, Msi—*M. sinensis*)

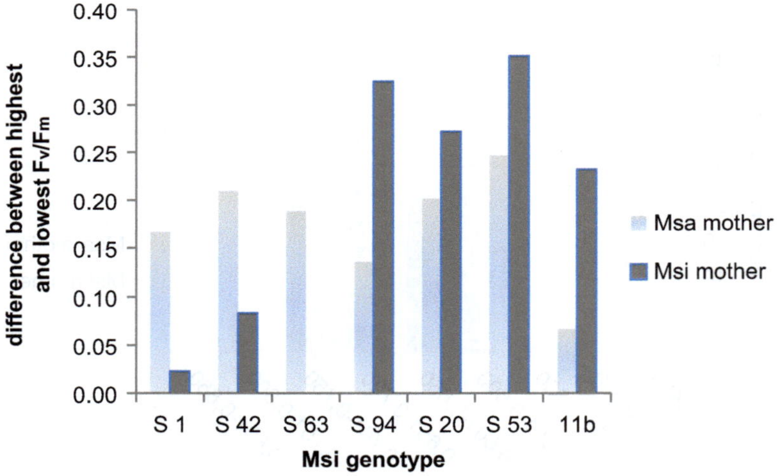

Fig. 13.4 Difference between the highest and the lowest F_v/F_m value measured with CF-Imager at 12 °C and separated according to Msi genotypes and to the mother from whom the seeds were harvested, $n=881$ (Msa—*M. sacchariflorus*, Msi—*M. sinensis*)

Selection of Genotypes with High and low Cold Tolerance After Chl a Fluorescence Measurements

Plants that had high F_v/F_m values when measured both with the CF-Imager and Mini-PAM systems were selected as cold tolerant genotypes. For comparison, plants with low F_v/F_m based on both methods were selected and expected to be less

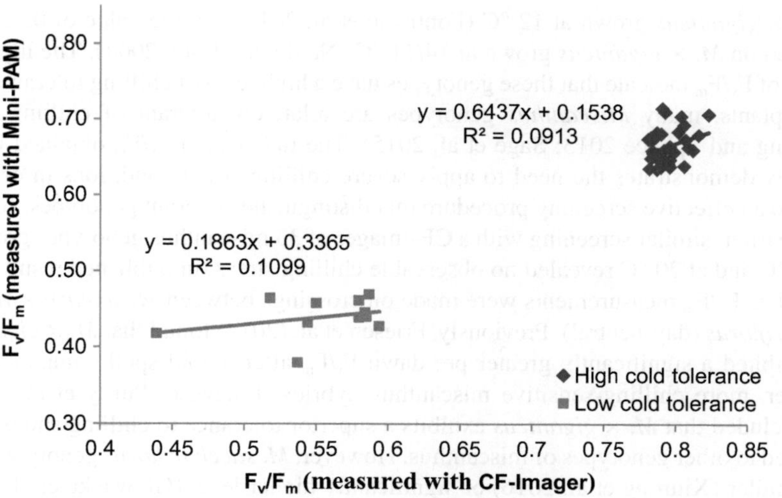

Fig. 13.5 Genotypes with high ($n = 33$) and low ($n = 11$) cold tolerance selected on the basis of high and low F_v/F_m measured with a CF-Imager at 12 °C and a Mini-PAM at 8 °C

cold tolerant genotypes. The F_v/F_m values were determined on the basis of measurements of the intermediate part of the leaves as the leaf tips showed a high variation in the Chl *a* fluorescence values. Within the 50 genotypes with the highest F_v/F_m (values between 0.781 and 0.823) when measured with the CF-Imager, 33 genotypes were selected because they also had high F_v/F_m (values between 0.635 and 0.741) when they were measured with the Mini-PAM (Fig. 13.5). Likewise, 11 genotypes of the 36 plants with the lowest F_v/F_m (values between 0.439 and 0.594) when measured with the CF-Imager that also had low values (between 0.315 and 0.470) when measured with the Mini-PAM were selected (Fig. 13.5). The selected genotypes were planted in the field for further studies.

Discussion

Low temperatures can reduce the growth of plants (Venema et al. 2000; Laing et al. 1995) and limit miscanthus production (Jones et al. 2015). Overall cold tolerance in miscanthus is the result of the chilling and frost tolerance of the shoots and the ability of rhizomes to overwinter successfully (Friesen et al. 2014). To extend the potential growing area of miscanthus to cooler climates it is necessary to find cold adapted genotypes. The screened offspring from the crossings in the present experiment indicated a broad genetic variability in cold tolerance determined by F_v/F_m measurements. Some of the genotypes had very high values at 12 °C (<0.825) and even at 8 °C (<0.741), which are higher than values of ≤0.655 under field conditions after a severe cold spell (Głowacka et al. 2015b), the value of 0.631 measured on

M. × *giganteus* grown at 12 °C (Fonteyne et al. 2015) and the value of 0.72 measured on *M.* × *giganteus* grown at 14/11 °C (Naidu and Long 2004). The high values of F_v/F_m indicate that these genotypes have a high level of chilling tolerance. For C_4 plants, many *Miscanthus* genotypes are relatively tolerant of chilling stress (Long and Spence 2013; Sage et al. 2015). The rather high F_v/F_m obtained in this study demonstrates the need to apply severe chilling stress conditions in order to have an effective screening procedure that distinguishes tolerant genotypes. Indeed, an earlier, similar screening with a CF-Imager of 33 miscanthus genotypes grown at 14 °C and at 20 °C revealed no observable chilling stress (unpublished results).

The F_v/F_m measurements were made on crossings between *M. sinensis* × *M. sacchariflorus* (day-neutral). Previously, Friesen et al. (2014) found that *M.* × *giganteus* exhibited a significantly greater pre-dawn F_v/F_m after a cold spell compared with other, more chilling-sensitive miscanthus hybrids. Likewise, Purdy et al. (2013) concluded that *M.* × *giganteus* exhibits a superior tolerance to chilling shock compared to other genotypes of miscanthus. However, *M. sacchariflorus* genotypes with a similar (Xiurong et al. 2016) or significantly higher level (Głowacka et al. 2014) of photosynthesis at low temperatures compared to *M.* × *giganteus* can be found. Furthermore, Farrell et al. (2006) reported that a *M. sinensis* hybrid exhibited higher cold tolerance than *M.* × *giganteus*. These results and our F_v/F_m measurements show that there is still value and potential in searching for superior tolerance. Though, for a definitive conclusion regarding our own F_v/F_m measurements a direct comparison of these cold tolerant genotypes with *M.* × *giganteus* and other new hybrids should be a target for future work.

The present results showed that the offspring from the Msa mother had higher F_v/F_m than the reciprocal offspring from the Msi mother. Contribution of maternal inheritance has been demonstrated for biomass related traits in miscanthus (Gifford et al. 2014), but not yet for stress responses. The crosses were performed with one Msa genotype collected in the Russian Far East (Sergey Peltek, pers. com.). It can be assumed that this Msa genotype is better adapted to the severe winter in this region than *M. sinensis,* which is uncommon there. Such geographically derived strong cold tolerance has been observed previously in *M. sacchariflorus*, for example northern populations had the highest tolerance of those studied by Yan et al. (2012). Our results indicate that the cytoplasm inherited from the mother contributes to cold tolerance features of the offspring. The variation of F_v/F_m values between and within populations is genotype dependent and a sign of genetic variation. Other studies have also shown the potential in searching for more chilling tolerance in miscanthus (Głowacka et al. 2014, 2015b; Clifton-Brown and Jones 1997).

Survival in the field after winter was recorded at 91 % in Denmark, while at the same site *M.* × *giganteus* can suffer near complete mortality in some winters (Clifton-Brown et al. 2001). The survival rates of the new crosses was higher than those of wild Japanese *M. sinensis* seedlings planted in northern Japan or wild Chinese accessions of *M. sinensis, M. sacchariflorus* and *M. lutarioriparius* planted in northern China (Anzoua et al. 2015; Yan et al. 2011). While the offspring showed heterogeneity in overwintering, the poor soil conditions additionally influenced the field survival in Ireland. Generally, genetic heterogeneity in the offspring can be estimated (Deuter 2000 and Sacks et al. 2013) to assess the potential for improvement. Also Hastings et al. (2009) and Jones et al. (2015) suggested that new miscanthus varieties should be superior to parent types concerning drought and winter frost tolerance. New hybrids give the opportunity to apply the genetic variability for adaptation to different and challenging environment conditions.

Measurement of Chl *a* fluorescence is a non-destructive method useful for screening plants for abiotic stress (Murchie and Lawson 2013). Both the CF-Imager and Mini-PAM can be used on miscanthus at seedling or an early growth stage, so that their application is fast and cost efficient. The CF-Imager measures the Chl *a* fluorescence on the basis of a larger leaf area than the Mini-PAM, which only provides point measurements and is thus more sensitive to within leaf variability. Furthermore, the CF-Imager does not require leaves of a certain width, which the Mini-PAM does, and variation over the whole leaf can be investigated. However, the Mini-PAM is suitable for application in the field, and in addition it is cheaper, faster, and easier to use than the CF-Imager. The measurements of approximately 900 plants in this experiment indicated a good correlation of F_v/F_m between CF-Imager and Mini-PAM. Previously measurements of Chl *a* fluorescence have been used as a small-scale screening tool for cold tolerance in miscanthus (Głowacka et al. 2015b) Here we find that both applied methods are useful for large-scale screening of cold tolerance in miscanthus to select the most promising genotypes for further analysis.

Acknowledgements This work was supported by FP7 KBBE.2011.3.1-02 Grant number 289461 (GrassMargins).

References

Anzoua KG, Suzuki K, Fujita S, Toma Y, Yamada T (2015) Evaluation of morphological traits, winter survival and biomass potential in wild Japanese *Miscanthus sinensis* Anderss. Populations in northern Japan. Grassl Sci 61:83–91

Beale CV, Long SP (1995) Can perennial C4 grasses attain high efficiencies of radiant energy conversion in cool climates? Plant Cell Environ 18:641–650

Clifton-Brown JC, Jones MB (1997) The thermal response of leaf extension rate in genotypes of the C4-grass Miscanthus: an important factor in determining the potential productivity of different genotypes. J Exp Bot 48:1573–1581

Clifton-Brown JC, Lewandowski I, Andersson B, Basch G, Christian DG, Kjeldsen JB, Jørgensen U, Mortensen JV, Riche AB, Schwarz K-U, Tayebi K, Teixeira F (2001) Performance of 15 Miscanthus genotypes at five sites in Europe. Agron J 93:1013–1019

Deuter M (2000) Breeding approaches to improvement of yield and quality in *Miscanthus* grown in Europe, in: European Miscanthus Improvement, Final Report, FAIR CT-96-1392, pp 28–52

Farrell AD, Clifton-Brown JC, Lewandowski I, Jones MB (2006) Genotypic variation in cold tolerance influences the yield of *Miscanthus*. Ann Appl Biol 149:337–345

Fonteyne S, Lootens P, Muylle H, Van den Ende W, De Swaef T, Reheul D, Roldán-Ruiz I (2015) Chilling tolerance and early vigor related characteristics evaluated in two Miscanthus genotypes. Photosynthetica. doi:10.1007/s11099-016-0193-y

Fracheboud Y, Haldimann P, Leipner J, Stamp P (1999) Chlorophyll fluorescence as a selection tool for cold tolerance of photosynthesis in maize (*Zea mays* L.). J Exp Bot 50(388): 1533–1540

Friesen PC, Peixoto MM, Busch FA, Johnson DC, Sage RF (2014) Chilling and frost tolerance in *Miscanthus* and *Saccharum* genotypes bred for cool temperate climes. J Exp Bot. doi:10.1093/jxb/eru105

Gifford JM, Chae WB, Swaminathan K, Moose SP, Juvik JA (2014) Mapping the genome of *Miscanthus sinensis* for QTL associated with biomass productivity. GCB Bioenergy. doi:10.1111/gcbb.12201

Głowacka K, Adhikari S, Peng J, Gifford J, Juvik JA, Long SP, Sacks EJ (2014) Variation in chilling tolerance for photosynthesis and leaf extension growth among genotypes related to the C4 grass *Miscanthus* × *giganteus*. J Exp Bot. doi:10.1093/jxb/eru287

Głowacka K, Clark LV, Adhikari S, Peng J, Stewart JR, Nishiwaki A, Yamada T, Jørgensen U, Hodkinson TR, Gifford J, Juvik JA, Sacks EJ (2015a) Genetic variation in *Miscanthus* × *giganteus* and the importance of estimating genetic distance thresholds for differentiating clones. GCB Bioenergy 7:386–404

Głowacka K, Jorgensen U, Kjeldsen JB, Kørup K, Spitz I, Sacks EJ, Long SP (2015b) Can the exceptional chilling tolerance of C4 photosynthesis found in *Miscanthus* × *giganteus* be exceeded? Screening of a novel Miscanthus Japanese germplasm collection. Ann Bot. doi:10.1093/aob/mcv035

Hastings A, Clifton-Brown J, Wattenbach M, Mitchell CP, Stampfl P, Smith P (2009) Future energy potential of *Miscanthus* in Europe. GCB Bioenergy 1:180–196

Heaton EA, Clifton-Brown J, Voigt TB, Jones MB, Long SP (2004) *Miscanthus* for renewable energy generation: European Union experience and projections for Illinois. Mitig Adapt Strat Glob Chang 9:433–451

Hodkinson TR, Klaas M, Jones MB, Prickett R, Barth S (2015) Miscanthus: a case study for the utilization of natural genetic variation. Plant Genet Resour 13:219–237

Jones MB, Finnan J, Hodkinson TR (2015) Morphological and physiological traits for higher biomass production in perennial rhizomatous grasses grown on marginal land. GCB Bioenergy 7:375–385

Kao WY, Tsai TT, Chen WH (1998) Response of photosynthetic gas exchange and chlorophyll a fluorescence of *Miscanthus floridulus* (Labill) Warb. to temperature and irradiance. J. Plant Physiol 152:407–412

Laing WA, Greer DH, Schnell TA (1995) Photoinhibition of photosynthesis causes a reduction in vegetative growth rates of dwarf bean (*Phaseolus vulgaris*) plants. Aust J Plant Physiol 22:511–520

Lewandowski I, Clifton-Brown J, Scurlock JMO, Huisman W (2000) Miscanthus: European experience with a novel energy crop. Biomass and Bioenergy 19:209–227

Long SP, Spence AK (2013) Toward cool C4 crops. Annu Rev Plant Physiol Plant Mol Biol 64:701–722

Maxwell K, Johnson GN (2000) Chlorophyll fluorescence – a practical guide. J Exp Bot 51: 659–668

Murchie EH, Lawson T (2013) Chlorophyll fluorescence analysis: a guide to good practice and understanding some new applications. J Exp Bot. doi:10.1093/jxb/ert208

Naidu SL, Long SP (2004) Potential mechanisms of low-temperature tolerance of C4 photosynthesis in Miscanthus x giganteus: an in vivo analysis. Planta 220:145–155

Naidu SL, Moose SP, Al-Shoaibi AK, Raines CA, Long SP (2003) Cold tolerance of C4 photosynthesis in *Miscanthus* × *giganteus*: adaptation in amounts and sequence of C4 photosynthetic enzymes. Plant Physiol 132:1688–1697

Purdy SJ, Maddison AL, Jones LE, Webster RJ, Andralojoc J, Donnision I, Clifton-Brown J (2013) Characterization of chilling-shock responses in four genotypes of *Miscanthus* reveals the superior tolerance of *M.* × *giganteus* compared with *M. sinensis* and *M. sacchariflorus*. Ann Botany. doi:10.1093/aob/mct059

Sacks E, Juvik JA, Lin Q, Stewart R, Yamada T (2013) The gene pool of *Miscanthus* species and its improvement. In: Paterson AH (ed), Genomics, plant genetics and genomics: crops and models 11. Springer Science+Business Media, New York, pp 73–101

Sage RF, de Melo Peixoto M, Friesen P, Deen B (2015) C4 bioenergy crops for cool climates, with special emphasis on perennial C 4 grasses. J Exp Bot 66:4195–4212

Schalitz G, Behrendt A (2012) Mehrjähriger Anbau von *Triarrhena* in Lysimetern und Ergebnisse zu Wasser- und Stoffbilanzen. In: Energetische Nutzung von Grünlandaufwüchsen: Referate und Poster der 56. Jahrestagung der Arbeitsgemeinschaft Grünland und Futterbau in der Gesellschaft für Pflanzenbauwissenschaften 2012 in Witzenhausen. pp 213–217

Venema JH, Eekhof M, van Hasselt PR (2000) Analysis of low-temperature tolerance of a tomato (*Lycopersicon esculentum*) cybrid with chloroplasts from a more chilling-tolerant *L. hirsutum* accession. Ann Bot 85:799–807

Xiurong J, Kørup K, Andersen MN, Sacks EJ, Zhu X-G, Lærke PE, Jørgensen U (2016) Can miscanthus C_4 photosynthesis compete with festulolium C_3 photosynthesis in a temperate climate? GCB Bioenergy, Accepted

Yan J, Chen W, Luo F, Ma H, Meng A, Li X, Zhu M, Li S, Zhou H, Zhu W, Han B, Ge S, Li J, Sang T (2011) Variability and adaptability of Miscanthus species evaluated for energy crop domestication. GCB Bioenergy 4:49–60

Yan J, Chen W, Luo F, Ma H, Meng A, Li X, Zhu M, Li S, Zhou H, Zhu W, Han B, Ge S, Li J, Sang T (2012) Variability and adaptability of Miscanthus species evaluated for energy crop domestication. GCB Bioenergy 4:49–60

Chapter 14
Comparison of Different Miscanthus Genotypes for Ash Melting Behaviour at Different Locations

Y. Iqbal and I. Lewandowski

Introduction

The issue of climatic change and scarcity of fossil fuels has diverted the focus of research towards renewable energy resources. Among the renewable energy resources, biomass-based energy production is one of the options with great potential. The major biomass resources include agricultural crops and residues, forest wood and residues, short rotation coppice crops and dedicated energy crops. Biomass can be converted through different pathways such as combustion to produce heat and electricity, fermentation to produce ethanol.

The end use of biomass is defined based on the composition of biomass. For example, biomass with low lignin content is preferred for fermentation process to produce ethanol. In recent years, the focus is on exploiting such biomass resources, which have no conflict with food supply. Therefore, dedicated energy crops emerged as an attractive option. Dedicated energy crops can contribute significantly towards sustainable biomass supply by exploiting those areas which are less suitable for agricultural crops.

There are different forms of biomass-based solid fuels such as wood chips, wood pellet and briquettes which are being used to produce heat and electricity. However, in Europe, the use of dedicated energy crops such as miscanthus has increased significantly in recent years, mainly for direct combustion to produce heat and electricity. The main obstacle to use such crops for combustion is high mineral, ash and moisture content. In biomass composition, the contents of potassium (K) and chloride (Cl) play a key role in defining suitability of biomass for combustion process (Lewandowski and Heinz 2003). Generally, in field crops, the Cl content is high in comparison to wood as well as coal. The high mineral content especially K and Cl

Y. Iqbal (✉) • I. Lewandowski
Department of Biobased Products and Energy Crops (340b), Institute of Crop Science, University of Hohenheim, Fruwirthstr 23, Stuttgart 70599, Germany
e-mail: Iqbal_Yasir@uni-hohenheim.de

S. Barth et al. (eds.), *Perennial Biomass Crops for a Resource-Constrained World*, DOI 10.1007/978-3-319-44530-4_14

participate in many chemical reactions which lead to ash melting at low temperature (Baxter et al. 2014). The low ash melting temperature can lead to deposition and damages the boilers which reduce the conversion efficiency and increases the maintenance cost. Therefore, it is important to keep the content of aforementioned elements low to carry out efficient combustion process. The biomass composition can be improved through selection of appropriate genotype and improved management practices such as appropriate harvesting and fertilisation (Baxter et al. 2014; Kludze et al. 2013).

Therefore, the focus of this study is on comparing biomass quality of different miscanthus genotypes for ash melting behaviour. In addition, the effect of different factors such as mineral and ash content on ash melting behaviour will also be part of this study.

The field trials were established with 15 miscanthus genotypes as a part of OPTIMISC project. The samples were collected, processed and analysed in laboratory for mineral, ash and moisture content as well as for ash melting behaviour. For this study some of the genotypes are selected to compare the ash melting behaviour.

Materials and Methods

Biomass Samples

The biomass samples for this analysis were collected from a field trial with 15 miscanthus genotypes (Table 14.1) established in 2012 as a part of OPTIMISC at 4 sites (Aberystwyth, Moscow, Stuttgart, and Wageningen). Out of these 15 miscanthus genotypes, 11 are in vitro derived clones and 4 are seed based. The 15 genotypes were planted in plots in a randomised block design with three replications, resulting in 45 experimental plots.

The biomass samples analysed in this study were collected between January and March, 2014–2015 (depending on location) by harvesting the middle 5 m^2 with manual cutters. Samples were chopped, dried (60 °C for 48 h), ground and then passed through a 1-mm sieve. The samples were used for mineral analysis and preparation of ash samples to monitor ash melting behaviour.

Analytical Methods

Dried biomass samples were analysed in the laboratory for N, P, K, Na, Cl, Si, Ca, Mg and ash content. N analysis was carried out by using Vario Macro cube, Elementaranalysensysteme (GmbH, Hanau, Germany) by following the Dumus principle (Naumann and Bassler (1976/2012) VDLUFA Methods Book III). The K, Ca, Na analyses was carried out after microwave (Ethos.Lab MLS GmbH) digestion

Table 14.1 Descriptions of miscanthus genotypes used in field trials at Ihinger Hof research station

Genotype ID	Species	Propagation method	Supplier
OPM-1	*M. sacchariflorus*	In vitro	IBERS
OPM-2	*M. sacchariflorus*	In vitro	IBERS
OPM-3	*M. sacchariflorus*	In vitro	IBERS
OPM-4	*M. sacchariflorus*	In vitro	IBERS
OPM-5	*M. sinensis* × *M. sacchariflorus hybrid*	In vitro	IBERS
OPM-6	*M. sinensis* × *M. sacchariflorus hybrid*	In vitro	IBERS
OPM-7	*M. sinensis* × *M. sacchariflorus hybrid*	In vitro	IBERS
OPM-8	*M. sinensis* × *M. sacchariflorus hybrid*	In vitro	IBERS
OPM-9	*M* × *giganteus*	In vitro	IBERS
OPM-10	*M. sinensis* × *M. sacchariflorus hybrid*	In vitro	Schwarz
OPM-11	*M. sinensis*	In vitro	IBERS
OPM-12	*M. sinensis*	Seedling	IBERS
OPM-13	*M. sinensis*	Seedling	WUR
OPM-14	*M. sinensis*	Seedling	WUR
OPM-15	*M. sinensis*	Seedling	IBERS

at 120–180 °C and pressure (24.16 bar) for 40 min with HNO_3 (65 %) and addition of H_2O_2 and measuring by using Flame photometer (ELEX 6361, Eppendorf AG, Hamburg, Germany) (Naumann and Bassler (1976/2012) VDLUFA Methods Book III). After digestion, lanthanum solution was added to the extracts and analysed for Mg content by using Atomic absorptions spectrometer (220 FS, Varian) (Naumann and Bassler (1976/2012) VDLUFA Methods Book III). For determination of Cl, extracts were prepared through hot water extraction and by using CARREZ solutions. The extracts were measured by using HPLC (high-performance liquid chromatography) (ICS 2000, Dionex Corporation, Sunnyvale, California, US). The microwave was used to digest the samples with HNO_3 and HF and Si content was determined by using ICP-OES (Vista Pro, Varian Inc., Palo Alto, California, US). To determine the P content, extracts were prepared through microwave digestion of biomass samples with HNO_3 and addition of H_2O_2, and then extracts were amended by ammonium vanadate and ammonium molybdate. The extracts were analysed for P content by using Atomic absorption spectrometry (Spectrophotometer PM6 W, Carl Zeiss AG, Oberkochen, Germany). For determination of ash content, samples were kept in Muffle furnace at 550 °C for 4 h (Naumann and Bassler (1976/2012) VDLUFA Methods Book III).

Preparation of ash samples and heating treatments: To prepare the ash samples, 20 g of biomass samples was taken in ceramic crucibles and heated in an electric muffle furnace at 550 °C for 4 h. For ash melting behaviour, approximately 100 mg of ash samples was taken and transferred to separate ceramic combustion boats. The ceramic combustion boats were placed in an electric muffle furnace, which was heated at an average rate of 10 °C min^{-1} until the required heating temperature was

Table 14.2 Ash-fusion temperature and ash-fusion classes

Ash-fusion temperature	Ash-fusion classes	Microscopic observations
Initial temperature	Loosening (1–2)	Particles are arranged in loose layers, spatula can move through without any resistance, shiny surfaces with tiny molten vesicles
Softening temperature	Partially sintered (2–3)	Particles start becoming compact through strong adhesive forces, still easy to disintegrate, produces crispy sound when spatula passes through, larger molten vesicles on the surface
Hemisphere temperature	Highly sintered (3–4)	Difficult to disintegrate, most of the area covered with larger molten vesicles. Organogenic material also visible in some parts
Flow temperature	Molten (4–5)	Particles are completely molten, manual disintegration is not possible, no organogenic material visible

achieved. Each of the ash samples was subjected to 4 different heating treatments of 800 °C, 900 °C, 1000 °C and 1100 °C for 2 h. After 2 h, the combustion boats were removed and placed into desiccators to allow them to cool before further analysis.

Monitoring of Ash Melting Behaviour

After different heating treatments and letting the samples to cool down in desiccators, the samples were analysed under a stereo microscope (Zeiss Stemi 2000-C, Carl Zeiss AG, Oberkochen, Germany) at magnifications up to 40×. Based on microscopic observations, the ash samples were graded and classified into 4 ash-fusion classes from 1 (no sintering) to 5 (completely molten). The ash-fusion classes with a description of microscopic observations are listed in Table 14.2 (Tonn et al. 2012).

Results

Ash Melting Behaviour at Different Locations

To estimate the ash melting behaviour, the genotypes OPM-3, OPM-6, OPM-9, OPM-10, OPM-11 and OPM-15 were selected based on dry matter yield potential except for OPM-15. These genotypes were tested at selected locations (Aberystwyth, Moscow, Stuttgart and Wageningen). The results indicate that at Aberystwyth, no strong sintering was observed at 800 °C–900 °C heating temperature for all genotypes except for OPM-6 and OPM-11. OPM-10 performed better at all heating treatments in comparison to other genotypes, whereas OPM-6 showed strong sintering at all heating treatments except for 800 °C (Fig. 14.1). OPM-15 also performed

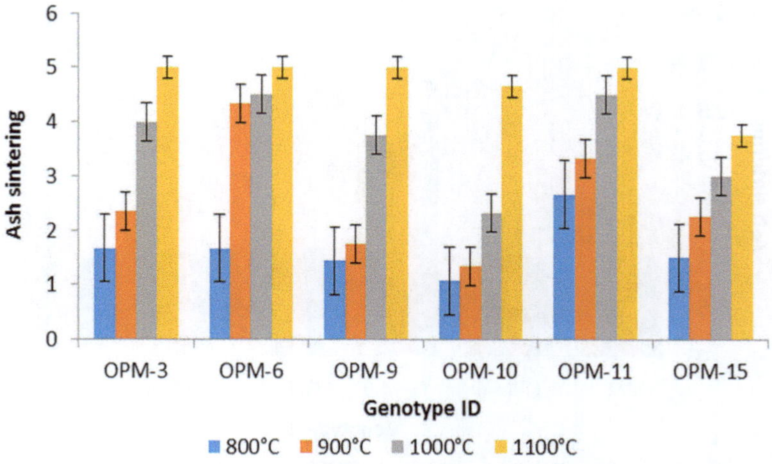

Fig. 14.1 comparison of ash melting behaviour for selected miscanthus genotypes at Aberystwyth

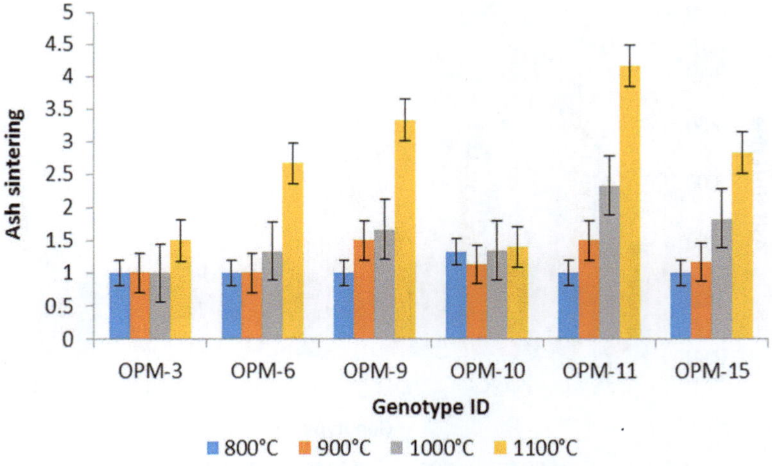

Fig. 14.2 Comparison of ash melting behaviour for selected miscanthus genotypes at Moscow

better in comparison to other genotypes except OPM-10 but this genotype is among the low yielding.

At Moscow, most of the genotypes performed better than Aberystwyth. At all heating treatments, no sintering was observed for OPM-3 and OPM-10. For other genotypes, partial sintering to strong sintering was recorded at 1000 °C and 1100 °C, respectively (Fig. 14.2). The high yielding genotype (OPM-6) has shown no sintering at all heating treatments except for 1100 °C, where partial sintering was observed.

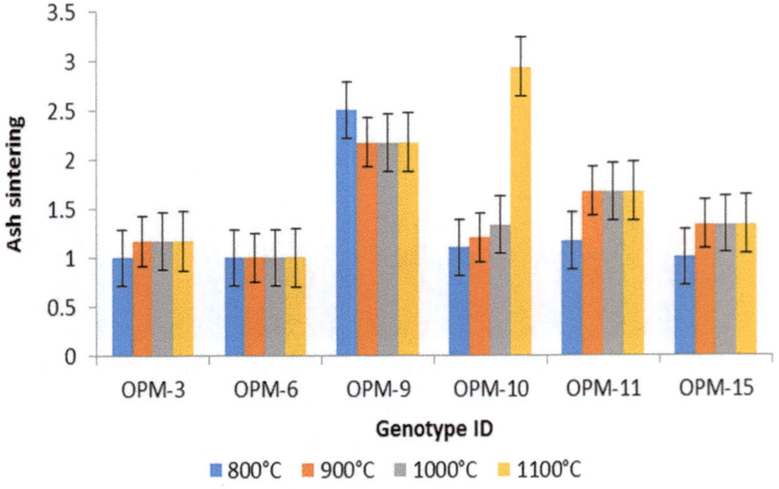

Fig. 14.3 Comparison of ash melting behaviour for selected miscanthus genotypes at Wageningen

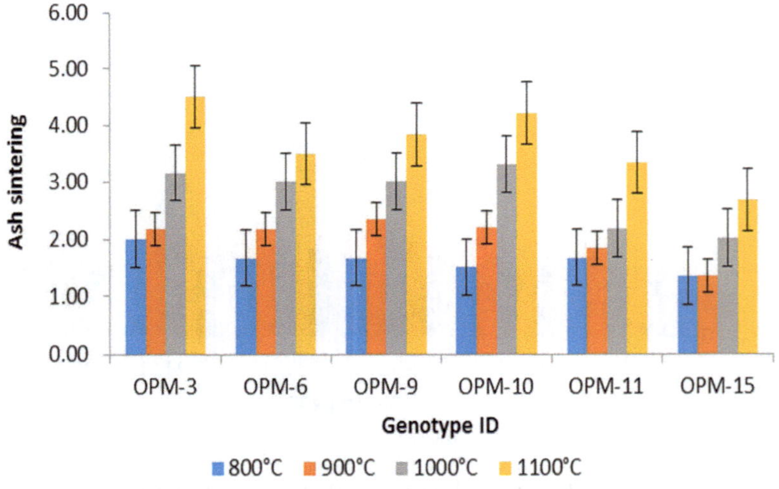

Fig. 14.4 Comparison of ash melting behaviour for selected miscanthus genotypes at Stuttgart

At Wageningen, most of the genotypes performed better especially at high temperatures in comparison to all other locations. OPM-3, OPM-6, OPM-11 and OPM-15 showed no sintering to slight sintering. At this location, OPM-9 showed partial sintering at all temperatures, there was no significant effect of heating treatments on sintering (Fig. 14.3). OPM-10 showed no sintering from 800 to 1000 °C and partial sintering was recorded at 1100 °C.

At Stuttgart, OPM-11 and OPM-15 performed better with low sintering tendencies in comparison to other genotypes. For all genotypes, increase in temperature led to significant increase in ash sintering. At 1100 °C, all the genotypes showed strong sintering except for OPM-15 (Fig. 14.4).

Relationship Between Biomass Composition and Ash Melting Behaviour

The correlation analysis indicates that there is strong relationship between Cl content in biomass and ash sintering. With the increase in Cl content, the tendency of ash sintering also increased significantly. The correlation coefficient for Cl and sintering was $r=0.70$.

The correlation analysis was also performed between K content in biomass and ash sintering. It also showed the same trend as for Cl content. Any increase in K content in biomass leads to high sintering tendencies which subsequently led to low ash melting temperatures. The value for correlation coefficient for K and ash sintering tendency was $r=0.65$. It signifies the importance of K and Cl content in biomass and its impact on ash melting behaviour.

Discussion

The combustion quality of biomass is determined by a) composites that determine the heating value of the biomass, e.g., ash, moisture and lignin content; b) composites that lead to harmful emissions, e.g., nitrogen (N), sulphur (S), chloride (Cl) and heavy metals; c) composites that have an impact on ash fouling, slagging and corrosion, e.g., chloride (Cl), potassium (K), phosphorus (P), magnesium (Mg), silicon (Si), calcium (Ca) and sodium (Na). High contents of chloride and potassium in the dry matter lead to ash melting at low temperatures and cause mechanical problems in the combustion processes such as corrosion and fouling (Baxter et al. 2012; Lewandowski and Kicherer 1997). Ash and moisture contents are relevant for the combustion process mainly due to their simultaneous effect on operating costs and heating value. High contents of nitrogen and sulphur in the plant biomass contribute to high emissions of primary pollutants such as nitrogen oxides (NO_x) and sulphur oxides (SO_x).

There are a number of factors responsible for the variation in the chemical composition of miscanthus genotypes. These include genetic differences (Jezowski et al. 2011), harvesting time (Miguez et al. 2008), climatic aspects, mainly temperature and rainfall (Woli et al. 2011) and soil conditions. There are several reports on the anatomical and morphological differences among the miscanthus genotypes, which lead to variation in chemical constituents of the biomass (Kaack et al. 2003; Lewandowski et al. 2003). Regarding climatic conditions, rainfall plays a key role by facilitating the leaching process such as Cl leaching which subsequently improves combustion quality. However, leaching efficiency depends on morphology especially stem thickness. Therefore, the variation in quality characteristics among the genotypes can be partially explained by morphological differences especially thickness of the stem, and leaf to stem ratio. Based on the current study, genotypes can be categorised into two main groups by considering their morphological characteristics a) thick stemmed genotypes with long stalks; b) thin stemmed genotypes with short stalks. The thick stemmed genotypes such as OPM-9 had low leaching rate compared to thin stemmed genotypes such as OPM-15. It indicates that the thickness of the stem is highly relevant to leaching of the minerals mainly for K and Cl. The passive leaching of minerals was observed in thick stemmed genotypes in comparison to thin stemmed genotypes (Jørgensen 1997: Iqbal and Lewandowski 2014), which could be the partial reason for difference in ash melting behaviour among the genotypes. The ash melting behaviour is directly linked with K and Cl contents of biomass, if there is low content of K and Cl, there is less chance of ash melting at low temperature because Cl participates in most of the chemical reactions taking place during combustion process. Therefore, the low K and Cl contents in the biomass composition make it favourable for combustion process. In addition, keeping K and Cl contents low through crop management practices and genotype selection is highly critical because it is not possible to keep these contents low through combustion technology as it is possible in case of N, where emissions can be reduced.

The other important factor to optimise biomass quality through crop management practices is optimal harvesting time. The time of harvesting is very important to define the biomass composition and biomass qualities for combustion. Presently, delayed harvest is being used effectively as a tool to improve the combustion quality characteristics, although it is at the expense of dry matter yield (Lewandowski and Heinz 2003). Delayed harvest provides ample time for the relocation of nutrients and increases the proportion of standing dead biomass with more broken cuticles. The high proportion of broken cuticles facilitates the leaching process. Therefore, the differences among the genotypes for ash melting behaviour at different locations can also be due to differences in harvesting time for each location. It indicates that the effect of harvesting time is highly significant for biomass qualities, which is in general agreement with those observed for miscanthus genotypes in previous studies (Lewandowski and Kicherer 1997; Lewandowski et al. 2000; Clifton-Brown et al. 2001 and Lewandowski and Heinz 2003).

Conclusions

Mineral composition especially K and Cl play key role in defining the ash melting behaviour of a specific genotype. The leaching process is very crucial for improvement of biomass quality especially through decreasing of Cl content in biomass. For locations, where ash fusion temperature was low, further delay in harvest could be an option to improve quality.

Acknowledgement This manuscript is part of OPTIMISC (optimising miscanthus biomass production) project, which is an international research project funded by the European Union's Seventh Programme for research, technological development and demonstration under the grant agreement FP7-289159.

References

Baxter XC, Darvell LI, Jones JM, Barraclough T, Yates NE, Shield I (2012) Study of *Miscanthus x giganteus* ash composition—variation with agronomy and assessment method. Fuel 95: 50–62

Baxter XC, Darvell LI, Jones JM et al (2014) Miscanthus combustion properties and variations with Miscanthus agronomy. Fuel 117:851–869

Clifton-Brown JC, Lewandowski I, Andersson B, Basch G, Christian DG, Kjeldsen JB, Jørgensen U, Mortensen J, Riche AB, Schwarz KU, Tayebi K, Teixeira F (2001) Performance of 15 *Miscanthus* genotypes at five sites in Europe. Agron J 93:1013–1019

Iqbal Y, Lewandowski I (2014) Inter-annual variation in biomass combustion quality traits over five years in fifteen Miscanthus genotypes in south Germany. Fuel Process Technol 121:47–55. doi:10.1016/j.fuproc.2014.01.003

Jezowski S, Glowacka K, Kaczmark Z (2011) Variation on biomass yield and morphological traits of energy grasses from the genus *Miscanthus* during the first years of crop establishment. Biomass and Bioenergy 35:814–821

Jørgensen U (1997) Genotype variation in dry matter accumulation and content of N. K and Cl in *Miscanthus* in Denmark. Biomass and Bioenergy 12:155–169

Kaack K, Schwarz KU, Brander PE (2003) Variation in morphology, anatomy and chemistry of stems of *Miscanthus* genotypes differing in mechanical properties. Ind Crops Prod 17: 131–142

Kludze H, Deen B, Dutta A (2013) Impact of agronomic treatments on fuel characteristics of herbaceous biomass for combustion. Fuel Process Technol 109:96–102

Lewandowski I, Heinz A (2003) Delayed harvest of *Miscanthus*—influences on biomass quantity and quality and environmental impacts of energy production. Eur J Agron 19:45–63

Lewandowski I, Kicherer A (1997) Combustion quality of biomass: practical relevance and experiments to modify the biomass quality of *Miscanthus giganteus*. Eur J Agron 6:163–177

Lewandowski I, Clifton-Brown JC, Scurlock JMO, Huisman W (2000) *Miscanthus*: European experience with a novel energy crop. Biomass and Bioenergy 19:209–227

Lewandowski I, Clifton-Brown JC, Andersson B, Basch G, Christian DG, Jørgensen U, Jones MB, Riche AB, Schwarz KU, Tayebi K, Teixeira F (2003) Environment and harvest time affects the combustion qualities of *Miscanthus* genotypes. Agron J 95:1274–1280

Miguez FE, Villamil MB, Long SP, Bollero GA (2008) Meta-analysis of the effects of management factors on *Miscanthus x giganteus* growth and biomass production. Agric For Meteorol 148:1280–1292

Naumann C, Bassler R (1976/2012) Die chemische Untersuchung von Futtermitteln. Methodenbuch/ Verband Deutscher Landwirtschaftlicher Untersuchungs—und Forschungsanstalten, Bd. 3. VDLUFA -Verl., Darmstadt

Tonn B, Thumm U, Lewandowski I, Claupein W (2012) Leaching of biomass from semi-natural grasslands—effects on chemical composition and ash high-temperature behaviour. Biomass and Bioenergy 36:390–403. doi:10.1016/j.biombioe.2011.11.014

Woli KP, David MB, Tasai J, Voigt TB, Darmody RG, Mitchell CA (2011) Evaluating silicon concentrations in biofuel feedstock crops *Miscanthus* and switchgrass. Biomass and Bioenergy 35:2807–2813

Chapter 15
Bioenergy Trees: Genetic and Genomic Strategies to Improve Yield

G. Taylor, M.R. Allwright, H.K. Smith, A. Polle, H. Wildhagen, M. Hertzberg, R. Bhalerao, J.J.B. Keurentjes, S. Scalabrin, D. Scaglione, and M. Morgante

Global Drivers for Increasing Bioenergy from Trees

The recently concluded COP21 climate negotiations committed UN signatories to restrict global temperature increases to 'well below' 2 °C above pre-industrial levels and to aspire towards 1.5 °C (United Nations 2015). If such ambitious climate and GHG emissions targets are to be met and future energy security assured, it is essential that renewable and sustainable alternative energy sources are developed and utilised on a global scale. Bioenergy and bioenergy with CCS will be a central part of that commitment, since these technologies feature strongly in many forward scenarios for reduced and negative carbon emissions, as described in the work leading

G. Taylor (✉) • M.R. Allwright • H.K. Smith
Centre for Biological Sciences, Life Sciences Building, University of Southampton, Southampton SO17 1BJ, UK
e-mail: G.Taylor@soton.ac.uk

A. Polle • H. Wildhagen
Department of Forest Botany and Tree Physiology, Büsgen Institute, Georg-August-University, Göttingen 37077, Germany

M. Hertzberg
SweTree Technologies AB, P.O. Box 4095, SE-904 03 Umeå, Sweden

R. Bhalerao
Department of Forest Genetics and Plant Physiology, Umeå Plant Science Center, Swedish University of Agricultural Sciences, 901 83 Umeå, Sweden

J.J.B. Keurentjes
Laboratory of Genetics, Wageningen University, 6708PB Wageningen, The Netherlands

S. Scalabrin • D. Scaglione
IGA Technology Services, Via J.Linussio, 51—Z.I.U., Udine 33100, Italy

M. Morgante
Università Di Udine, Istituto die Genomica Applicata, Udine 33100, Italy

S. Barth et al. (eds.), *Perennial Biomass Crops for a Resource-Constrained World*, DOI 10.1007/978-3-319-44530-4_15

to COP21 (Fuss et al. 2014). Many outstanding issues remain on delivery however, the concept of 'land sparing' (increasing agricultural yields and reducing farmland area to allow the spare land to be used for climate change mitigation and biodiversity conservation) could enable high-yield feedstocks to produce significant, low-impact lignocellulosic resources (Lamb et al. 2016). Sources of lignocellulosic biomass under active consideration over the past decade include energy grasses (David and Ragauskas 2010; Jørgensen 2011), crop residues (Gomez et al. 2008) and fast growing trees cultivated under short rotation coppice or short rotation forestry (Hinchee et al. 2009; Tullus et al. 2012).

Prioritisation of the Sustainable Intensification of Biomass-Tree Cultivation

Bioenergy trees are prime examples of second-generation (2G) bioenergy feedstocks (as defined by Manning et al. 2015) in which lignocellulosic biomass is harvested from dedicated perennial species (Somerville et al. 2010). Such crops must be appropriate to the climate and region in which they are grown and able to grow on marginal lands thus minimising competition with food crops or the destruction of high-nature-value ecosystems. They should require few inputs; both to minimise the economic and energy costs of their cultivation and management and to reduce the environmental impacts associated with the fertilisers and pesticides necessary for intensive farming. A review by Manning et al. (2015) suggested that farmland biodiversity and the provision of regulating and cultural ecosystem services could even be improved by perennial biomass crop cultivation if appropriately managed across the landscape. Benefits (in addition to provisioning services such as low carbon energy) include the provision of habitat corridors between and within intensively farmed land areas; harbouring biodiversity including pollinators and other insect populations; preventing soil erosion and buffering water sources against nutrient run-off and sedimentation (Manning et al. 2015). Land use change modelling reported by Milner et al. (2015) supports this idea and proposes that planting perennial lignocellulosics, in temperate landscapes at least, can enhance the provision of a basket of ecosystem services. A review by Don et al. (2012) reported that soils under dedicated perennial crops emit significantly less N_2O than soils under conventional arable cultivation. They also have the potential to sequester more carbon, though the effect of transition from grassland to perennial biomass cropping can be neutral or even slightly negative (Don et al. 2012; Harris et al. 2015). The net greenhouse gas balance of bioenergy crops depends strongly upon good land management practices (Davis et al. 2013) and maintaining soil-based ecosystems services is an essential aspect of sustainable agricultural intensification (Schulte et al. 2014). Figure 15.1 provides an overview of the 'more-from-less', paradigm for such sustainable intensification; increasing yield in a given area without degrading the land or resorting to energy intensive cultivation practices (Allwright and Taylor

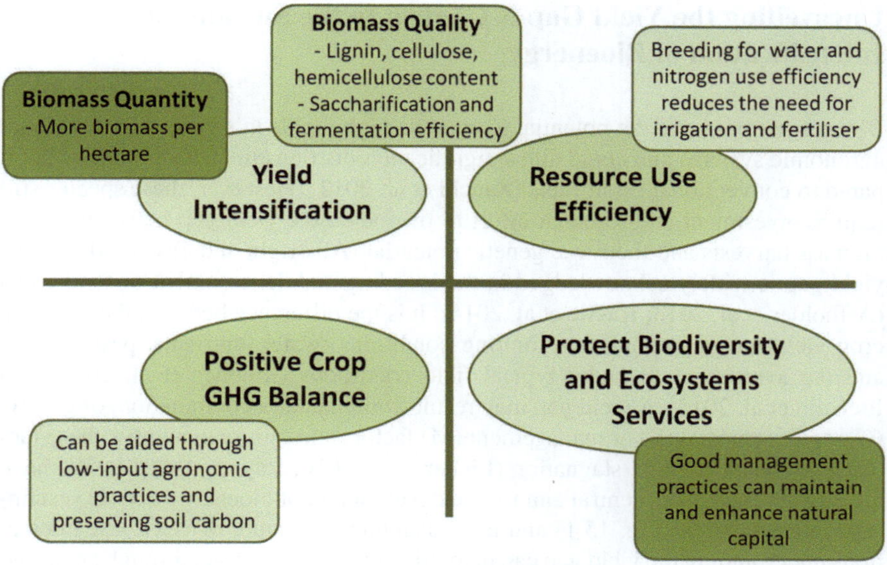

Fig. 15.1 More from less—sustainable intensification. The figure illustrates how breeding targets to fill the yield gap through yield intensification should be a high priority. However, this yield intensification must make full consideration of the wider inputs required to achieve high yield and that more efficient plants with respect to resources such as water and nitrogen are required. In addition, yield intensification must occur alongside assessment of the GHG balance of the crop system such that lower GHG emissions can be targeted; perhaps through management practices related to the preservation of soil carbon. Finally, a basket of ecosystem services that includes biodiversity protection must be delivered from future multi-functional landscapes

2016). This essential principle drives bioenergy tree breeding and development and thus the remainder of this chapter is focussed on research in three important species: poplar (Verlinden et al. 2015), willow (Stolarski et al. 2013) and eucalyptus (Freeman et al. 2013).

The primary goal of bioenergy tree breeding and development is the sustainable intensification of biomass production. In addition to breeding for yield traits, this also means targeting feedstock quality to increase the efficiency of conversion to liquid fuels and decrease emissions across the whole life cycle of the system. Resource use efficiency with respect to water and nutrients is also a priority to ensure a low-input crop without high irrigation or fertigation needs. Good agronomic practices are also an essential aspect of making high yielding bioenergy trees sustainable. The crop GHG balance can be improved through the preservation of soil carbon stocks and biodiversity and ecosystems services may be preserved and enhanced by good land management.

Unravelling the Yield Gap is Central to the Sustainable Intensification of Bioenergy

Bioenergy trees have the potential to achieve high yields in sustainable, low input agronomic systems and could allow significant reductions in GHG emissions compared to conventional fossil fuels (Zanchi et al. 2012). However, these species still require investment and research effort to overcome the yield gap between typical biomass harvests and their true genetic potential (Allwright and Taylor 2016). The yield gap is widely acknowledged in the breeding and development of food crops (Affholder et al. 2013; Kassie et al. 2014). It is the difference between the potential crop yield under optimal, non-limiting conditions (water, nutrients, pest control) and the average yield under typical field conditions (Mueller et al. 2012; Van Ittersum et al. 2013). A yield gap may result from one, or a combination, of genetic (G), environmental (E) or management (M) factors and failure to address these factors may result in yield stagnation (Licker et al. 2010; Ray et al. 2012). We have already outlined that a central aim for the development of bioenergy trees is sustainable intensification (Fig. 15.1) and understanding the nature of G×E×M interactions underpinning the yield gap can help drive this. For example, through enhanced water and nutrient resource management (Mueller et al. 2012; Bredemeier et al. 2015); the optimisation of soil pH (Tilman et al. 2011); protecting and increasing soil organic carbon stocks (Powlson et al. 2011) or improving soil aeration which is important to fine root development and growth (Weltecke and Gaertig 2012). Figure 15.2 (modified from Allwright and Taylor 2016) demonstrates significant yield gaps for all three tree species; although this figure generally draws on data from small research-scale yield plots since those are the only data widely available at present for these crops for such an analysis. With this caveat, poplar and eucalyptus show a greater range of values than willow with larger maximum biomass yields reported. Poplar trials range from 3 to 35 $t^{-1}ha^{-1}y^{-1}$ (mean 16.1), eucalyptus trials from 10.5 to 34 $t^{-1}ha^{-1}y^{-1}$ (mean 22.4) and willow trials from 11.6 to 27.5 $t^{-1}ha^{-1}y^{-1}$ (mean 17.3). It can be seen that the highest yields are generally achieved in trials in which irrigation or fertigation are supplied, yet potential yields are rarely reached. There is however, an exception in the case of the highest yielding eucalyptus trial whose small experimental plot size means, in the words of the authors, 'commercial yields are likely to be considerably lower' (Sims et al. 1999). In practice therefore, the yield gap may be greater than indicated by the shaded region of the chart as commercial yields fall short of those reported in experimental plots and trials (Nonhebel 2002) where only climatic conditions may be limiting. In the remainder of this chapter, we explore the potential for high-throughput phenotyping and transcriptomics, forward genetics (association studies and genomic selection), reverse genetics and genetic modification/genome editing as tools to help close the yield gap and drive the sustainable intensification of the cultivation of biomass trees as part of the molecular breeding pipeline.

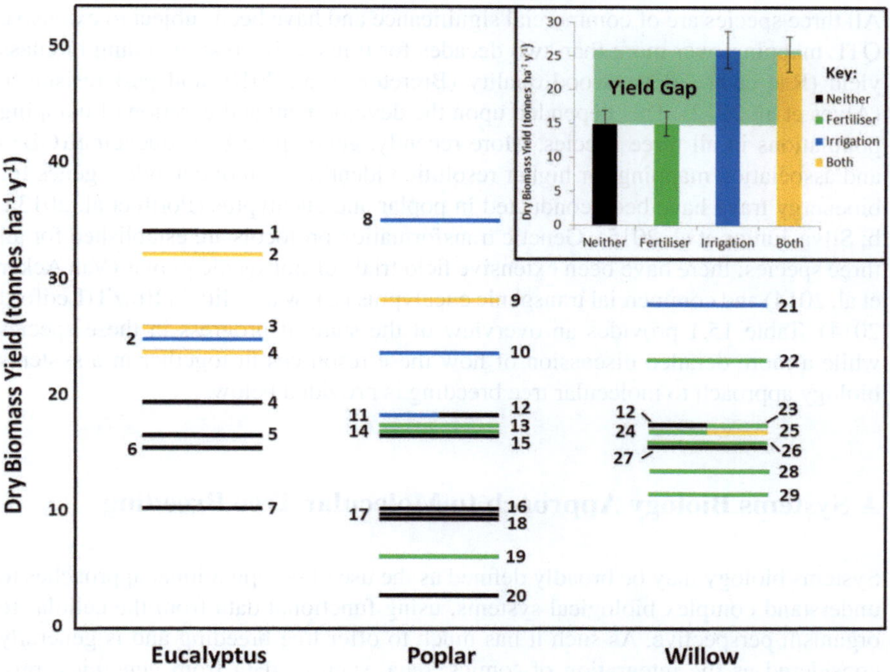

Fig. 15.2 Reported biomass yields reveal a yield gap for biomass trees. Poplar, willow and eucalyptus show wide variation in their biomass yields. Where more than one value is reported in a publication those given here are the maximum reported oven-dry biomass yields (t ha^{-1} y^{-1}) for the best performing sites, genotypes and years or coppice cycles within each study. The inset bar chart displays the mean yield and standard error for all the trials shown, pooled across feedstocks for each management practice. Numerical citations adjacent to each bar correspond to a single published field trial: **1**. (Sims et al. 1999) **2**. (Shankhwar and Srivastava 2015) **3**. (Minhas et al. 2015) **4**. (Guo et al. 2006) **5**. (de Andrade et al. 2013) **6**. (Herrero et al. 2014) **7**. (Müller *et al.* 2005) **8**. (Scaracia-Mugnozza et al. 1997) **9**. (Pontailler et al. 1999) **10**. (Carmona et al. 2015) **11**. (Rae et al. 2007) **12**. (Labrecque and Teodorescu 2005) **13**. (Fortier et al. 2010) **14**. (Nassi O Di Nasso N et al. 2010) **15**. (Verlinden et al. 2015) **16**. (Dillen et al. 2013) **17**. (Truax et al. 2012) **18**. (Nielsen et al. 2014) **19**. (Bungart and Hüttl 2004) **20**. (Bungart 1999) **21**. (Adegbidi et al. 2001) **22**. (Labrecque and Teodorescu 2003) **23**. (Volk et al. 2011) **24**. (Stolarski et al. 2013) **25**. (Kopp et al. 2001) **26**. (Stolarski et al. 2011) **27**. (McElroy and Dawson 1986) **28**. (Serapiglia et al. 2013) **29**. (Adegbidi et al. 2003). (Modified from Allwright and Taylor 2016)

Three Tree Species for Biomass Production

There is now an extensive knowledge and technology foundation for the improvement of poplar, willow and eucalyptus. These include phenotyping facilities, genetic mapping, genetic modification and advanced molecular breeding. The publication of the poplar genome in 2006 (Tuskan et al. 2006) was followed by that of eucalyptus in 2011 (Myburg et al. 2011) while the willow genome is still in progress.

All three species are of commercial significance and have been subject to extensive QTL mapping over more than two decades for traits of interest including biomass yield (Rae et al. 2009), wood quality (Brereton et al. 2010) and pest resistance (Alves et al. 2012). This depended upon the development and curation of mapping populations in all three species. More recently, genotyping-by-sequencing (GBS) and association mapping for higher resolution identification of candidate genes for bioenergy traits have been conducted in poplar and eucalyptus (Porth et al. 2013a, b; Silva-Junior et al. 2015). Genetic transformation protocols are established for all three species; there have been extensive field trials of transgenic poplar (Van Acker et al. 2014) and commercial transgenic eucalyptus is now a reality in Brazil (Ledford 2014). Table 15.1 provides an overview of the state of progress in these species while a more detailed discussion of how these resources fit together in a systems biology approach to molecular tree breeding is provided below.

A Systems Biology Approach to Molecular Tree Breeding

Systems biology may be broadly defined as the use of computational approaches to understand complex biological systems, using functional data from the cellular to organism perspective. As such it has much to offer tree breeding and is generally considered as the integration of 'omics' data, such as data from genomics, proteomics and metabolomics with data from the phenotyping. Figure 15.3 is an illustrated overview of how a systems biology approach might aid the discovery of links between genes and traits. Conducting the quantity of phenotyping required now represents a significant challenge and a bottleneck relative to the ability to obtain molecular data for genotyping. An exception to this is the ability to procure high-throughput RNA-Seq data which is now revolutionising eQTL approaches. Both reverse and forward genetics can be of value to the acceleration of the breeding pipeline. Reverse genetics seeks to elucidate a specific gene's function through mutagenising its DNA sequence and observing the phenotypic outcome. By contrast, forward genetics seeks to map the genetic basis of a specific trait of interest by seeking a statistical relationship between genetic markers and that phenotype. In general, reverse genetics approaches are valuable for understanding the basis of traits controlled by a small number of genetic loci of large effect while forward genetics approaches are better suited to understanding polygenetic traits with multiple small effect loci. One powerful forward genetic technique is genomic selection. Here the complex nature of a polygenetic trait such as yield is explicitly recognised, with 100s or 1000s of SNP molecular markers used together to establish breeding values and obviate significance testing in association and linkage studies (Beaulieu et al. 2014). In contrast, there is also now powerful evidence that CRISPR/Cas genome editing (a cutting edge reverse genetics approach) may be deployed in tree crops such as poplar (Zhou et al. 2015) where mutation breeding could be precisely deployed, again accelerating the breeding pipeline. The following sections give more detail of progress made to date for our target bioenergy trees.

Table 15.1 Biological and technical resources for the development of poplar, willow and eucalyptus

Resource	Poplar	Willow	Eucalyptus
Genome sequence	Fully sequenced genome (Tuskan et al. 2006) and some next generation sequencing (NGS).	In progress.	Fully sequenced genome (Myburg et al. 2011).
Genomic mapping	Genetic linkage maps for several Populus species (Cervera et al. 2001; Yin et al. 2002; Wang et al. 2011).	Extensive linkage mapping (Berlin et al. 2010; Hanley and Karp 2014).	Extensive linkage mapping (Grattapaglia and Sederoff 1994; Brondani et al. 2006; Hudson et al. 2012).
QTL mapping	Extensive QTL mapping including for biomass distribution (Wullschleger et al. 2005), biomass yield (Rae et al. 2007, 2009) and wood quality (Novaes et al. 2009).	QTL mapping for rust resistance (Hanley et al. 2011; Samils et al. 2011); enzymatic saccharification (Brereton et al. 2010) and coppicing response (Salmon et al. 2014).	QTL mapped for growth (Freeman et al. 2009, 2013); flowering time (Missiaggia et al. 2005); wood quality (Rocha et al. 2007) and rust resistance (Alves et al. 2012).
Advanced genotyping	Multiple reports of GBS in poplar utilising restriction enzyme (Schilling et al. 2014), random sheering (Slavov et al. 2012) and transcriptome resequencing (Geraldes et al. 2011) based methods. In the latter instance more than 500,000 single nucleotide polymorphisms (SNPs) were called and assisted in developing a 34 K genotyping array for *P. trichocarpa* (Geraldes et al. 2013).	Candidate gene cloning (Serapiglia et al. 2011) and transcriptome resequencing (Liu et al. 2013) but no GBS reported at this time.	Population genetic analysis and phylogeny reconstruction with 8000 markers across several taxa (Steane et al. 2011). A 60 K SNP chip was recently constructed from whole genome resequencing of 240 genotypes from 12 species (Silva-Junior et al. 2015) and employed for analysis of linkage disequilibrium, recombination and nucleotide diversity (Silva-Junior and Grattapaglia 2015).

(continued)

Table 15.1 (continued)

Resource	Poplar	Willow	Eucalyptus
Genome wide association studies (GWAS)	Candidate gene approaches have been employed in poplar (Wegrzyn et al. 2010). A 34 K genotyping array for *P. trichocarpa* has permitted a number of GWAS to be conducted in a widely utilised association population for this species. Large numbers of associations have been identified for wood quality (Porth et al. 2013a, b); rust resistance (La Mantia et al. 2013) and phenology, ecophysiology and biomass traits (McKown et al. 2014a, b).	Association populations have been established (Karp et al. 2011). In one diverse population, SSR markers were employed to assess genetic diversity and population structure to investigate the potential for future GWAS (Berlin et al. 2014).	Candidate gene approaches for growth and wood quality traits (Thavamanikumar et al. 2011, 2014). 60 K SNP chip available as of this year (Silva-Junior et al. 2015).
Genetic modification (GM)	Extensive glasshouse (Stewart et al. 2009; Wilkerson et al. 2014) and some field trials (Van Acker et al. 2014), predominantly addressing lignin content/structure or other aspects of wood chemistry.	Genetic transformation reported (Xing and Maynard 1995) but no trials of transgenic material published.	Extensive research with transgenic lines (Navarro et al. 2011; Yu et al. 2013) and the world's first commercial GM tree deployment in Brazil (Ledford 2014; FuturaGene 2015).
Deployment	Widely cultivated in temperate regions.	Limited deployment primarily in Canada and northern Europe.	Widely cultivated in southern hemisphere.

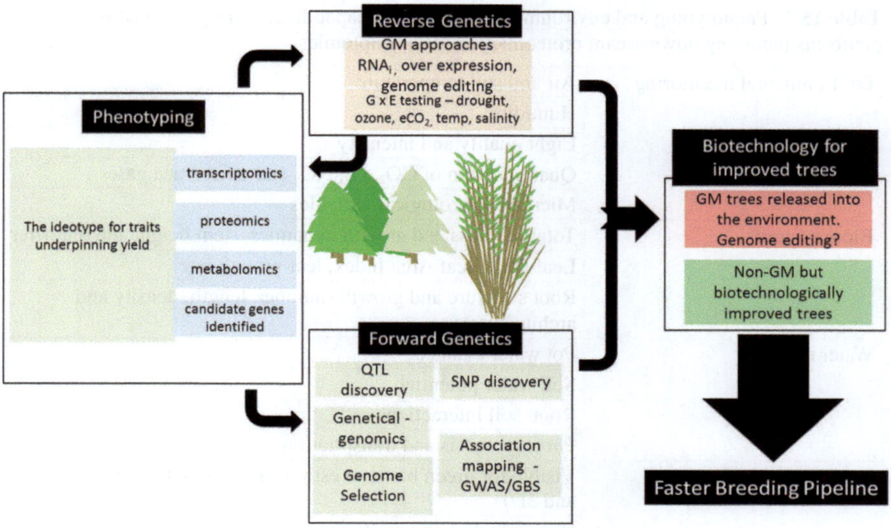

Fig. 15.3 Systems biology for optimised biomass tree breeding. Phenotyping and 'omics' technologies, linked to the development of both forward and reverse genetic approaches, are proposed as a mechanism to deliver the yield improvement required for sustainable intensification. (Modified from Sims et al. 2006)

Phenotyping: The Bottle Neck for Molecular Breeding

In order to make genetic gains to increase the productivity of biomass trees, it is necessary to thoroughly assess the phenotypes of large numbers of existing and emerging genotypes. Given the latest approaches use association rather than mapping populations, the number of individual genotypes and replicates can soon lead to very large and unwieldy experiments with several thousand plants (Porth et al. 2013a, b; McKown et al. 2014a, b). Phenotyping throughput is still limited and now stands as the major bottleneck for breeding programs. To this end, there is increased interest in developing high-throughput phenotyping platforms such as those which make up the International and European Plant Phenotyping Networks (http://www.plant-phenotyping.org/ and http://www.plant-phenotyping-network.eu/). These facilities include both controlled environment and field set-ups and generally increased throughput is reliant on the utilisation of imaging and remote sensing technologies (Table 15.2).

The facility at IPK Gatersleben (Leibniz-Institut für Pflanzengenetik und Kulturpflanzenforschung, Germany) combines a high-throughput controlled environment phenotyping platform (IPK LemnaTec Scanlayzer) with GC-MS for metabolite profiling. This accommodates plants of small–large size with a capacity for up to 4608 plants to be grown in parallel. The system has enabled the detailed evaluation of stress-related metabolic and phenotypic traits in crops such as lentil. In this case, drought and salinity stress were the focus. Four accessions were int-

Table 15.2 Phenotyping and environmental monitoring capabilities of major global phenotyping platforms including downstream proteomics and transcriptomics

Environmental monitoring	Air and soil temperature
	Humidity
	Light quality and intensity
	Quantification of CO_2, O_3, NO_x and other trace gases
	Micro-meteorological variables
Biomass traits	Total biomass and growth dynamics, stem height and diameter
	Leaf area, Leaf Area Index, leaf growth rate
	Root structure and growth (number, length, density and architecture)
Water relations	Pot water balance
	Soil water potential
	Root–soil interactions
	Photosynthesis and transpiration
Remote sensing/imaging	Visible for green biomass estimates and plant architecture (2D and 3D)
	Hyperspectral and multi-spectral
	Near infra-red (NIR) for moisture content
	Far infra-red (FIR) for transpiration rates and apex temperature
	Fluorescence—bulk and chlorophyll
	Raman spectrometry
Biochemistry and 'omics'	Cell/tissue processes
	Nitrogen content
	Metabolite profiling
	Thermogravimetric profiling
	GC/MS analysis of pyrolytic gas composition
	Proteomics profiling
	Transcriptomics

ensively phenotyped in order to link drought and salinity tolerance to observed metabolic differences (Muscolo et al. 2015). Phenotyping platforms which utilise controlled environments allow high-precision plant phenotyping and support the study of the genetic basis of these traits. However, for traits that are subject to high GxE interaction, field phenotyping to measure plants in conditions more similar to the target commercial environment is required. For example, DIAPHEN is a field-based platform, comprising imaging tools carried by drones and phenomobiles, managed by the Institut National de la Recherche Agronomique (INRA, France). These systems are GPS-equipped and have the capacity to frequently crop physiological parameters, for example green area cover and canopy activity, in field plots of medium–large plants with a throughput of up to 100 plots h^{-1}. The platform has been successfully used to identify genetic determinants of the drought response in an apple tree hybrid population consisting of 122 genotypes (Virlet et al. 2015).

Although substantial advances have been made in plant phenotyping in recent years (Grobkinsky et al. 2015), the gap between the relatively low-throughput of high accuracy, controlled environment platforms and the higher throughput, lower precision phenotyping which can be achieved in the field remains a challenge and several ongoing, large-scale projects are currently working on the problem of up-scaling phenotyping approaches to crop scale. High-precision, high-throughput phenotyping is necessary to support crop breeding and management.

Metabolomics, Proteomics and Transcriptomics

In addition to traditional morpho-physiological phenotyping, individuals and populations can be characterised based on their metabolome, proteome and transcriptome. Using these 'omics' approaches allows physiological phenotypes to be linked to the underlying metabolome, proteome and transcriptome. This is a valuable tool to elucidate the molecular and genetic basis of yield and underpin the breeding effort.

In eucalyptus, the drought stress response of two contrasting genotypes was examined at a metabolomic and physiological level (Shvaleva et al. 2006). These *Eucalyptus globulus* genotypes were found to differ in physiological drought avoidance mechanisms. The metabolomic analyses indicating that glutathione reductase plays a central role in response to drought. Similarly, proteomics can be applied alongside traditional phenotyping to unravel the molecular basis of traits of interest. For example, xylem development is important as it impacts donwstream bioethanol production. In *Populus* a proteomic method was used to identify co-expressed proteins in the secondary xylem and generate transgenic trees based on this analysis for field evaluation (Jia et al. 2011). In this way, proteomic and genomic-informed breeding strategies can be developed which utilise these rapidly advancing technologies to support breeding.

It is now possible to carry out whole-transcriptome sequencing in which expressed mRNA sequences are reverse transcribed and the cDNA (complementary DNA) sequenced to provide the entire coding region of the genome. This is a powerful and increasingly achievable tool to characterise the genetic control of traits of interest; such as yield under stress conditions. For example, a comparative transcriptomic approach has been used to identify genes with conserved expression patterns in the woody tissues of *Populus trichocarpa* and *Eucalyptus grandis* (Hefer et al. 2015). This identified conserved multi-gene orthologous gene clusters involved in secondary cell wall biosynthesis as well as species-specific gene regulation which allows xylem specialisation. Similar transcriptomic approaches have been employed in a number of plant species and many of these have been able to construct transcript correlation networks which can be linked to phenotypic traits (Porth et al. 2013a, b; Gehan et al. 2015; Vining et al. 2015). In *P. balsamifera* a high level of network module preservation was again present however organisation within modules and the central hub genes (highly interconnected genes at the centre of a network responsible for modulating a trait of interest) was found to vary between genotypes (Hamanishi et al. 2015). Through this transcriptome analysis, one of the six genotypes was found to have a large and distinct transcriptomic drought response while also exhibiting the smallest metabolomic response. The transcripts in this hub are likely to play a central role in regulating the drought response. This shows the power of these transcriptome-based strategies to determine critical gene hubs and gene connectivity at the genotype, organ and tissue levels. Furthermore, modelling approaches can be taken based on these identified gene hubs whereby phenotypic predictions can be made based on the alteration of genes in the hub network. The combination of transcriptomic network analysis and predictive modelling has further scope to be extended to other bioenergy tree species and to inform breeding programs for industrially-important bioenergy traits.

In the past, 'omics' approaches have been costly and constrained by both technologies, for example for protein identification or high-throughput sequencing, and the bioinformatics pipelines that must deal efficiently with large amounts of data. However, these are now rapidly decreasing in cost and accessibility. While it is not always straightforward; it is now possible to use these approaches as part of a powerful multi-omics strategy which can be linked with traditional phenotypic data to underpin breeding efforts.

Reverse Genetics: Proof of Concept or Direct Release of Biotechnologically Enhanced Trees?

Reverse genetic approaches such as gene knockouts or overexpression are required for proof of concept studies to confirm the function of putative candidate genes from QTL or GWAS analyses (Prado et al. 2014) but may themselves also produce new trees of valuable and distinctive phenotypes for direct commercialisation (Fig. 15.3). An example of this is seen in the successful launch of the world's first commercial, transgenic forest tree in Brazil this year (FuturaGene 2015). FuturaGene's GM eucalyptus has shown 20 % increases in biomass yield in field trials over a 10-year period and is ready for harvest after 5½ years instead of the usual 7 (Ledford 2014). The overexpressed protein (derived from a gene sequence identified from the model plant, *Arabidopsis*) accelerates growth by enhancing cell wall expansion; however, its identity remains a commercial secret. Overexpression of stress-responsive genes in eucalyptus can improve salt and cold tolerance (Navarro et al. 2011; Yu et al. 2013). The overexpression of several stress-related genes by gene stacking resulted in poplar with increased salt and pest tolerance (Polle and Chen 2015). Cold tolerance is a key breeding priority for eucalyptus to extend its growth range, sustain consistent yields and be commercially competitive in emerging bioenergy markets (Yu and Gallagher 2015). Reverse genetics can also assist in systems biology approaches to understand more complex pathways. For example, Vanholme et al. (2012) used Arabidopsis loss-of-function mutants for each of the ten genes in the lignin biosynthesis pathway to understand the responses to perturbations in this pathway. Low-lignin transgenic trees (generally knockouts or RNAi knockdowns) are of great research interest because they have the potential to yield a feedstock that is less recalcitrant to enzymatic saccharification (Studer et al. 2011). Van Acker et al. (2014) demonstrated improved saccharification and ethanol yield from field-grown GM poplar deficient in the lignin biosynthetic enzyme cinnamoyl-CoA reductase. Unfortunately, blunt reductions in lignin content can have negative consequences for yield (Van Acker et al. 2014) and pest resistance (Polle et al. 2013) and other, more novel, transgenic approaches are being investigated. One route is the heterologous expression (i.e. stimulating gene expression in cells that do not normally express the gene) of thermophilic, cell wall degrading enzymes *in planta* (Jung et al. 2012). These enzymes can be activated by mild temperature increases post-harvest and can decrease the energy and financial costs of the conversion of wood to ethanol. Poplar especially is known as an efficient bioreactor for the expression of foreign enzymes (Kim et al. 2012). Another exciting GM approach for reduced recalcitrance without impacting fitness has been reported by Wilkerson et al. (2014). They successfully incorporated a transferase gene into poplar which introduced ester linkages into the lignin backbone. These ester bonds can be readily hydrolysed by a mild, alkaline pre-treatment, aiding processing. In recent years, highly targeted, sequence-specific genome editing has become more feasible in eukaryotes through the development of CRISPR/Cas technology (Gaj et al. 2013). In a key development for bioenergy trees this technology has now been successfully used in poplar to target a lignin biosynthetic enzyme (4-coumarate:CoA ligase) and further innovations using this method are likely to follow (Zhou et al. 2015). Finally, tilling is a powerful and high-throughput reverse genetic approach to elucidate gene function in a mutagenised population using a mismatch endonuclease to detect the induced mutations (SNPs or indels). Ecotilling is closely related but seeks to identify polymorphisms in natural populations and evaluate their effects on genes of interest and phenotypic significance (methodology and development reviewed by Barkley and Wang 2008). Ecotilling has been successfully employed in food crops (Yu et al. 2012) and has potential for accelerating the domestication of forest trees (Harfouche et al. 2012). Marroni et al. (2011) reported the detection of rare alleles in poplar using NGS and believe that this methodology could drive next generation ecotilling in this species; allowing function to be ascribed to these low frequency variants.

From Trees to Genes: Forward Genetics

Forward genetic techniques seek to understand the genetic basis of a phenotype and identify genomic regions, markers and/or candidate genes linked with the trait of interest (Fig. 15.3). Forward genetic approaches are of particular value for elucidating quantitative, polygenic traits. Before the revolution in cost-efficient, high-throughput, next-generation sequencing (Mardis 2011), genetic marker density was generally limited and research was focussed on broad QTL mapping which can be achieved with only a few hundred SSLR or microsatellite markers. QTL have been mapped for yield in both eucalyptus (Freeman et al. 2013) and poplar (Wullschleger et al. 2005; Rae et al. 2007). Rae et al. (2009) identified five robust QTL hotspots for yield in short rotation coppice (SRC) poplar explaining 20 % of final biomass yield in the mapping population. In willow, QTL have been mapped for rust resistance; a major willow pathogen and responsible for commercial losses (Hanley et al. 2011; Samils et al. 2011). Expression QTL (eQTL) mapping is a more recent development, also known as 'genetical genomics' (Joosen et al. 2009). This approach considers gene expression (quantified levels of given mRNA transcripts) as a quantitative trait and maps this expression data as QTL (Ingvarsson and Street 2010). This can permit the identification of causal genes underpinning the phenotype of interest. Genetical genomics has been widely employed for several years in a number of plant and animal species including model organisms such as *Drosophila*, yeast and mice (Joosen et al. 2009). In the model plant *Arabidopsis* the technique has been useful for understanding the genetic basis of complex responses such as genotype-by-environment interactions (Joosen et al. 2013) and genetic regulatory networks (Terpstra et al. 2010). In bioenergy research genetical genomics has been applied in the biodiesel crop jatropha for oil production traits (Liu et al. 2011) and in poplar for leaf shape variation (Drost et al. 2015). Another genetical genomics approach that has been employed in poplar is a form of bulk segregant analysis with microarray expression data. Street et al. (2006) identified extreme genotypes for drought tolerance traits in response to soil drying and used microarrays to identify differentially expressed genes between these groups. They were able to identify promising candidate genes whose differential expression co-located with traditionally mapped QTL for these drought-specific traits. More recently, with the availability of NGS approaches to provide high-throughput DNA marker data, eQTL mapping has become relatively cheap and much more tractable to elucidate the link between phenotypes and their underlying resolution at the genomic level (Majewski and Pastinen 2011). These technologies are now being applied to plant improvement and combined with QTL-Seq approaches (Takagi et al. 2013). Using NGS for RNA-Seq can offer significant new potential to resolve traits in trees in future in a more time and cost-effective manner.

Association mapping is a more powerful forward genetic approach for elucidating the genetic basis of qualitative and quantitative traits in species of interest; seeking statistical associations between SNPs and phenotypes of interest within a population (Ingvarsson and Street 2010). The finesse with which a trait can be mapped is dependent on the rate of decay of linkage disequilibrium (LD), i.e. the non-random association of alleles at different loci. Since linkage is a major contributor to LD, LD declines with physical distance (Flint-Garcia et al. 2003). Outbreeding species (including poplar, eucalyptus and willow), which have a higher effective degree of recombination than inbreeders (Gaut and Long 2003), can achieve higher resolution association mapping but concomitantly require a higher marker frequency (Neale and Kremer 2011). The need for high marker density meant that initial association studies in bioenergy trees tended to take a candidate

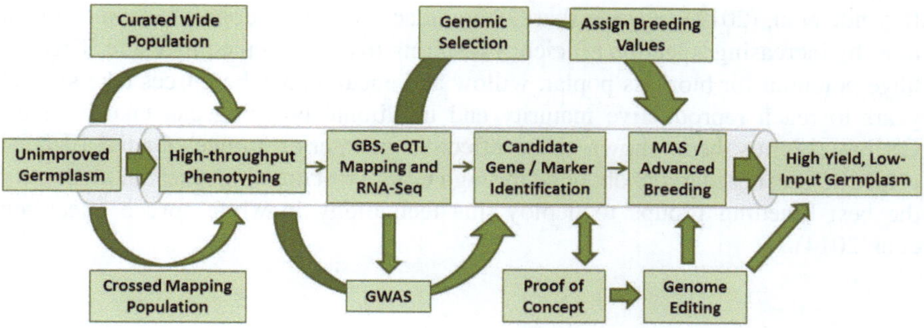

Fig. 15.4 The advanced molecular breeding pipeline

gene approach which are not genome-wide but useful for narrowing down genes of interest within a broader QTL region or identifying candidates within a group of genes of putatively similar function (Teare 2011). In both poplar (Wegrzyn et al. 2010; Guerra et al. 2013) and eucalyptus (Thavamanikumar et al. 2011, 2014) candidate gene approaches have been employed for wood quality traits with robust trait-marker associations identified. In poplar these have been superseded by the development of a 34,000 SNP array for *P. trichocarpa* with SNPs drawn from 3543 candidate genes for a variety of valuable bioenergy traits (Geraldes et al. 2013). This 'chip' has been employed in a number of GWAS in poplar in the past 2 years identifying hundreds of trait-marker associations for key traits including biomass yield (McKown et al. 2014a, b), wood quality (Porth et al. 2013a, b) and rust tolerance (La Mantia et al. 2013). Eucalyptus researchers are now pursuing a similar path with the recent publication of a 60,000 SNP chip that will permit GWAS in this species (Silva-Junior et al. 2015). Associations can then feed into the molecular breeding pipeline (Fig. 15.4) and marker-assisted selection (MAS), as seen in many crop plants (Miedaner and Korzun 2012), for the advanced breeding of superior bioenergy trees.

Beginning with unimproved germplasm curated in a natural, wide or mapping population, advanced molecular breeding may proceed through high-throughput phenotyping for traits of interest. In parallel; GBS, GWAS, genomic selection, transcriptome sequencing and/or eQTL mapping can allow the identification of candidate genes or markers for the phenotyped traits which may also serve as high value targets for GM proof of concept studies and genome editing. Collectively these techniques feed directly into advanced, marker-assisted selection and breeding programmes for novel, high yield, low-input feedstocks.

As NGS reduces genotyping costs and marker numbers and density increase there is the potential to move towards GS in bioenergy trees. GS assigns breeding values to individuals based on genome-wide markers of sufficient density to permit the assumption or knowledge that all relevant genomic regions are in LD with some of the genotyped SNPs (single nucleotide polymorphisms or single base changes in the DNA sequence) (Grattapaglia and Resende 2010). A modelling study from

Resende et al. (2012) suggests that GS could accelerate the domestication of forest trees by increasing selection efficiency resulting in a faster breeding cycle. This has huge potential for biomass poplar, willow and eucalyptus where trees take several years to reach reproductive maturity and traditional breeding can take decades. GS has recently been shown to be effective in interior spruce (Gamal El-Dien et al. 2015) using markers obtained through GBS; with ongoing research to identify the best breeding groups to deploy this technology in white spruce (Beaulieu et al. 2014).

The Breeding Pipeline

Trees are long-lived and largely out-breeding species and it is therefore difficult to make rapid improvements through breeding and selection (Harfouche et al. 2011; Allwright and Taylor 2016). In addition many tree species are dioecious (single sexed), making the selection of specific crosses difficult and genetic research complex. These lifecycle limitations have major impacts on the breeding cycles for woody plants and have been partially overcome in the past by the extensive use of vegetatively propagated or clonal material, as in the three species considered here (Liesebach and Naujoks 2004; Meilan et al. 2002; Stape et al. 2008). Recently, protoplast fusion has been introduced as a novel technique for the production of enhanced poplar germplasm (Hennig et al. 2015); however, the technique is still in its infancy.

This short review has highlighted several approaches that are combining next-generation DNA sequencing technologies with high-throughput phenotyping approaches to overcome this bottleneck in the next decade with accelerated breeding cycles possible. All pipelines begin with the collection and curation of novel germplasm material (Fig. 15.4) and future efforts to fulfil the necessity for sustainable intensification (Fig. 15.1) are likely to involve collection from extreme climate sites. The value of wild germplasm cannot be overestimated and has proved to be of central importance in recent breeding efforts in both rice (Arbelaez et al. 2015) and tomato (Blanca et al. 2015). Recent advances now mean that this material is tractable with large GWAS studies enabling the rapid development of links between traits and genes and the development of molecular markers with which to pursue MAS. The difficulty with this approach for trees is their outbreeding nature, although rare variants have been identified using a modified pooled multiplexing (the simultaneous sequencing of many DNA samples tagged for their identification, thus speeding DNA sequencing whilst reducing cost) approach that identifies rare variants of functional genes underpinning lignin production in poplar (Marroni et al. 2011). More promising are the genomic selection tools where training and validation populations are used to calculate genotypes' breeding values from multiple markers in relation to traits of interest. Such techniques offer significant potential to reduce breeding time since selections can be made in a fraction of the time required to follow the growth and performance of a breeding population using routine harvest and assessment methods. Alongside genomic selection, genome editing has also been shown as a proven technology for poplar (Fan et al. 2015; Zhou et al. 2015) and offers a route for the rapid assessment of individual genes that might emerge from the breeding pipeline and high-throughput phenotyping. In many respects, genomic selection and genome editing offer two contrasting routes to the production of improved, high yielding biomass material for the future bioenergy landscape and both should be considered over the coming decades.

Conclusions

Tree breeding for bioenergy is important as woody lignocellulosics crops can contribute to efforts to fulfil global commitments to reduced emissions and the move towards a low carbon economy. Most future energy scenarios highlight a significant role for energy from biomass, including through co-firing; biomass burning with CCS and biomass for liquid biofuels. However, the supply of high yielding, sustainable feedstock cultivars of biomass tree species is hampered by the biology of trees. This review has highlighted the importance of high-throughput phenotyping and new molecular technologies that can be deployed to significantly accelerate the breeding pipeline, without the necessity to produce a 'GM' tree; helping to address the current yield gap and increase potential yields in these important lignocellulosic crops.

Acknowledgement We acknowledge support from the Seventh Framework for Research of the European Commission, for the project WATBIO (www.watbio.eu), project number, FP7-311929.

References

Adegbidi HG, Volk TA, White EH et al (2001) Biomass and nutrient removal by willow clones in experimental bioenergy plantations in New York State. Biomass and Bioenergy 20:399–411

Adegbidi HG, Briggs RD, Volk TA et al (2003) Effect of organic amendments and slow-release nitrogen fertilizer on willow biomass production and soil chemical characteristics. Biomass and Bioenergy 25:389–398

Affholder F, Poeydebat C, Corbeels M et al (2013) The yield gap of major food crops in family agriculture in the tropics: assessment and analysis through field surveys and modelling. F Crop Res 143:106–118

Allwright MR, Taylor G (2016) Molecular breeding for improved second generation bioenergy crops. Trends Plant Sci 21:43–54

Alves AA, Rosado CCG, Faria DA et al (2012) Genetic mapping provides evidence for the role of additive and non-additive QTLs in the response of inter-specific hybrids of Eucalyptus to Puccinia psidii rust infection. Euphytica 183:27–38

Arbelaez JD, Moreno LT, Singh N et al (2015) Development and GBS-genotyping of introgression lines (ILs) using two wild species of rice, O. meridionalis and O. rufipogon, in a common recurrent parent, O.sativa cv. Curinga. Mol Breed 35:81
Barkley NA, Wang ML (2008) Application of TILLING and EcoTILLING as reverse genetic approaches to elucidate the function of genes in plants and animals. Curr Genomics 9: 212–226
Beaulieu J, Doerksen TK, MacKay J et al (2014) Genomic selection accuracies within and between environments and small breeding groups in white spruce. BMC Genomics 15:1048
Berlin S, Lagercrantz U, von Arnold S et al (2010) High-density linkage mapping and evolution of paralogs and orthologs in Salix and Populus. BMC Genomics 11:129
Berlin S, Trybush SO, Fogelqvist J et al (2014) Genetic diversity, population structure and phenotypic variation in European Salix viminalis L. (Salicaceae). Tree Genet Genomes 10: 1595–1610
Blanca J, Montero-Pau J, Sauvage C et al (2015) Genomic variation in tomato, from wild ancestors to contemporary breeding accessions. BMC Genomics 16:257
Bredemeier M, Busch G, Hartmann L et al (2015) Fast growing plantations for wood production - integration of ecological effects and economic perspectives. Front Bioeng Biotechnol 3:72
Brereton NJB, Pitre FE, Hanley SJ et al (2010) QTL mapping of enzymatic saccharification in short rotation coppice willow and its independence from biomass yield. Bioenergy Res 3:251–261
Brondani RPV, Williams ER, Brondani C, Grattapaglia D (2006) A microsatellite-based consensus linkage map for species of Eucalyptus and a novel set of 230 microsatellite markers for the genus. BMC Plant Biol 6:20
Bungart R (1999) Erzeugung von Biomasse zur energetischen Nutzung durch den Anbau schnellwachsender Baumarten auf Kippsubstraten des Lausitzer Braunkohlereviers unter besonderer Beru¨ cksichtigung der Na¨ hrelementversorgung und des Wasserhaushaltes. Cottbuser Schr Bodenschutz Rekult 7:159
Bungart R, Hüttl RF (2004) Growth dynamics and biomass accumulation of 8-year-old hybrid poplar clones in a short-rotation plantation on a clayey-sandy mining substrate with respect to plant nutrition and water budget. Eur J For Res 123:105–115
Carmona R, Nuñez T, Alonso MF (2015) Biomass yield and quality of an energy dedicated crop of poplar (Populus spp.) clones in the Mediterranean zone of Chile. Biomass and Bioenergy 74:96–102
Cervera M, Ivens B, Gusma J et al (2001) Dense genetic linkage maps of three populus species (populus deltoides, P. Nigra and P. Trichocarpa) based on AFLP and microsatellite markers. Genetics 158:787–809
David K, Ragauskas AJ (2010) Switchgrass as an energy crop for biofuel production: A review of its ligno-cellulosic chemical properties. Energy Environ Sci 3:1182
Davis SC, Boddey RM, Alves BJR et al (2013) Management swing potential for bioenergy crops. GCB Bioenergy 5:623–638
de Andrade TCGR, de Barros NF, Dias LE, Azevedo MIR (2013) Biomass yield and calorific value of six clonal stands of Eucalyptus urophylla ST Blake cultivated in Northeastern Brazil. Cern Lavras 19:467–472
Dillen SY, Djomo SN, Al Afas N et al (2013) Biomass yield and energy balance of a short-rotation poplar coppice with multiple clones on degraded land during 16 years. Biomass and Bioenergy 56:157–165
Don A, Osborne B, Hastings A et al (2012) Land-use change to bioenergy production in Europe: implications for the greenhouse gas balance and soil carbon. GCB Bioenergy 4:372–391
Drost DR, Puranik S, Novaes E et al (2015) Genetical genomics of Populus leaf shape variation. BMC Plant Biol 15:166
Fan D, Liu T, Li C et al (2015) Efficient CRISPR/Cas9-mediated targeted mutagenesis in populus in the first generation. Sci Rep 5:12217
Flint-Garcia SA, Thornsberry JM, Buckler ES (2003) Structure of linkage disequilibrium in plants. Annu Rev Plant Biol 54:357–374

Fortier J, Gagnon D, Truax B, Lambert F (2010) Biomass and volume yield after 6 years in multiclonal hybrid poplar riparian buffer strips. Biomass and Bioenergy 34:1028–1040
Freeman JS, Whittock SP, Potts BM, Vaillancourt RE (2009) QTL influencing growth and wood properties in Eucalyptus globulus. Tree Genet Genomes 5:713–722
Freeman JS, Potts BM, Downes GM et al (2013) Stability of quantitative trait loci for growth and wood properties across multiple pedigrees and environments in Eucalyptus globulus. New Phytol 198:1121–1134
Fuss S, Canadell JG, Peters GP et al (2014) Betting on negative emissions. Nat Clim Chang 4:850–853
FuturaGene (2015) FuturaGene's eucalyptus is approved for commercial use in Brazil
Gaj T, Gersbach CA, Barbas CF (2013) ZFN, TALEN and CRISPR/Cas-based methods for genome engineering. Trends Biotechnol 31:397–405
Gamal El-Dien O, Ratcliffe B, Klápště J et al (2015) Prediction accuracies for growth and wood attributes of interior spruce in space using genotyping-by-sequencing. BMC Genomics 16:1–16
Gaut BS, Long AD (2003) The lowdown on linkage disequilibrium. Plant Cell 15:1502–1506
Gehan MA, Greenham K, Mockler TC, McClung CR (2015) Transcriptional networks-crops, clocks, and abiotic stress. Curr Opin Plant Biol 24:39–46
Geraldes A, Pang J, Thiessen N et al (2011) SNP discovery in black cottonwood (Populus trichocarpa) by population transcriptome resequencing. Mol Ecol Resour 11:81–92
Geraldes A, Difazio SP, Slavov GT et al (2013) A 34K SNP genotyping array for populus trichocarpa: design, application to the study of natural populations and transferability to other populus species. Mol Ecol Resour 13:306–323
Gomez LD, Steele-King CG, McQueen-Mason SJ (2008) Sustainable liquid biofuels from biomass: the writing's on the walls. New Phytol 178:473–485
Grattapaglia D, Resende MDV (2010) Genomic selection in forest tree breeding. Tree Genet Genomes 7:241–255
Grattapaglia D, Sederoff R (1994) Genetic linkage maps of Eucalyptus grandis and Eucalyptus urophylla using a pseudo-testcross: mapping strategy and RAPD markers. Genetics 137: 1121–1137
Grobkinsky DK, Pieruschka R, Svensgaard J et al (2015) Phenotyping in the fields: dissecting the genetics of quantitative traits and digital farming. New Phytol 207:950–952
Guerra FP, Wegrzyn JL, Sykes R et al (2013) Association genetics of chemical wood properties in black poplar (Populus nigra). New Phytol 197:162–176
Guo LB, Sims REH, Horne DJ (2006) Biomass production and nutrient cycling in Eucalyptus short rotation energy forests in New Zealand: II. Litter fall and nutrient return. Biomass and Bioenergy 30:393–404
Hamanishi ET, Barchet GL, Dauwe R et al (2015) Poplar trees reconfigure the transcriptome and metabolome in response to drought in a genotype- and time-of-day-dependent manner. BMC Genomics 16:329
Hanley SJ, Karp A (2014) Genetic strategies for dissecting complex traits in biomass willows (Salix spp.). Tree Physiol 34:1167–1180
Hanley SJ, Pei MH, Powers SJ et al (2011) Genetic mapping of rust resistance loci in biomass willow. Tree Genet Genomes 7:597–608
Harfouche A, Meilan R, Altman A (2011) Tree genetic engineering and applications to sustainable forestry and biomass production. Trends Biotechnol 29:9–17
Harfouche A, Meilan R, Kirst M et al (2012) Accelerating the domestication of forest trees in a changing world. Trends Plant Sci 17:64–72
Harris ZM, Spake R, Taylor G (2015) Land use change to bioenergy: a meta-analysis of soil carbon and GHG emissions. Biomass and Bioenergy 82:27–39
Hefer CA, Mizrachi E, Myburg AA et al (2015) Comparative interrogation of the developing xylem transcriptomes of two wood-forming species: Populus trichocarpa and Eucalyptus grandis. New Phytol 206:1391–1405

Hennig A, Kleinschmit JRG, Schoneberg S et al (2015) Water consumption and biomass production of protoplast fusion lines of poplar hybrids under drought stress. Front Plant Sci 6:330
Herrero C, Juez L, Tejedor C et al (2014) Importance of root system in total biomass for Eucalyptus globulus in northern Spain. Biomass and Bioenergy 67:212–222
Hinchee M, Rottmann W, Mullinax L et al (2009) Short-rotation woody crops for bioenergy and biofuels applications. In Vitro Cell Dev Biol Plant 45:619–629
Hudson CJ, Freeman JS, Kullan AR et al (2012) A reference linkage map for Eucalyptus. BMC Genomics 13:240
Ingvarsson PK, Street NR (2010) Association genetics of complex traits in plants. New Phytol 189:909–922
Jia X, Zhao M, Zhao C et al (2011) Populus biomass protein-protein interactions and their functions. BMC Proc
Joosen RVL, Ligterink W, Hilhorst HWM, Keurentjes JJB (2009) Advances in genetical genomics of plants. Curr Genomics 10:540–549
Joosen RVL, Arends D, Li Y et al (2013) Identifying genotype-by-environment interactions in the metabolism of germinating Arabidopsis seeds using generalized genetical genomics. Plant Physiol 162:553–566
Jørgensen U (2011) Benefits versus risks of growing biofuel crops: the case of Miscanthus. Curr Opin Environ Sustain 3:24–30
Jung SK, Parisutham V, Jeong SH, Lee SK (2012) Heterologous expression of plant cell wall degrading enzymes for effective production of cellulosic biofuels. J Biomed Biotechnol 2012: 405842
Karp A, Hanley SJ, Trybush SO et al (2011) Genetic improvement of willow for bioenergy and biofuels. J Integr Plant Biol 53:151–165
Kassie BT, Van Ittersum MK, Hengsdijk H et al (2014) Climate-induced yield variability and yield gaps of maize (Zea mays L.) in the Central Rift Valley of Ethiopia. F Crop Res 160:41–53
Kim S, Kim Y, Ee YL et al (2012) The transgenic poplar as an efficient bioreactor system for the production of xylanase. Biosci Biotechnol Biochem 76:1140–1145
Kopp RF, Abrahamson LP, White EH et al (2001) Willow biomass production during ten successive annual harvests. Biomass and Bioenergy 20:1–7
La Mantia J, Klápště J, El-Kassaby YA et al (2013) Association analysis identifies melampsora × columbiana poplar leaf rust resistance SNPs. PLoS One 8, e78423
Labrecque M, Teodorescu TI (2003) High biomass yield achieved by Salix clones in SRIC following two 3-year coppice rotations on abandoned farmland in southern Quebec, Canada. Biomass and Bioenergy 25:135–146
Labrecque M, Teodorescu T (2005) Field performance and biomass production of 12 willow and poplar clones in short-rotation coppice in southern Quebec (Canada). Biomass and Bioenergy 29:1–9
Lamb A, Green R, Bateman I et al (2016) The potential for land sparing to offset greenhouse gas emissions from agriculture. Nat Clim Chang 1
Ledford H (2014) Brazil considers transgenic trees. Nature 512:357
Licker R, Johnston M, Foley JA et al (2010) Mind the gap: how do climate and agricultural management explain the "yield gap" of croplands around the world? Glob Ecol Biogeogr 19: 769–782
Liesebach M, Naujoks G (2004) Approaches on vegetative propagation of difficult-to-root Salix caprea. Plant Cell Tissue Organ Cult 79:239–247
Liu P, Wang C, Li L et al (2011) Mapping QTLs for oil traits and eQTLs for oleosin genes in jatropha. BMC Plant Biol 11:132
Liu J, Yin T, Ye N et al (2013) Transcriptome analysis of the differentially expressed genes in the male and female shrub willows (Salix suchowensis). PLoS One 8, e60181
Majewski J, Pastinen T (2011) The study of eQTL variations by RNA-seq: From SNPs to phenotypes. Trends Genet 27:72–79
Manning P, Taylor GE, Hanley M (2015) Bioenergy, food production and biodiversity - an unlikely alliance? GCB Bioenergy 7:570–576
Mardis ER (2011) A decade's perspective on DNA sequencing technology. Nature 470:198–203

Marroni F, Pinosio S, Di Centa E et al (2011) Large-scale detection of rare variants via pooled multiplexed next-generation sequencing: towards next-generation Ecotilling. Plant J 67: 736–745

McElroy GH, Dawson WM (1986) Biomass from short-rotation coppice willow on marginal land. Biomass 10:225–240

McKown AD, Guy RD, Quamme L et al (2014a) Association genetics, geography, and ecophysiology link stomatal patterning in Populus trichocarpa with carbon gain and disease resistance trade-offs. Mol Ecol 23:5771–5790

McKown AD, Klápště J, Guy RD et al (2014b) Genome-wide association implicates numerous genes underlying ecological trait variation in natural populations of Populus trichocarpa. New Phytol 203:535–553

Meilan R, Auerbach DJ, Ma C et al (2002) Stability of herbicide resistance and GUS expression in transgenic hybrid poplars (Populus sp.) during four years of field trials and vegetative propagation. HortScience 37:277–280

Miedaner T, Korzun V (2012) Marker-assisted selection for disease resistance in wheat and barley breeding. Phytopathology 102:560–566

Milner S, Holland RA, Lovett A et al (2015) Potential impacts on ecosystem services of land use transitions to second generation bioenergy crops in GB. GCB Bioenergy 8:317–333

Minhas PS, Yadav RK, Lal K, Chaturvedi RK (2015) Effect of long-term irrigation with wastewater on growth, biomass production and water use by Eucalyptus (Eucalyptus tereticornis Sm.) planted at variable stocking density. Agric Water Manag 152:151–160

Missiaggia AA, Piacezzi AL, Grattapaglia D (2005) Genetic mapping of Eef1, a major effect QTL for early flowering in Eucalyptus grandis. Tree Genet Genomes 1:79–84

Mueller ND, Gerber JS, Johnston M et al (2012) Closing yield gaps through nutrient and water management. Nature 490:254–257

Müller MD, Filho AAT, Vale RS, do Couto L (2005) Biomass yield and energetic content in Agroforestry Systems with Eucalypt in Vazante-MG. Biomassa Energ 2:125–132

Muscolo A, Junker A, Klukas C et al (2015) Phenotypic and metabolic responses to drought and salinity of four contrasting lentil accessions. J Exp Bot 66:5467–5480

Myburg A, Grattapaglia D, Tuskan G et al (2011) The Eucalyptus grandis Genome Project: Genome and transcriptome resources for comparative analysis of woody plant biology. BMC Proc 5:I20

Nassi O Di Nasso N, Guidi W, Ragaglini G et al (2010) Biomass production and energy balance of a 12-year-old short-rotation coppice poplar stand under different cutting cycles. GCB Bioenergy 2:89–97

Navarro M, Ayax C, Martinez Y et al (2011) Two EguCBF1 genes overexpressed in Eucalyptus display a different impact on stress tolerance and plant development. Plant Biotechnol J 9:50–63

Neale DB, Kremer A (2011) Forest tree genomics: growing resources and applications. Nat Rev Genet 12:111–122

Nielsen UB, Madsen P, Hansen JK et al (2014) Production potential of 36 poplar clones grown at medium length rotation in Denmark. Biomass and Bioenergy 64:99–109

Nonhebel S (2002) Energy yields in intensive and extensive biomass production systems. Biomass and Bioenergy 22:159–167

Novaes E, Osorio L, Drost DR et al (2009) Quantitative genetic analysis of biomass and wood chemistry of Populus under different nitrogen levels. New Phytol 182:878–890

Polle A, Chen S (2015) On the salty side of life: molecular, physiological and anatomical adaptation and acclimation of trees to extreme habitats. Plant Cell Environ 38:1794–1816

Polle A, Janz D, Teichmann T, Lipka V (2013) Poplar genetic engineering: promoting desirable wood characteristics and pest resistance. Appl Microbiol Biotechnol 97:5669–5679

Pontailler JY, Ceulemans R, Guittet J (1999) Biomass yield of poplar after five 2-year coppice rotations. Forestry 72:157–163

Porth I, Klapšte J, Skyba O et al (2013a) Genome-wide association mapping for wood characteristics in Populus identifies an array of candidate single nucleotide polymorphisms. New Phytol 200:710–726

Porth I, Klápště J, Skyba O et al (2013b) Network analysis reveals the relationship among wood properties, gene expression levels and genotypes of natural Populus trichocarpa accessions. New Phytol 200:727–742

Powlson DS, Gregory PJ, Whalley WR et al (2011) Soil management in relation to sustainable agriculture and ecosystem services. Food Policy 36:S72–S87

Prado JR, Segers G, Voelker T et al (2014) Biotech crop development: from idea to product. Annu Rev Plant Biol 65:769–790

Rae AM, Pinel MPC, Bastien C et al (2007) QTL for yield in bioenergy Populus: identifying G×E interactions from growth at three contrasting sites. Tree Genet Genomes 4:97–112

Rae AM, Street NR, Robinson KM et al (2009) Five QTL hotspots for yield in short rotation coppice bioenergy poplar: the Poplar Biomass Loci. BMC Plant Biol 9:23

Ray DK, Ramankutty N, Mueller ND et al (2012) Recent patterns of crop yield growth and stagnation. Nat Commun 3:1293

Resende MFR, Muñoz P, Acosta JJ et al (2012) Accelerating the domestication of trees using genomic selection: accuracy of prediction models across ages and environments. New Phytol 193:617–624

Rocha R, Barros E, Cruz C et al (2007) Mapping of QTLS related with wood quality and developmental characteristics in hybrids (Eucalyptus Grandis X Eucalyptus mapping of QTLS related with wood quality and developmental Characteristics In Hybrids (Eucalyptus Grandis X Madeira E Crescimento EM. Rev Arvore 31:13–24

Salmon J, Ward SP, Hanley SJ et al (2014) Functional screening of willow alleles in Arabidopsis combined with QTL mapping in willow (*Salix*) identifies *SxMAX4* as a coppicing response gene. Plant Biotechnol J 12:480–491

Samils B, Rönnberg-Wästljung AC, Stenlid J (2011) QTL mapping of resistance to leaf rust in Salix. Tree Genet Genomes 7:1219–1235

Scaracia-Mugnozza GE, Ceulemans R, Heilman PE et al (1997) Species and their hybrids grown under short rotation. II. Biomass components and harvest index of hybrid and parental species clones. Can J For Res 27:285–294

Schilling MP, Wolf PG, Duffy AM et al (2014) Genotyping-by-sequencing for populus population genomics: an assessment of genome sampling patterns and filtering approaches. PLoS One 9, e95292

Schulte RPO, Creamer RE, Donnellan T et al (2014) Functional land management: a framework for managing soil-based ecosystem services for the sustainable intensification of agriculture. Environ Sci Policy 38:45–58

Serapiglia MJ, Cameron KD, Stipanovic AJ, Smart LB (2011) Correlations of expression of cell wall biosynthesis genes with variation in biomass composition in shrub willow (Salix spp.) biomass crops. Tree Genet Genomes 8:775–788

Serapiglia MJ, Cameron KD, Stipanovic AJ et al (2013) Yield and woody biomass traits of novel shrub willow hybrids at two contrasting sites. Bioenergy Res 6:533–546

Shankhwar AK, Srivastava RK (2015) Biomass production through grey water fertigation in eucalyptus hybrid and its economic significance. Environ Prog Sustain Energy 34:222–226

Shvaleva AL, Costa E, Silva F, Breia E et al (2006) Metabolic responses to water deficit in two Eucalyptus globulus clones with contrasting drought sensitivity. Tree Physiol 26:239–248

Silva-Junior OB, Grattapaglia D (2015) Genome-wide patterns of recombination, linkage disequilibrium and nucleotide diversity from pooled resequencing and single nucleotide polymorphism genotyping unlock the evolutionary history of Eucalyptus grandis. New Phytol 208: 830–845

Silva-Junior OB, Faria DA, Grattapaglia D (2015) A flexible multi-species genome-wide 60K SNP chip developed from pooled resequencing of 240 Eucalyptus tree genomes across 12 species. New Phytol 206:1527–1540

Sims REH, Senelwa K, Maiava T, Bullock BT (1999) Eucalyptus species for biomass energy in New Zealand - part II: Coppice performance. Biomass and Bioenergy 17:333–343

Sims REH, Hastings A, Schlamadinger B et al (2006) Energy crops: current status and future prospects. Glob Chang Biol 12:2054–2076
Slavov GT, Difazio SP, Martin J et al (2012) Genome resequencing reveals multiscale geographic structure and extensive linkage disequilibrium in the forest tree Populus trichocarpa. New Phytol 196:713–725
Somerville C, Youngs H, Taylor C et al (2010) Feedstocks for lignocellulosic biofuels. Science 329:790–792
Stape JL, Binkley D, Ryan MG (2008) Production and carbon allocation in a clonal Eucalyptus plantation with water and nutrient manipulations. For Ecol Manage 255:920–930
Steane DA, Nicolle D, Sansaloni CP et al (2011) Population genetic analysis and phylogeny reconstruction in Eucalyptus (Myrtaceae) using high-throughput, genome-wide genotyping. Mol Phylogenet Evol 59:206–224
Stewart JJ, Akiyama T, Chapple C et al (2009) The effects on lignin structure of overexpression of ferulate 5-hydroxylase in hybrid poplar. Plant Physiol 150:621–635
Stolarski MJ, Szczukowski S, Tworkowski J, Klasa A (2011) Willow biomass production under conditions of low-input agriculture on marginal soils. For Ecol Manage 262:1558–1566
Stolarski MJ, Szczukowski S, Tworkowski J, Klasa A (2013) Yield, energy parameters and chemical composition of short-rotation willow biomass. Ind Crops Prod 46:60–65
Street NR, Skogström O, Sjödin A et al (2006) The genetics and genomics of the drought response in Populus. Plant J 48:321–341
Studer MH, Demartini JD, Davis MF et al (2011) Lignin content in natural Populus variants affects sugar release. Proc Natl Acad Sci U S A 108:6300–6305
Takagi H, Abe A, Yoshida K et al (2013) QTL-seq: Rapid mapping of quantitative trait loci in rice by whole genome resequencing of DNA from two bulked populations. Plant J 74:174–183
Teare MD (2011) Candidate gene association studies. In: Teare MD (ed) Genetic epidemeology. Humana Press, Totowa, NJ, pp 105–117
Terpstra IR, Snoek LB, Keurentjes JJB et al (2010) Regulatory network identification by genetical genomics: signaling downstream of the Arabidopsis receptor-like kinase ERECTA. Plant Physiol 154:1067–1078
Thavamanikumar S, Tibbits J, McManus L et al (2011) Candidate gene-based association mapping of growth and wood quality traits in Eucalyptus globulus Labill. BMC Proc 5:O15
Thavamanikumar S, McManus LJ, Ades PK et al (2014) Association mapping for wood quality and growth traits in Eucalyptus globulus ssp. globulus Labill identifies nine stable marker-trait associations for seven traits. Tree Genet Genomes 10:1661–1678
Tilman D, Balzer C, Hill J, Befort BL (2011) Global food demand and the sustainable intensification of agriculture. Proc Natl Acad Sci U S A 108:20260–20264
Truax B, Gagnon D, Fortier J, Lambert F (2012) Yield in 8 year-old hybrid poplar plantations on abandoned farmland along climatic and soil fertility gradients. For Ecol Manage 267:228–239
Tullus A, Rytter L, Tullus T et al (2012) Short-rotation forestry with hybrid aspen (Populus tremula L. × P. tremuloides Michx.) in Northern Europe. Scand J Forensic Res 27:10–29
Tuskan GA, Difazio S, Jansson S et al (2006) The genome of black cottonwood, Populus trichocarpa (Torr. & Gray). Science 313:1596–1604
United Nations (2015) Adoption of the Paris Agreement
Van Acker R, Leplé JC, Aerts D et al (2014) Improved saccharification and ethanol yield from field-grown transgenic poplar deficient in cinnamoyl-CoA reductase. Proc Natl Acad Sci U S A 111:845–850
Van Ittersum MK, Cassman KG, Grassini P et al (2013) Yield gap analysis with local to global relevance: a review. Field Crop Res 143:4–17
Vanholme R, Storme V, Vanholme B et al (2012) A systems biology view of responses to lignin biosynthesis perturbations in Arabidopsis. Plant Cell 24:3506–3529
Verlinden MS, Broeckx LS, Ceulemans R (2015) First vs. Second rotation of a poplar short rotation coppice: above-ground biomass productivity and shoot dynamics. Biomass and Bioenergy 73:174–185

Vining K, Romanel E, Jones R et al (2015) The floral transcriptome of Eucalyptus grandis. New Phytol 206:1406–1422
Virlet N, Costes E, Martinez S et al (2015) Multispectral airborne imagery in the field reveals genetic determinisms of morphological and transpiration traits of an apple tree hybrid population in response to water deficit. J Exp Bot 66:5453–5465
Volk TA, Abrahamson LP, Cameron KD et al (2011) Yields of willow biomass crops across a range of sites in North America. Asp Appl Biol 112:67–74
Wang Y, Zhang B, Sun X et al (2011) Comparative genome mapping among Populus adenopoda, P. alba, P. deltoides, P. euramericana and P. trichocarpa. Genes Genet Syst 86:257–268
Wegrzyn JL, Eckert AJ, Choi M et al (2010) Association genetics of traits controlling lignin and cellulose biosynthesis in black cottonwood (Populus trichocarpa, Salicaceae) secondary xylem. New Phytol 188:515–532
Weltecke K, Gaertig T (2012) Influence of soil aeration on rooting and growth of the Beuys-trees in Kassel, Germany. Urban For Urban Green 11:329–338
Wilkerson CG, Mansfield SD, Lu F et al (2014) Monolignol ferulate transferase introduces chemically labile linkages into the lignin backbone. Science 344(80):90–93
Wullschleger SD, Yin TM, Difazio SP et al (2005) Phenotypic variation in growth and biomass distribution for two advanced-generation pedigrees of hybrid poplar. Can J For Res 35: 1779–1789
Xing Z, Maynard C (1995) Producing transgenic shining willow (Salix lucida Muhl.) shoots from stem segments via Agrobacterium tumefaciens transformation. Vitr Cell Dev Biol Plant 31:221–226
Yin T, Zhang X, Huang M et al (2002) Molecular linkage maps of the Populus genome. Genome 45:541–555
Yu A, Gallagher T (2015) Analysis on the growth rhythm and cold tolerance of five-year old eucalyptus benthamii plantation for bioenergy. Open J For 5:585–592
Yu S, Liao F, Wang F et al (2012) Identification of rice transcription factors associated with drought tolerance using the ecotilling method. PLoS One 7:1–9
Yu X, Kikuchi A, Matsunaga E et al (2013) Environmental biosafety assessment on transgenic Eucalyptus globulus harboring the choline oxidase (codA) gene in semi-confined condition. Plant Biotechnol 30:73–76
Zanchi G, Pena N, Bird N (2012) Is woody bioenergy carbon neutral? A comparative assessment of emissions from consumption of woody bioenergy and fossil fuel. GCB Bioenergy 4: 761–772
Zhou X, Jacobs T, Xue L et al (2015) Exploiting SNPs for biallelic CRISPR mutations in the outcrossing woody perennial Populus reveals 4-coumarate : CoA ligase specificity and redundancy. New Phytol 208:298–301

Chapter 16
Screening of Giant Reed Clones for Phytoremediation of Lead Contaminated Soils

S. Sidella, B. Barbosa, J. Costa, S. L. Cosentino, and A.L. Fernando

Introduction

Soil pollution by heavy metals is a global problem causing vast areas of agricultural land to become nonarable and hazardous for both wildlife and human populations (Alloway 1995). In particular, the amount of soils contaminated with anthropogenic lead (Pb) is high, given its widespread deposition over the course of the nineteenth and twentieth centuries from a range of point and non-point sources, including industrial waste and pollution, leaded paint, and automobile exhaust (McClintock 2015). Thanks to the gradual phase-out of leaded fuels and paints, anthropogenic deposition of Pb began to decline in the late 1970s. However, historic deposition continues to play an active role and soil levels in certain spots are significantly higher than geogenic "background" soil Pb levels (McClintock 2015), causing quality and ecosystems services loss (Dauber et al. 2012). Phytoremediation is an emerging technology that involves the use of plants and their associated microbes to remediate soils, water, air and groundwater contaminated with toxic metals, organics, and radionuclides (Bañuelos et al. 2000; Barbosa et al. 2016). Perennial energy crops, such as giant reed, have been recognized as tolerant to contaminated matrices

S. Sidella • S.L. Cosentino
Dipartimento di Agricoltura, Alimentazione e Ambiente—Di3A, University of Catania, via Valdisavoia 5, 95123 Catania, Italy

B. Barbosa
MEtRiCS, Departamento de Ciências e Tecnologia da Biomassa, Faculdade de Ciências e Tecnologia, FCT, Universidade Nova de Lisboa, 2829-516 Caparica, Portugal

Universidade Federal do Oeste da Bahia, Barreiras, Brazil

J. Costa • A.L. Fernando (✉)
MEtRiCS, Departamento de Ciências e Tecnologia da Biomassa, Faculdade de Ciências e Tecnologia, FCT, Universidade Nova de Lisboa, 2829-516 Caparica, Portugal
e-mail: ala@fct.unl.pt

S. Barth et al. (eds.), *Perennial Biomass Crops for a Resource-Constrained World*, DOI 10.1007/978-3-319-44530-4_16

and the high yielded biomass can be used for the production of energy, paper pulp, and biomaterials (Barbosa et al. 2015a; Fernando 2005; Nsanganwimana et al. 2014; Sidella 2014). Moreover, perennial energy grasses show high water and nitrogen efficiencies, low susceptibility to pests and diseases, and a deep, dense, and extensive belowground apparatus that can prevent soil erosion and groundwater and surface water contamination (Barbosa et al. 2015b; Fernando 2005; Fernando et al. 2010). In this context, this research works aimed to study the phytoremediation response of different *Arundo donax* L. clones to soils contaminated with lead.

Methodology

The work was carried out in the Department of Biomass Science and Technology of the Faculty of Science and Technology at the Universidade NOVA de Lisboa—Caparica, Portugal—with the aim to study the potential of some clones of giant reed in the phytoremediation of soils contaminated with lead. The trial was carried out in pots containing 12 kg of soil (loam soil previously analyzed, collected from the *Campus* area) artificially contaminated with a lead rich sludge (waste product derived from a battery manufacturing company, "Sociedade Portuguesa do Acumulador Tudor," located in Castanheira do Ribatejo, near Lisbon), containing 14 % Pb (dry weight basis). In April 2013, two rhizomes were placed in each pot (10 cm deep) and fertilization was applied: 3 g N m^{-2} (urea, 46 % N), 3 g N m^{-2} (nitrolusal, mixture of $NH_4NO_3 + CaCO_3$, 27 % N), 17 g K_2O m^{-2} (potassium sulfate, 51 % K_2O), 23 g P_2O_5 m^{-2} (superphosphate, 18 % P_2O_5). The two studied factors, taken into account in this work were: (1) giant reed genotypes and (2) soil lead content. Four different giant reed clones (19, 22, 27, and 30), belonging to the University of Catania collection (Cosentino et al. 2006) (Table 16.1 and Fig. 16.1) were tested in this experiment.

These clones were selected from a previous screening concerning the salinity resistance. Three different Pb soil lead content were tested: control pots without contamination (Pb_0); Pb_I, with 450 mg Pb kg^{-1} dry matter (DM), and Pb_{II}, with 900 mg Pb kg^{-1} DM, corresponding, respectively, to the maximum quantity of lead in soil and to twice this limit value, permitted by the Portuguese Law and reported in Annex I of the Decree Law No. 276/09 (2009). The pots were placed outside and full irrigation was applied with tap water to overcome water stress (*ca.* 950 mm along the growing cycle, in each pot). Each combination of genotype/level of

Table 16.1 Geographical coordinates of the giant reed clones origin

Genotype	Place	Geographical Coordinates		Altitude
		Latitude N	Longitude E	m a.s.l.
19	Tortorici	38°01′	14°49′	450
22	Cefalù (PA)	38°01′	14°00′	16
27	Agrigento (AG)	37°19′	13°35′	230
30	Licata (AG)	37°06′	13°56′	8

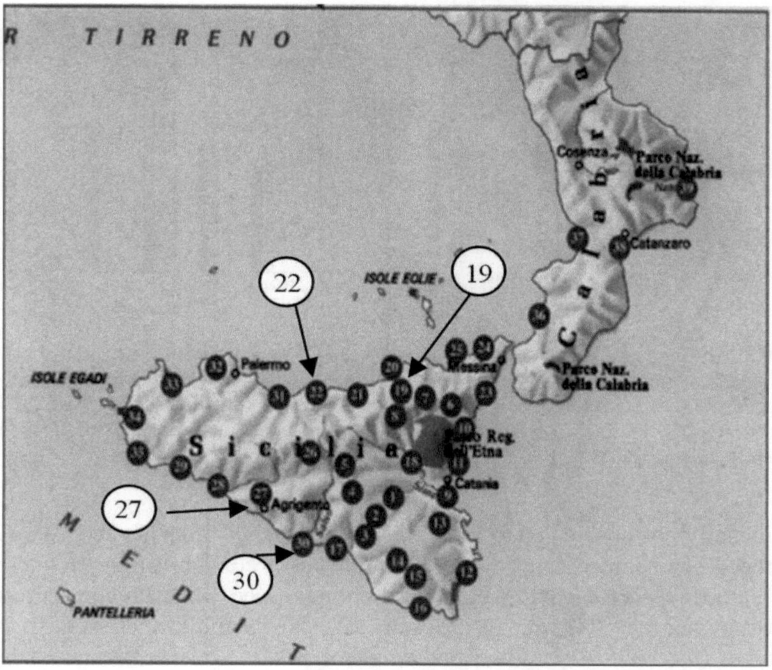

Fig. 16.1 Map indicating giant reed clones origin in the island of Sicily in Italy

contamination was independently evaluated in triplicate. In September 2013, after a period of growth of 5 months, the plants were harvested and biomass yield (above and belowground) and biomass quality (i.e., Pb content) were determined. Biomass was separated into the different organs (stems, leaves, roots, rhizomes), dried at 70 °C and dry weight determined. Plant tissues were mineralized by dry combustion (550 °C) and the Pb content was determined by atomic absorption spectrometry after dissolving the ash residue with nitric acid (Vandecasteele and Block 1993).

The statistical interpretation of the results was performed using analysis of variance (two-way ANOVA) by means of CoStat software (version 6.0) and the means were separated according to the test of Student-Newman-Keuls (SNK) when ANOVA revealed significant differences ($p \leq 0.05$).

Results and Discussion

Biomass Yield

Yield performance of *A. donax* biomass in lead contaminated soils is showed in Fig. 16.2. Results indicated that the increased Pb content in the soil did not influence the production of aboveground biomass of *Arundo donax* L. ($p > 0.05$) and increased

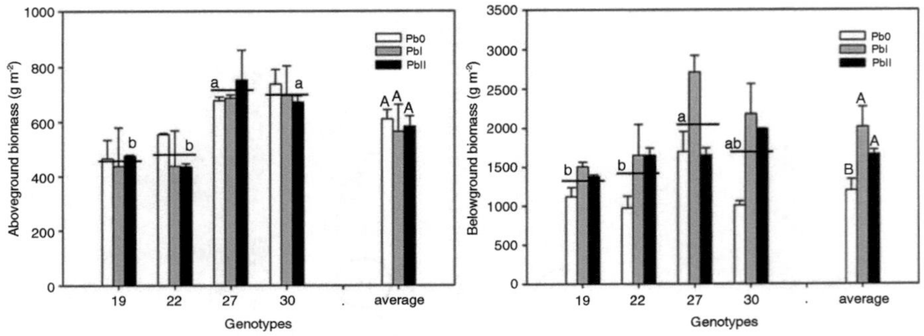

Fig. 16.2 Aboveground and belowground biomass (g DM m^{-2}) of the different studied genotypes under lead contaminated soils. The *horizontal line* represents the average value associated with each genotype. Different *lower-case letters* indicate statistical significance ($p<0.05$, SNK test) among genotypes. Different *capital letters* indicate statistical significance ($p<0.05$, SNK test) between Pb treatments

(significantly) the production of rhizomes and roots ($p<0.05$). Guo and Miao (2010) also reported tolerance to Pb for the same level of contamination. However, the same authors reported phytotoxic effects on giant reed when Pb concentration in soil was higher than 1000 mg kg^{-1}. Results also show significant differences among genotypes ($p<0.05$). Among the studied genotypes, the results showed that genotype 27 and genotype 30 were the most productive. Higher production was attained with the belowground organs by comparison with the aerial fraction. This is consistent with the common behavior of perennial crops that spend more energy on the development and establishment of the belowground organs on the first years of cultivation (Fernando 2005).

Biomass Lead Content

Results indicated that for all the fractions of the plants (stems, leaves, roots and rhizomes), differences were statistically significant among treatments ($p<0.05$), and higher Pb content was obtained with increasing Pb level in the soil (Figs. 16.3 and 16.4). The same was also reported by other authors for giant reed biomass harvested from Pb or other heavy metals contaminated soils (Nsanganwimana et al. 2014). As Pb concentration in the biomass reflects the level of soil Pb concentration, *A. donax* was considered an indicator plant (Leung et al. 2007). Highest content of Pb was verified in the belowground organs (rhizomes but especially in roots), while stems and leaves showed much lower lead content. These results are consistent with other studies performed with giant reed in matrices contaminated with heavy metals (Nsanganwimana et al. 2014). Moreover, Kabata-Pendias (2011) indicated that the translocation of Pb from belowground organs to the aerial fraction is limited because Pb pyrophosphate bind to cell walls.

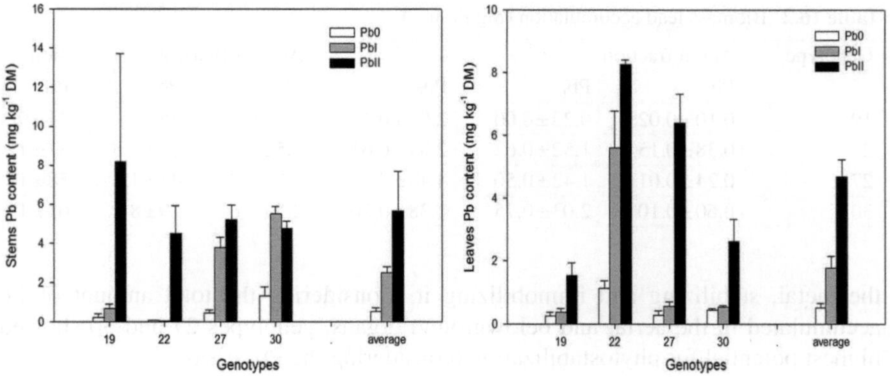

Fig. 16.3 Lead content of stems and leaves (mg Pb kg^{-1} DM) of the different studied clones

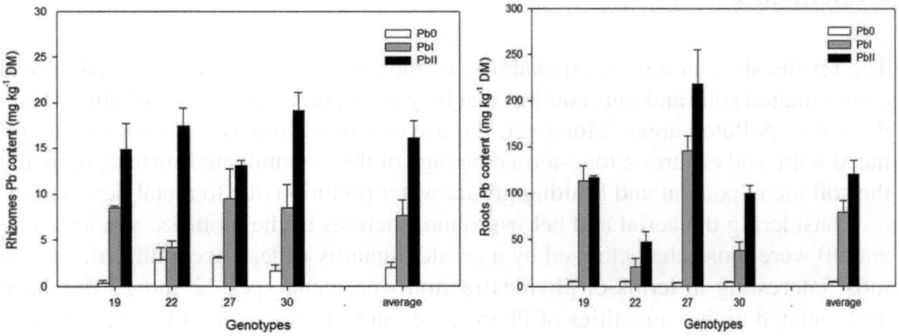

Fig. 16.4 Lead content of rhizomes and roots (mg Pb kg^{-1} DM) of the different studied clones

Biomass Lead Accumulation

Table 16.2 presents the biomass lead accumulation in the biomass. Pb accumulation in the biomass was higher in the contaminated pots compared to the control pots, meaning that this grass is able to take up and accumulate Pb in higher amounts than habitual, thus showing phytoextraction potential. According to Mirza et al. (2010), the accumulation of the metals in the aerial fraction of the plant can be considered an indicator of the capacity of the plant to serve as a phytoextractor, once the metal will be harvested. According to the results presented in Table 16.2, genotypes 22 and 27 showed higher potential to phytoextract Pb than genotypes 19 and 30.

This is because the lead accumulation in the aerial fraction was higher in those genotypes (and in the Pb_{II} essay). However, the Pb transfer to the aerial fraction is limited and highest Pb remained accumulated in the belowground organs of the plant. Therefore, the phytoremediation potential of this plant, to Pb contaminated soils, is linked with the capacity of the plants, to absorb on the belowground fraction

Table 16.2 Biomass lead accumulation (mg Pb m^{-2})

Genotype	Aerial fraction			Aerial + Belowground fraction		
	Pb_0	Pb_I	Pb_{II}	Pb_0	Pb_I	Pb_{II}
19	0.10±0.02	0.22±0.00	2.08±0.63	1.3±0.4	35±11	34±9
22	0.38±0.15	1.52±0.62	2.89±0.61	3.5±1.0	11±3	37±12
27	0.24±0.01	1.42±0.50	4.48±0.72	5.6±2.2	41±12	58±15
30	0.60±0.10	2.03±0.75	2.38±0.20	2.5±0.6	29±8	61±17

the metal, stabilizing and immobilizing it. Considering the total amount of Pb accumulated in the aerial and belowground organs, genotypes 27 and 30 showed highest potential for phytostabilization (considering the Pb_{II} essay).

Conclusion

The results showed a good adaptability of the studied clones of giant reed to lead contaminated soils and thus confirm that they have potential in terms of phytoremediation of polluted areas. Moreover, the use of *Arundo donax* L. in soils contaminated with lead ensures a long-term coverage of the contaminated surface, reducing the soil metal content and limiting groundwater pollution due to metal percolation.

Considering the aerial and belowground fractions of the biomass, genotypes 27 and 30 were those characterized by a greater quantity of lead accumulated. But the most interesting in terms of phytoextraction were genotypes 22 and 27 that have accumulated higher quantities of Pb in leaves and stems which will be harvested.

References

Alloway BJ (1995) Heavy metals in soils, 2nd edn. Blackie Academic and Professional, Glasgow, UK

Bañuelos G, Zambrzuski S, Mackey B (2000) Phytoextraction of Se from soils irrigated with selenium-laden effluent. Plant and Soil 224:251–258

Barbosa B, Boléo S, Sidella S, Costa J, Duarte MP, Mendes B, Cosentino SL, Fernando AL (2015a) Phytoremediation of Heavy Metal- Contaminated Soils Using the Perennial Energy Crops Miscanthus spp. and Arundo donax L. Bioenerg Res 8: 1500–1511

Barbosa B, Costa J, Fernando AL, Papazoglou EG (2015b) Wastewater reuse for fiber crops cultivation as a strategy to mitigate desertification. Ind Crop Prod 68:17–23

Barbosa B, Costa J, Boléo S, Duarte MP, Fernando AL (2016) Phytoremediation of inorganic compounds. In: Ribeiro AB, Mateus EP, Couto N (eds) Electrokinetics across disciplines and continents—new strategies for sustainable development. Springer International Publishing, Switzerland, pp 373–400

Cosentino SL, Copani V, D'Agosta GM, Sanzone E, Mantineo M (2006) First results on evaluation of Arundo donax L. clones collected in Southern Italy. Ind Crop Prod 23:212–222

Dauber J, Brown C, Fernando A, Finnan J, Krasuska E, Ponitka J, Styles D, Thrän D, Groenigen K, Weih M, Zah R (2012) Bioenergy from "surplus" land: environmental and socio-economic implications. BioRisk 7:5–50

Decreto-Lei (Decree-Law) n° 276/09 (2009) Anexo I, Valores limite de concentração relativos a metais pesados, compostos orgânicos e dioxinas e microrganismos, Diário da República, 192, I Série, 7154–7165 (in Portuguese)

Fernando ALAC (2005) Fitorremediação por *Miscanthus* x *giganteus* de solos contaminados com metais pesados, PhD Thesis, Universidade Nova de Lisboa (Lisbon, Portugal)

Fernando AL, Duarte MP, Almeida J, Boléo S, Mendes B (2010) Environmental impact assessment of energy crops cultivation in Europe. Biofuels, Bioprod Bior 4:594–604

Guo ZH, Miao XF (2010) Growth changes and tissues anatomical characteristics of giant reed (Arundo donax L.) in soil contaminated with arsenic, cadmium and lead. J Cent South Univ Technol 17:770–777

Kabata-Pendias A (2011) Trace elements in soils and plants, 4th edn. CRC, Boca Raton

Leung HM, Ye ZH, Wong MH (2007) Survival strategies of plants associated with arbuscular mycorrhizal fungi on toxic mine tailings. Chemosphere 66:905–915

McClintock N (2015) A critical physical geography of urban soil contamination. Geologija 65:69–85

Mirza N, Mahmood Q, Pervez A, Ahmad R, Farooq R, Shah MM, Azim MR (2010) Phytoremediation potencial of *Arundo donax* in arsenic-contaminated synthetic wastewater. Bioresour Technol 101:5815–5819

Nsanganwimana F, Marchland L, Douay F, Mench M (2014) *Arundo donax* L., a candidate for phytomanaging water and soils contaminated by trace elements and producing plant-based feedstock, a review. Int J Phytorem 16:982–1017

Sidella S (2014) Adaptability, biomass yield and phytoremediation of *Arundo donax* L. on marginal lands: salt, dry and lead-contaminated soils. PhD Thesis, University of Catania (Catania, Italy)

Vandecasteele C, Block CB (1993) Modern methods for trace element determination. John Wiley & Sons, Chichester, UK

Chapter 17
Moisture Loss Rate in Grass Cut at Anthesis: Variation Among Selected Traditional Species

S. Jeżowski, S. Ornatowski, J. Finnan, Z. Kaczmarek, and J. Cerazy

Introduction

There is growing interest in the use of marginal land to grow feedstocks for bioenergy production (Dauber et al. 2012). In Europe, including Poland, such marginal areas comprise large sections of mountainous, foothill and lowland regions (e.g. adjacent to marshes along large rivers). In such areas perennial traditional grasses may successfully be grown and used as renewable energy feedstocks (Ceotto 2008; Prochnow et al. 2009a,b). Grass harvested from such areas can be used as a feedstock for anaerobic digestion (Prochnow et al. 2009a) or as a feedstock for combustion (Van Loo and Koppejan 2008). Grass is used as a feedstock for combustion both to generate heat in small and medium scale applications (Cherney 2010) or to generate electricity in larger scale power plants (Pahkala et al. 2008). The favoured method for managing grass grown as a solid biofuel involves extensive grassland management practices with one late biomass cut and low levels of fertilisation (Prochnow et al. 2009b). Solid biofuels derived from such low input systems can provide more usable energy, greater greenhouse gas reductions and less pollution per hectare compared to first generation liquid biofuel production (Tilman et al. 2006). For safe storage of grass, it is necessary to remove a substantial proportion of the water from the crop when mown (Jones 1979). Also, the efficiency of the combustion system increases as the moisture content of the fuel decreases (Van Loo and Koppejan 2008). Thus, high rates of drying would be advantageous in grass grown for biofuel. Previously, it has been shown that the rate of moisture loss differs

S. Jeżowski (✉) • S. Ornatowski • Z. Kaczmarek • J. Cerazy
Institute of Plant Genetics, Polish Academy of Sciences, ul. Strzeszyńska 34, 60-479 Poznań, Poland
e-mail: ajez@igr.poznan.pl

J. Finnan
Teagasc, Crops, Environment and Land Use Programme, Oak Park Crops Research Centre, Carlow, Ireland

S. Barth et al. (eds.), *Perennial Biomass Crops for a Resource-Constrained World*, DOI 10.1007/978-3-319-44530-4_17

between grass species (Jones and Prickett 1981). In order to identify grass species with high rates of drying, studies were initiated at the Institute of Plant Genetics PAS in Poznań. The objective of the study was to identify genetic variation in moisture content loss in cut biomass in selected species of traditional grasses, taking into consideration the effect of the environment, in which moisture loss takes place during the drying process.

Plant Material and Methods

Plant material comprised four species of traditional grasses, each represented by two cultivars. These cultivars included Emeraude and Jordane (*Festuca)*, Ambassador and Donata (*Dactylis)*, Felina and Hykor (*Festulolium)* and Chieftain and Bamse (*Phalaris)*.

The field trial was established as a randomised complete block design with three replications and a plot size of 20 m^2. Biomass (1 m^2) was harvested from the centre of each plot at anthesis . Half of the cut biomass was left in the field (the natural environment) and half was stored in a barn (a stable environment). The moisture content of the cut biomass was measured at weekly intervals over a four week period after harvest by collecting samples from the two environments (natural—the field, and stable—the barn). Moisture content in biomass samples collected from each environment and at respective dates was determined by the gravimetric method. The experiment was carried out in the years 2013, 2014 and 2015 on the experimental field of the Institute of Plant Genetics PAS in Poznań (lat. 16 41 E, long. 52 25 N). Prior to the establishment of the experiment in the spring of 2013 the field to be used in the experiment was fertilised with a mineral Azofoska fertiliser (13 % nitrogen, 19 % phosphorus, 16 % potassium, 0.18 % copper, 0.045 % zinc, 0.27 % manganese and 0.09 % boron) at 250 kg/ha before deep tillage was performed. During the growing season, i.e. from April to September, rainfall was 356 mm (2013), 261 mm (2014) and 248 mm (2015). After the completion of the field trial and compilation of recorded data a four-factor analysis of variance was performed. These factors included the years of the study, tested cultivars, dates of biomass sample collection for analyses of moisture content and the environment, in which the moisture loss process took place (i.e. the field and the barn). To compare means for these factors and means for the combinations of factors we applied the LSD test (Comparison between the means of treatments: Least Significance Difference (LSD) test used at the 5 % level of significance (Gomez and Gomez 1984).

Results and Discussion

Analysis of variance showed statistically significant variations in all investigated factors, i.e. the tested cultivars, years of the study, dates of cut biomass sample collection for the analyses of moisture content, and the environment, in which cut biomass was drying (Figs. 17.1, 17.2, 17.3, 17.4 and 17.5 and Table 17.1).

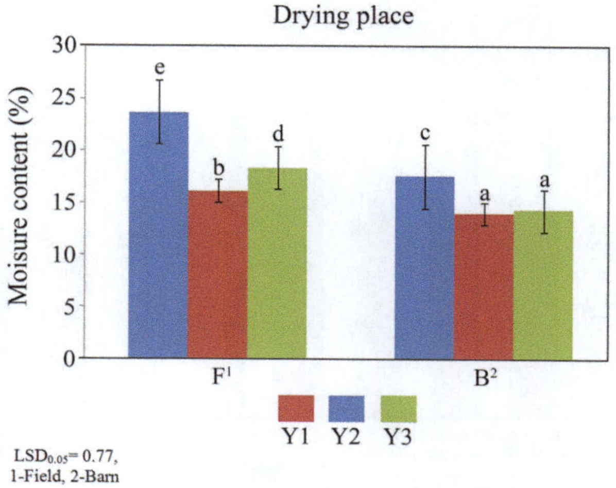

Fig. 17.1 Variation in moisture content (%) of tested cultivars depending on the year of the study and the environment of biomass drying. Data for each year represents an average of all four sampling times

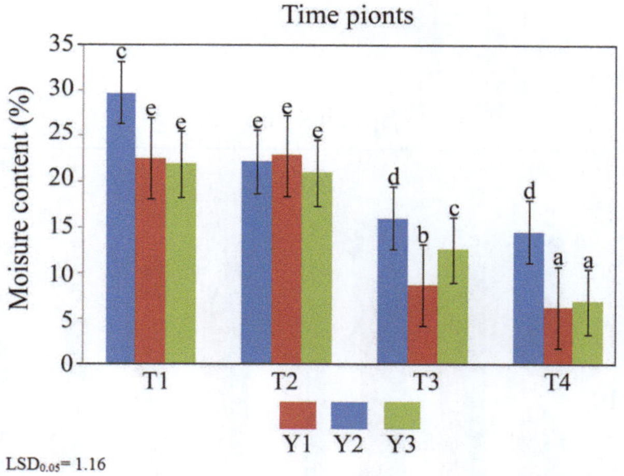

Fig. 17.2 Variation in moisture content (%) of tested cultivars depending on the dates of biomass sampling

Figure 17.1 presents results which show variation in moisture content in all of the tested cultivars jointly averaged over the four drying times in each year based on the year of the study and the environment in which drying took place. The greatest biomass moisture content in the tested cultivars in both drying environments was recorded in the first year, while it was lowest in the second year. For biomass drying in the field, there were significant differences between each of the 3 years of the study. There were significant differences in the moisture content of biomass dried in the barn between the 1st and the 2nd, and between the 1st and the 3rd year.

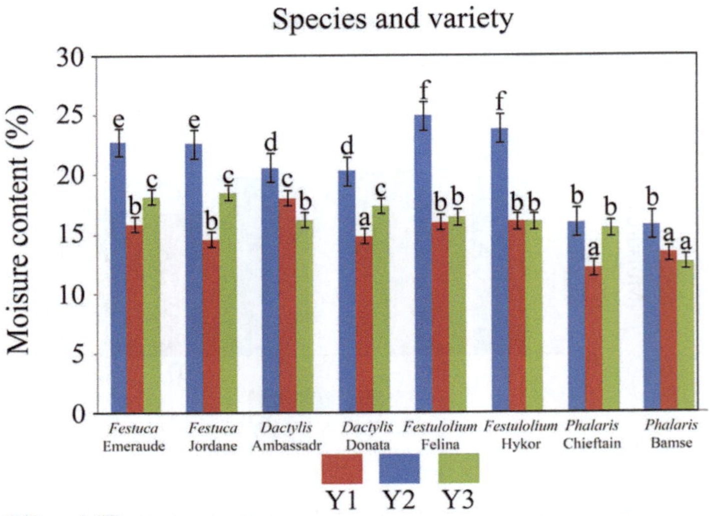

LSD$_{0.05}$= 1.56

Fig. 17.3 Mean values of moisture contents (%) in biomass of tested grass cultivars in successive years of the study

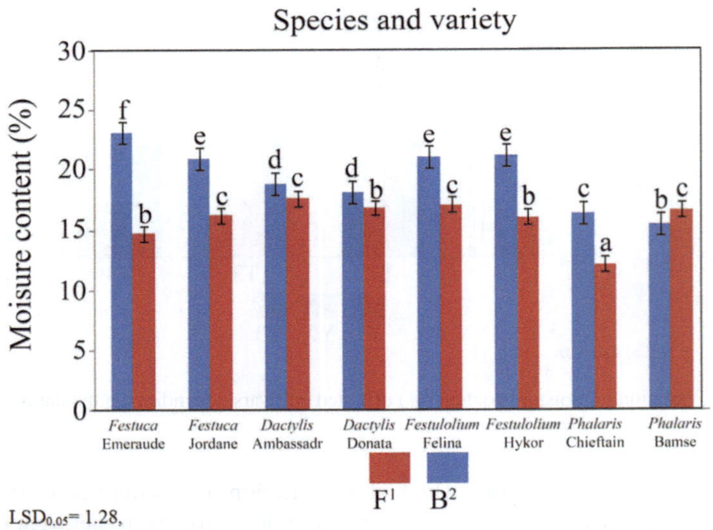

LSD$_{0.05}$= 1.28,
1-Field, 2-Barn

Fig. 17.4 Mean values of moisture contents (%) in biomass of tested grass cultivars in the 3 years of the study

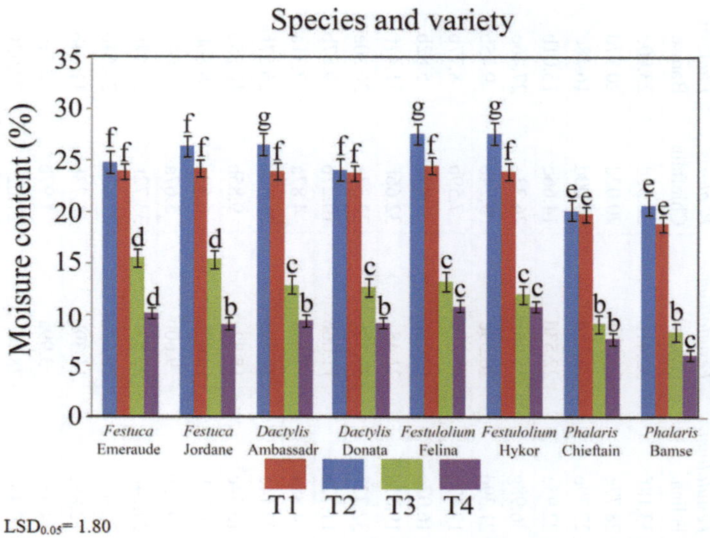

Fig. 17.5 Mean values of moisture contents (%) in biomass of tested grass cultivars in the 3 years of the study in relation to moisture content measurement dates

Differences in moisture loss between the time points of biomass sampling was tested in each individual year for all cultivars and for both drying environments (Fig. 17.2). There were significant differences between the biomass moisture content at each of the sampling points in year 1 although there was no significant difference between the first two sampling points in years 2 and 3. In year 1, the greatest drop in moisture content occurred in the week between the first and second timing whereas the greatest drop in moisture content occurred between the second and third timings in years 2 and 3.

Figure 17.3 presents mean values of moisture content in biomass of the cultivars of the tested grass species in successive years averaged over both drying environments. There were significant differences in moisture content between cultivars in all years of the study. The greatest moisture content was found in the first year of the study in *Festulolium* cv. Felina and Hykor as well as *Festuca* cv. Emeraude and Jordane. The lowest moisture content was recorded in *Phalaris* cv. Chieftain and Bamse in all the years of the study. Mean moisture contents presented in Fig. 17.4 for biomass of the tested grass cultivars over the 3-year period of the study in relation to the environment of biomass drying showed that the greatest biomass moisture content was recorded for both cultivars of the species *Festuca* and *Festulolium*, while it was lowest for the species *Phalaris*. There were significant differences between the moisture content of different cultivars in both drying environments. The environment most advantageous for moisture loss was found to be the shed, which protected the cut biomass against excess moisture from rainfall.

Table 17.1 Variation in moisture contents (%) in biomass of tested grass cultivars taking onto consideration the years of analyses, the environment of biomass drying and the dates of moisture content analyses

Years	Drying place	Time points	Species and variety							
			Festuca Emeraude	*Festuca* Jordane	*Dactylis* Ambassador	*Dactylis* Donata	*Festulolium* Felina	*Festulolium* Hykor	*Phalaris* Chieftain	*Phalaris* Bamse
Y1	F	T1	31.50f	31.07f	25.32e	26.39e	33.18f	32.41f	22.52d	23.90d
		T2	29.99f	29.81f	24.68e	26.54e	28.77e	29.86e	20.02d	20.97d
		T3	24.54e	24.01e	23.37d	22.76d	27.79e	25.15d	18.00d	16.48c
		T4	19.76d	17.47c	15.05c	16.45c	22.95e	22.53d	14.66c	13.01b
	B	T1	36.22g	33.93f	30.49f	31.96f	36.98g	35.06g	26.33e	27.46e
		T2	22.34d	21.83d	20.76d	19.03d	21.39d	20.59e	9.24b	9.45b
		T3	17.38c	12.73b	11.11b	9.97b	17.47c	12.87b	7.59b	8.77b
		T4	10.58b	12.07b	13.25b	9.36b	16.02c	12.23b	4.87a	5.82b
Y2	F	T1	29.24f	29.41e	24.22e	25.07e	19.61d	21.48	32.66f	31.47f
		T2	26.97e	24.32d	27.28e	21.33d	32.72f	30.33f	22.37e	26.80e
		T3	12.31b	11.81b	12.18b	10.71b	11.72b	11.06b	10.27b	9.87b
		T4	6.37a	3.17a	5.57a	4.28a	4.88a	7.63a	4.87a	2.31a
	B	T1	26.91e	26.44e	27.81e	26.34e	27.14e	27.39f	21.60d	26.57f
		T2	14.75c	19,74d	23.32d	15.32c	15.54c	15.82c	9.85b	10.25b
		T3	5.60a	4,84a	11.95b	7.65b	9.71b	10.03b	6.53a	8.80b
		T4	7.75a	6,92a	11.88b	8.20b	9.33b	9.60b	5.67a	5.67a
Y3	F	T1	25.50d	28.07e	23.07d	29.16f	27.24e	32.53f	33.72f	29.47f
		T2	22.30d	26.12e	23.02d	27.70e	22.62d	25.62e	22.62d	23.64d
		T3	11.54b	13.21c	13.21c	12.37c	12.88b	11.39b	12.73b	11.04b
		T4	12.03b	4.05a	4.05a	5.99a	4.43a	3.99a	7.97b	2.51a
	B	T1	28.17f	28.59	28.59	28.27	27.46	28.14	27.62	28.93f
		T2	20.70d	19.87d	19.87d	22.26d	16.23c	14.66c	15.27c	16.59c
		T3	10.89b	9.82b	8.82b	11.77b	8.39b	6.75a	4.90a	4.28a
		T4	4.42a	6.78b	6.88a	10.37b	7.88a	9.14b	9.04b	5.42a

$LSD_{0.05}=4.42$

Figure 17.5 presents mean moisture contents in biomass of cultivars of the tested species across both drying environments for the 3 years of the study. Results show that losses in biomass moisture contents in most cultivars between successive measurements were significant. Only losses in moisture between the first and the second dates for *Festuca* cv. Emeraude and Jordane as well as *Phalaris* cv. Chieftain and Bamse were non-significant. In turn, between the other dates these differences, as well as those for most other cultivars of the investigated grass species, were significant We clearly see that *Phalaris* cv. Chieftain and Bamse lost moisture the fastest, while the moisture loss process was slowest in *Festulolium* cultivars.

A complete picture of the observed genetic variability in the tested grass species in terms of the rate and magnitude of changes in moisture content during the drying process of cut biomass in the tested grass species is provided by Table 17.1, which presents a summary of the entire study. This table presents differences in moisture content of the biomass of selected grass cultivars depending on the years of the study, locations (the environment) of biomass drying and the dates of moisture content determination. Results presented in the table clearly show significant differences, particularly between the years of the study, between the tested cultivars in terms of moisture contents in their cut biomass and between moisture losses over time. The same dates of moisture content measurements in cut biomass of the analysed species as well as biomass drying locations in each year of the study had a significantly different effect on the level and rate of moisture loss in individual species. In this respect *Phalaris* proved to be the most stable, as it lost moisture from biomass at the fastest rate of all species and the differences in the loss of moisture between individual years of the study were least evident in comparison to the other tested species. In turn, these considerable fluctuations in moisture content in the cut biomass in the other tested species during the drying process were caused mainly by the volume of precipitation during the vegetation period and primarily at anthesis, when these measurements were conducted. This was most evident in 2014 and 2015, when in the vegetation period in those years much lower precipitation was recorded (261 and 248 mm) than in the first year of the study (2013), when precipitation total was 356 mm.

Similar to Jones and Prickett (1981), we found considerable variation in moisture loss between grass species. Field drying was highly dependent on atmospheric drying conditions but, in nearly all instances, sufficient drying took place to ensure that the biomass was safe to conserve and suitable for combustion 2 weeks after cutting.

The moisture loss process from biomass of energy crops species, e.g. *Salix viminalis*, *Populus*, *Miscanthus* sp., *Panicum virgatum* has been described in detail by El-Bassam (1998). Our study illustrates the moisture loss mechanism from traditional grasses, which can also be used as feedstocks for solid fuel combustion.

Conclusions

1. Considerable genetic variation between cultivars was observed in terms of the rate of moisture loss in dried biomass.
2. Differences were observed between the drying rates of different cultivars depending on the environment of biomass drying and the year of the study.

References

Ceotto E (2008) Grasslands for bioenergy production. A review. Agron Sustain Dev 28:47–55
Cherney JH (2010) Grass combustion update. http://www.cce.cornell.edu/EnergyClimateChange/Documents/Cherney%20Grass%20for%20Combustion%202010%20update%20_3_%20_3_.pdf. Accessed 8 Jan 2015
Dauber J, Brown C, Fernando A, Finnan J, Krasuska E, Ponitka J, Sytles D, Thran D, Van Groeningen KJ, Weih M, Zah R (2012) Bioenergy from "surplus" land: environmental and socio-economic implications. Biorisk 7:5–50
El-Bassam N (1998) Energy plant species. James & James (Science Publishers) Ltd, 1–320
Gomez KA, Gomez AA (1984) Statistical procedures for agricultural research. Wiley, New York
Jones L (1979) The effect of stage of growth on the rate of drying of cut grass at 20 °C. Grass Forage Sci 34:139–144
Jones L, Prickett J (1981) The rate of water loss from cut grass of different species dired at 200C. Grass Forage Sci 36:17–23
Pahkala K, Aalto M, Isolahti M, Poikola J, Jauhiainen L (2008) Large-scale energy grass farming for power plants—a case study from Ostrobothnia, Finland. Biomass Bioenerg 32:1009–1015
Prochnow A, Heiermann M, Plochl M, Linke B, Idler C, Amon T, Hobbs PJ (2009a) Bioenergy from permanent grassland—a review: 1. Biogas Bioresour Technol 100:4931–4944
Prochnow A, Heiermann M, Plochl M, Amon T, Hobbs PJ (2009b) Bioenergy from permanent grassland—a review: 2. Combust Bioresour Technol 100:4945–5954
Tilman D, Hill J, Lehman C (2006) Carbon-negative biofuels from low-input, high-diversity grassland mixtures. Science 314:1598–1600
Van Loo S, Koppejan J (2008) Biomass combustion and co-firing. Earthscan, London

Part IV
Ecophysiology of Perennial Biomass Crops

Chapter 18
A Systems Approach Guiding Future Biomass Crop Development on Marginal Land

X.-G. Zhu, T.-G. Chang, Q.-F. Song, J. Finnan, S. Barth, L.-M. Mårtensson, and M.B. Jones

Background

The use of fossil fuels as our current principal source of energy is leading to an unacceptable impact on the Earth's climate system (Marland and Obersteiner 2008). Additionally, rapidly increasing energy demand has become a serious challenge both in developed and developing countries. Exploitation of renewable and sustainable energy sources may represent an effective solution to this problem (Zhuang et al. 2011) as renewable sources of primary energy, such as biomass, can help stabilise atmospheric concentrations of greenhouse gases below dangerous levels (Fischer and Schrattenholzer 2001). Although biomass has the potential to become one of the primary global energy sources (Berndes et al. 2003), estimates for the potential contribution of biomass resources to global energy production have produced a wide range of results. Berndes et al. (2003) reviewed 17 studies on this subject and reported estimates from below 100 EJ $year^{-1}$ to above 400 EJ $year^{-1}$. More recently, Offermann et al. (2011) reported estimates of between 0 EJ $year^{-1}$ to more than 1550 EJ $year^{-1}$, the latter figure corresponding to approximately three times global primary energy consumption. The major difference in estimates has

X.-G. Zhu (✉) • T.-G. Chang • Q.-F. Song
CAS-MPG Partner Institute for Computational Biology, Shanghai Institutes for Biological Sciences, Chinese Academy of Sciences, Shanghai 200031, China
e-mail: zhuxinguang@picb.ac.cn

J. Finnan • S. Barth
Teagasc, Crops, Environment and Land Use Programme, Oak Park Crops Research Centre, Carlow, Ireland

L.-M. Mårtensson
Department of Biosystems and Technology, Swedish University of Agricultural Sciences, Alnarp, Sweden

M.B. Jones
School of Natural Sciences, Trinity College Dublin, The University of Dublin, Dublin, Ireland

S. Barth et al. (eds.), *Perennial Biomass Crops for a Resource-Constrained World*, DOI 10.1007/978-3-319-44530-4_18

been ascribed to different assumptions on land availability and yield levels (Berndes et al. 2003). Consequently, the potential of bioenergy for climate change mitigation remains unclear due to large uncertainties about future agricultural yield improvements and the land available for biomass production (Popp et al. 2011).

Although biomass could contribute significantly to the mitigation of global CO_2 emissions and to lessening reliance on petroleum resources, the use of large areas for biomass production raises important concerns about the implications of such a land use change on ecology, biodiversity and society (Marland and Obersteiner 2008). Land and resource conflicts are foreseen between bioenergy and food supply, water use and biodiversity conservation (Slade et al. 2014) which has led to a fuel vs. food debate (Cassman and Liska 2007). It is anticipated that competition for scarce land and water resources will further intensify during the twenty-first century due to rising global population and wealth on the demand side and due to changing climate conditions on the supply side (Lotze-Campen et al. 2010). Such competition will be increased if there is large-scale land use change for bioenergy production. It has been estimated that large-scale introduction of bioenergy would raise food prices (Bryngelsson and Lindgren 2013). Growing and conflicting demands are likely to require conversion of additional land to agricultural use and/or improvements in the productivity of existing land (Dauber et al. 2012).

Competition between food production and bioenergy could be potentially avoided by using abandoned agricultural land for bioenergy production (Campbell et al. 2008). Field et al. (2008) suggested that abandoned agricultural land offers the greatest potential for producing biomass energy which will reduce net global warming but will avoid competition with food production. The use of marginal lands, abandoned croplands and abandoned pasture lands for bioenergy production offers significant environmental and economic advantages as long as perennial crops and sustainable land management practises are employed (Dale et al. 2010). It has been estimated that there are 430–580 Mha of degraded land available for bioenergy production (Hoogwijk et al. 2003). More recently, the global area of abandoned agricultural land has been estimated to be between 385 and 580 million hectares, which can deliver biomass for less than 10 % of primary energy demand in developed countries, but more than the total energy demand in some African countries, where grasslands are relatively productive but current energy demand is low (Campbell et al. 2008). Ambiguity in the definition and characterisation of land which might be available for bioenergy production as well as uncertainty in assessments of land availability has caused confusion about the prospects and the environmental and socio-economic consequences of bioenergy development on land no longer used for food production (Dauber et al. 2012). However, recent estimates are that there are at least 100 million ha of marginal land available for bioenergy production (Fritz et al. 2013).

Large-scale exploitation of biomass resources entails a risk of sacrificing natural areas to managed monocultures (Field et al. 2013). Bryngelsson and Lindgren (2013) suggest that the use of low productivity land for bioenergy production represents a suboptimal land use allocation and is thus costly from a societal point of view. Furthermore, the cultivation of energy crops on marginal land may contribute to losses in soil carbon, indirect land use change and changes in biodiversity although such changes might be minimised by utilising the biomass already growing on marginal land (Dauber et al. 2012; Tilman et al. 2006). However, the necessity of achieving minimum quality standards may well entail the utilisation of specific species. The newly developed biotechnological methods combined with high-throughput plant phenotyping offer opportunities to rapidly select new genotypes that could achieve economic yields on large areas of marginal land. The use of such genotypes combined with the use of sustainable agricultural practises should allow marginal lands to be used for biomass production whilst avoiding any deleterious effects on the environment.

In the following, we first discuss the major physiological features required by biomass crops growing on marginal land, then we review the more widely desired traits associated with biomass crops. This is followed by a discussion of physiological features currently largely un-explored in biomass (or even food crop) breeding, and finally we discuss the potential of using crop systems model to guide future efforts of breeding biomass crops tailored for particular environments.

Desirable Physiological Traits for Biomass Crops

Tolerant plants need to possess combinations of traits that enable them to survive and thrive on marginal land. The objective in exploiting marginal land for bioenergy crops is to identify traits that plants use to avoid or tolerate stress and by using appropriate breeding technology to incorporate these traits in bioenergy crops grown on the marginal land. In broad terms, traits of particular importance on marginal land are tolerance of drought, flooding, frosts, low non-freezing temperatures, saline and contaminated soils.

Perennial rhizomatous grasses (PRGs) are a group of plants which have been widely supported as second-generation biofuel crops but very little genetic improvement has taken place and most emphasis has been placed on improved management to, for example, identify optimal fertiliser requirements and timely harvesting in good conditions to maximise the economic yield of biomass (Meehan et al. 2013). Here we review the major traits that could be incorporated into biofuel crops, and PRGs in particular, in order to utilise the vast areas of marginal land that are available worldwide. Examples of selection for stress tolerance are taken from the final report of the GrassMargins EU FP7 research project (GrassMargins 2015). Several stress-tolerant grass species have been identified and are listed in Table 18.1. In this session, we discuss features which are more generally desired for improved biomass productivity in both fertile and also the marginal land (Table 18.1). In fact, many of these features are equally desirable for food crops as well (Long et al. 2015).

Biomass crops need to have a high water use efficiency in general. In this respect, the C_4 crop species, such as *Miscanthus* and switchgrass, show in general higher water use efficiencies than C_3 species (Heaton et al. 2004). Also, some CAM species, such as agave, can maintain even higher water use efficiency and should be

Table 18.1 The potential features that are desired for biomass crops

Features	Known species with such features
Currently explored options	
High drought tolerance	Switchgrass, poplar, *Arundo donax*
High salinity tolerance	*Miscanthus x giganteus*, tall fescue, switchgrass
High temperature tolerance	Switchgrass
Low temperature tolerance	*Miscanthus x giganteus*, Festulolium,
High light use efficiency	*Miscanthus x giganteus*, sweet sorghum
High nutrient use efficiency	*Miscanthus x giganteus*
Environmental friendly (high pest and disease resistance)	*Miscanthus x giganteus*, switchgrass
Unexplored options	
Optimal root:shoot ratio	Not screened
Optimal biomass composition	Not screened
Optimal respiration	Not screened
Optimal leaf photosynthetic properties	Not screened

considered as potential options for biomass crops (Stewart 2015). Genetic screening and breeding for cultivars with enhanced water use efficiency will always be regarded as a major focus of the biomass crop breeding. However, the potential of the biomass crops to influence the water table also needs to be considered. As a reflection of the potential impact of this, the dramatic increase in agriculture irrigation in Northern China has resulted in a significant drop in the water table in a large area of China (Yuan and Shen 2013).

Biomass crops need to have high nutrient use efficiency, that is, a high biomass productivity per unit of nutrient input. Characteristics of crops with high nutrient efficiency are high nutrient recycling capacity at the end of the growing season or deep root systems for nutrient uptake. Furthermore, nitrogen fixation capacity of root/symbiont associations is a specially desired property for any biomass crops.

A high light use efficiency is a required feature for any successful biomass crop. The biomass productivity per ground area essentially represents the energy conversion efficiency of the crop. Long-term theoretical and experimental studies have identified a number of candidate approaches to improve light use efficiencies in plants, in particular in C_3 plants. These are summarised in a number of recent reviews, including Zhu et al. (2010) and Long et al. (2015). These approaches include optimisation of the Rubisco kinetic properties, increase of the RuBP regeneration rates, optimisation of the leaf chlorophyll concentration (or decreasing the photosystem antenna size), manipulation of the canopy architecture, increase of the speed of recovery of leaves from photoprotective states to the normal state, etc. Since these different features of photosynthesis are highly conservative, once they are proved to be effective, they can be directly either bred or engineered into bioenergy crops to increase the light use efficiencies. These features should benefit photosynthesis in both fertile soil and on marginal lands. Close interaction between the research communities of the bioenergy and food crops is therefore needed to ensure that the new options to improve photosynthesis can be rapidly transferred between the research communities.

One of the key elements in developing biomass-based energy is to create a sustainable and environmentally friendly source of energy, i.e. its large-scale adaptation should benefit rather than endanger the land growing them, and should not decrease the ecosystem service of the original landscape. In addition to high light, water and nitrogen use efficiencies any biomass crop need to have high pest or disease resistance to decrease the demand for pesticide and to ensure high sustainability and less potential damage to the environment.

Physiological Features Required by Biomass Crops on Marginal Land

Crops growing on marginal land are consistently subjected to a range of abiotic stresses, including shortage/excess of soil water, low nutrient availability, salinity and high and low temperatures, which reduce their yields below their potential. Stresses on marginal land are often multiple, combining extremes of temperature with water stress and possibly salinity stress and soil toxins. Marginal land by definition severely limits the productivity of plants which grow on it. Natural vegetation on marginal land consists predominantly of stress tolerators (Jones and Jones 1989) that have evolved functional traits that reflect their ecological strategies in stressful environments. The yield potential of crops grown on marginal land is reduced below the optimum by the abiotic stresses and in extreme cases yields that are so low that they are not economically viable.

Temperature Stress

Low- and high-temperature stress leads to a number of physiological responses in plants, mainly associated with direct effects on metabolic activity and the process of cell expansion which ultimately control the rate of growth. Extremes of temperature both low and high can be lethal but most attention to date has been given to rapid screening for frost tolerance by, for example, measuring electrolyte leakage following freezing (Perez-Harguindeguy et al. 2013). Clifton-Brown and Lewandowski (2000) demonstrated cold tolerance variation between species and genotypes of *Miscanthus* and more recently Yan et al. (2012) found considerable variation in a range of traits including cold tolerance and overwinter survival/establishment among 93 wild collected genotypes of *M. sacchariflorus* and *M. sinensis* grown in common garden experiments at three sites with contrasting climate in China, thus confirming that the selection of genotypes with enhanced cold-temperature tolerance represents an effective strategy for maximising production in environments

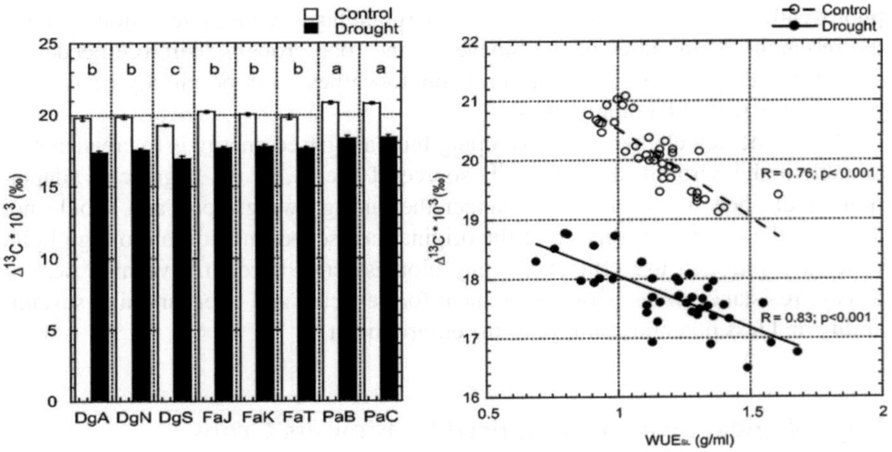

Fig. 18.1 *LEFT*: Discrimination of ^{13}C (‰) in shoot material of species and varieties under control (*white bars*) and drought treatment (*black bars*), where *DgA Dactylis glomerata* 'Amba', *DgN D. glomerata* 'Niva', *DgS D. glomerata* 'Sevenop', *FaJ Festuca arundinacea* 'Jordane', *F. arundinacea* 'Kora', *FaT F. arundinacea* 'Tower', *PaB Phalaris arundinacea* 'Bamse', *PaC P. arundinacea* 'Chieftain'. *RIGHT*: Relationship between $\Delta^{13}C$ discrimination (‰) and WUE_{SL} (g/mL)

subjected to cold stress (Yan et al. 2012). Low non-freezing temperatures, particularly at high light levels, can photo-inhibit photosynthesis in plants with C_4 photosynthesis but one genus that appears to be uniquely cold tolerant in this respect is *Miscanthus* (Naidu et al., 2013).

Water Stress

Plants can be stressed by shortage of water (drought) and excess of water (waterlogging) in the soil. Drought tolerance is the ability of plant tissue to maintain metabolism at low plant water potentials brought about by water stress. Selection for drought tolerance has been achieved by selecting for plants that have higher water use efficiencies measured as the amount of biomass produced per mm of water lost through evapotranspiration (Richards 2006). Screening for water use efficiency (WUE) is achieved by measuring the discrimination in photosynthesis between the two stable isotopes of carbon (Richards 2006) and it has been used successfully in selecting for water stress tolerance in cereals (Richards 2006). In GrassMargins the use of stable isotope ratios as an indicator of WUE was investigated in a greenhouse experiment, which showed that lower discrimination was induced by drought treatment and close correlations between discrimination and WUE was detected within the control treatment as well as within the drought treatment, confirming that isotope ratio measurements can be used for rapid screening of WUE under similar growth conditions (Fig. 18.1). Of the C_3 grasses grown under drought stress, Cocksfoot was the most stress tolerant while Reed Canary Grass was the most sensitive and Tall Fescue was intermediate. Within species, genotypic variation was most pronounced in Cocksfoot. A controlled drought experiment with 50 *Miscanthus* genotypes was used to compare their water use efficiency under optimal and low water supply. Significant genotypic variation in tolerance of water stress was identified by measuring the dynamic responses to water stress using novel time-lapse digital photography to measure process relevant traits such as leaf temperature and chlorophyll fluorescence.

In GrassMargins up to ten genotypes of each of five PRG species were screened under water-logged conditions and the most stress tolerant were Reed Canary Grass and *Miscanthus* while Cocksfoot was the most sensitive. Tall Fescue and Festulolium (a hybrid between *Festuca* and *Lolium*) were intermediate in sensitivity (Lærke et al. 2014).

Salinity Stress

Salt stress is an increasing problem for agriculture and is likely to become worse as summer rainfall declines under climate change and more irrigation is used. Salt tolerance by plants is achieved by either excluding salt from the roots, secreting excess salt uptake or tolerating salt in the tissues, usually by increasing succulence (Perez-Harguindeguy et al. 2013). Each of these forms of tolerance is associated with characteristic physiological traits. Extensive genetic diversity for salt tolerance exists in plant taxa which colonise saline environments. For example, salt tolerance has been shown to evolve within communities of Reed Canary Grass (*Phalaris arundinacea*) subjected to salt stress over a period of more than 20 years (Maeda et al. 2006) and in the *Miscanthus* genus large variations in salt tolerance between and within species has been demonstrated. In GrassMargins a greenhouse screening of four PRGs was performed by establishing plants without salt stress and then subjecting them to increasing salt concentrations from 0.5 to 2.5 %. There was variation in tolerance between species and the ranking from most to least tolerant was as follows: Tall fescue > Festulolium > Cocksfoot > Reed Canary Grass. It was also noted that the most susceptible accessions were the fastest growing under non-stress conditions. In a field trial on saline soils it became clear that salt tolerance during germination is an important and independent plant trait which determines a plant's ability to tolerate saline conditions.

Largely Unexplored Options to Increase Energy Conversion Efficiency of Biomass Crops

Improving photosynthetic efficiency is a major option for improving biomass yields but many other crop growth and development-related features can also potentially increase the energy conversion efficiency of crops and warrant systematic evaluation. Many of these have not been considered as relevant for biomass production so far (Table 18.1) but some are considered below:

Altered root:shoot ratio. Altering root:shoot ratio can theoretically greatly influence biomass productivity of crops by allocating more biomass above ground for harvest. However, the optimal root:shoot ratio is also influenced by the demand of recycling nutrients below ground at the end of a growing season as well as the needs to ensure supply of the water and nutrient required for rapid biomass accumulation in the biomass crops, both of which may mean that more biomass needs to be allocated below ground.

Altered biomass composition. Depending on the composition of biomass, the efficiency of converting the photosynthate into biomass can differ dramatically (Amthor 2010). Therefore, it is highly preferable to breed or design biomass crops to maintain an optimal biomass composition, which confer both a high photosynthetic efficiency and a high conversion efficiency from photosynthate into biomass.

Different rates of respiration. Respiration is another aspect that has been largely ignored in the current study of potential options to improve crop biomass productivity. Increased night temperature has been demonstrated to be a major factor that led to the decreased rice yields (Peng et al. 2004), due to the increased night respiration. Maintaining a proper respiratory flux under different environments can therefore potentially dramatically increase the biomass productivity of crops. In plants, there are large genetic variations of leaf respiration in tropical forest (Asao et al. 2015). We hence expect that there should also be large genetic variations in biomass crops that can be potential targets for selective breeding.

Genetic potential of photosynthesis in the field. Photosynthesis has large natural variations (Lawson et al. 2012). However, so far we lack a mechanistic understanding of what sets the photosynthetic rates of a plant. That is, why a particular accession of a species maintains a certain photosynthetic rates under a particular condition. For example, for *Miscanthus* grown in the field, its photosynthetic rate can reach 40 μmol CO_2 m^{-2} s^{-1}. But why is this the rate instead of 30 or 50 μmol m^{-2} s^{-1}. What are the genetic mechanisms underlying the variation of photosynthetic efficiency in the field? Systematic evaluation of genetic mechanisms controlling the natural variations of photosynthesis can potentially open the door to large-scale improvement of photosynthetic energy conversion efficiencies in corps.

Crop Systems Models: A Critical Tool to Support the Design of New Ideotype and Agronomic Practice to Support Biomass Crop Development

In contrast to the hundred or even thousands of years of selective breeding in major food crops, the breeding of biomass crops has only recently started as a response to global climate change and the shrinking supply of fossil fuel. It is also important to appreciate that the final breeding targets for biomass crops and food crops are very different because many of the features that are ideal for food crops are unimportant for biomass crops. One example is the semi-dwarf characteristics of many food crops currently used in the wheat and rice breeding program. The semi-dwarf characteristics were critical for the success of the first green revolution, since these characteristics led to an increased harvest index and the increased grain productivity in rice and wheat (Claeys et al. 2014). However, maintaining a short stature is certainly not desired for biomass crops. In fact, the harvest index, i.e. the ratio of grain to the total biomass, becomes irrelevant for biomass crop breeding. Therefore, completely new breeding selection criteria need to be established for biomass crops. But what are these ideal features for biomass crops and how are they defined?

The breeding of many crops has in recent years been guided by crop growth and development models. For example, rice breeding has been guided by the rice ideotype based on simulation studies (Peng et al. 1994, 2008) but equivalent models have not been developed for biomass crops. Here, we argue that systems models of biomass crops can play a similar role that systems models for food crops have done in the past. This is particularly important since humanity simply does not have the time to slowly go through the same trial and error process, as was done for the food crops, if the pending risk of global climate change is to be halted before too serious damage occurs. In the following two sections, we first briefly discuss major components needed for a crop systems model, which can support biomass crop design and breeding, and then, we briefly comment on the major applications of such crops systems models in biomass crop breeding and cultivation.

Major modules needed in a biomass crop growth and development model are shown in Fig. 18.2 and a basic schema of a biomass crop systems model is shown in Fig. 18.3. This model includes a leaf photosynthesis module, a canopy microclimate and photosynthesis module, soil process module, plant development module, photosynthate partitioning module and finally modules at the agro-ecosystem scale.

The Leaf Physiology Module

This module should be able to predict photosynthetic rates and transpiration rates under a range of light, CO_2, temperature and humidity conditions. The steady-state biochemical models of C_3 photosynthesis (Farquhar et al. 1980) and C_4 photosynthesis (von Caemmerer 2000) can be used to predict the photosynthetic rates. The stomatal conductance, and correspondingly the transpiration rate, can be predicted through the Ball and Berry model of stomatal conductance, which uses the photosynthetic rates together with atmospheric CO_2 and humidity to predict stomatal conductance (Ball et al. 1987). The potential impacts of nitrogen concentration on leaf photosynthetic efficiencies also need to be reflected in this module.

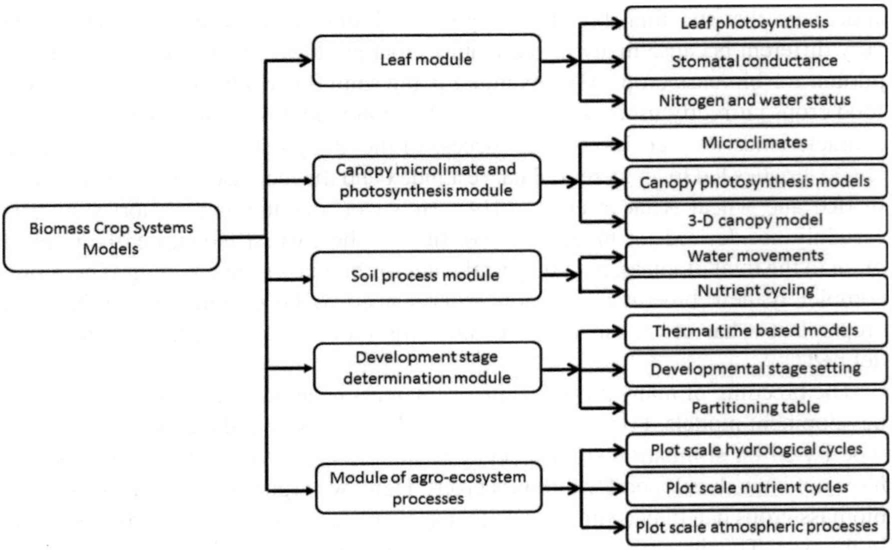

Fig. 18.2 The components needed to develop a mechanistic model of biomass crop growth and development. As a minimum, five modules need to be developed to form a basic biomass crop growth and development model. Adapted from Zhu et al. (2012a, b)

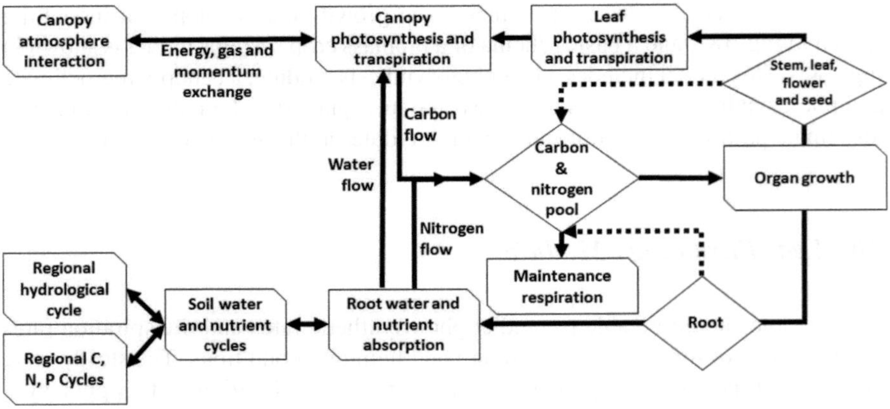

Fig. 18.3 A basic schema of a biomass crop systems model. A biomass crop systems model needs to include modules representing leaf photosynthesis, canopy microclimate and photosynthesis, soil process, plant development and photosynthate partitioning, and even agro-ecosystem level processes, i.e. regional hydrological and C, N, P cycles, and also interaction between canopy and atmosphere. The dotted lines represent the senescence of plant tissues, which contribute new carbon and nitrogen to the pool for partitioning

The Canopy Microclimate and Photosynthesis Module

The canopy microclimate includes information of the light, temperature, humidity and CO_2, inside the canopy which together determine the leaf level performances. An effective model should be able to predict accurately these different components and further enable accurate prediction of the photosynthetic rate of the whole canopy, i.e. all leaves in the canopy. Currently, the photosynthetic rate of a canopy can be calculated through models with different levels of details (see review in Zhu et al. 2012b), which include the big-leaf model which represents the whole canopy as one single leaf (dePury and Farquhar 1997), the sunlit-shaded model which represents leaves in the canopy as being either sunlit or shaded leaves (Norman 1980), and the latest 3-D canopy photosynthesis model where the 3D canopy architecture is represented and the light, temperature and humidity for each of the individual leaves are used to calculate the photosynthetic rate of the total canopy (Song et al. 2013, 2016a). During the calculation of the canopy photosynthesis, the energy balance on each leaf surface needs to be determined to ensure a proper estimate of the photosynthetic performance under different ambient temperatures.

Module Describing the Soil Related Processes

Leaf photosynthesis is heavily influenced by the leaf water potential, which is in turn influenced by the soil water status. Furthermore, the nitrogen availability in the soil will largely influence the leaf nitrogen concentration and hence physiological parameters of photosynthesis, such as the maximal RuBisco-limited RuBP carboxylation rate (V_{cmax}), and the maximal rate of electron transfer (J_{max}) (Evans 1989a, b). Therefore, to ensure an accurate prediction of the crop photosynthetic efficiency, the soil water and nutrient status need to be precisely simulated. This requires the simulation of the nitrogen, carbon, and water movement dynamics in the soil and their interaction with plants. A number of soil nutrient cycling models, such as the CENTURY model, can be used for this purpose (Parton et al. 1988, 1993). However, much work is needed to ensure an effective integration of the soil process models with the crop growth and development models.

Plant Development and Photosynthate Partitioning Modules

Being able to precisely predict the plant developmental process is required for any crop growth model. So far, this aspect of crop models is mainly managed through empirical approaches, such as the thermal time-based developmental stage determination, as mechanistic models of crop developmental processes are not yet available.

This is mainly due to the lack of mechanistic understanding of the molecular mechanisms controlling plant development. There are semi-mechanistic approaches to predict the flowering date of major crops such as rice and wheat by coupling the Quantitative Trait Loci (QTLs) related to flowering and the environmental conditions (Zheng et al. 2013). This approach is likely to be the major form of crop developmental model in the foreseeable future. At different developmental stages, a plant partitions different proportion of photosynthate into different organs for biomass development. So far, this aspect is usually modelled based on a partitioning table established through empirical experiments and altered based on the environmental conditions of the soil and atmosphere.

Modules of the Agro-Ecosystem Scale Processes

In order to use crop models at the landscape and global scale the biomass crop models need to simulate larger scale processes, such as the hydrological cycles, nutrient cycle, and the interaction with the atmosphere (Wang and Zhu 2008). These large-scale processes need to be integrated with the biomass crop models because the large-scale biomass production can potentially influence the hydrological, nitrogen, and carbon cycles and hence influence whether a biomass crop production scenario in a particular region is sustainable. A number of models, such as SiB (Sellers et al. 1986), SiB-2 (Sellers et al. 1996), can be used as the linkage between crop models and regional general circulation models.

The biomass crop systems models can be used in two major ways. First, they can be used to define the physiological and architectural features for an ideal biomass crop. In this respect, effective multivariable optimisation method needs to be developed. Because different regions have different environmental conditions, the ideal features for biomass crops should be region-specific (Song et al. 2016a). Once features for ideal biomass crops can be defined through a crop systems model, the ideal-biomass crops can be bred or engineered. Considering that most of the biomass crops are perennials, and hence require a long breeding cycle of possibly up to 15 years it is extremely important to develop effective crop systems models as a priority to better define the ideal type so that we can obtain the optimal breeding traits and avoid selecting inappropriate traits. Having a highly robust biomass model therefore represents one of the highest priority research areas to advance the future biomass-based bioenergy production.

The second major application of such crop systems model is to identify the optimal agronomic practices for maximising yield of biomass crops. Because biomass crops have not historically been a significant component of agriculture, many basic agronomic practices for biomass production have not been systematically evaluated. Examples of important knowledge to obtain are the optimal germplasm selection for a particular region, optimal water and nitrogen application strategies, the best harvesting time for both the biomass productivity and also biomass quality. In fact, without a good agronomic practice combined with good germplasms to ensure high productivity, the biomass-based bioenergy production might never be adopted by farmers at a large scale. Well-validated crop growth and development models can be used as an effective tool to identify the "best" or "better" agronomic practice, which can then be further tested infield trials. The crop models themselves need to be parameterised and validated using data collected through experiments performed on sites in diverse environments. Although in any individual experiment, only a limited set of parameters are collected in detail, if combined together, these experiments can form the basis to develop a robust crop systems model. In this respect, new algorithms to enable better data assimilation and model improvement are in high demand.

Conclusions

Biomass-based bioenergy production on marginal land is a major option to gain an environmentally friendly, sustainable source of energy production without jeopardising food security. After a few decades of research on biomass crops, many physiological features that would be advantageous for biomass crops have been identified, including high photosynthetic efficiency, increased recycling of nutrients at the end of the crop growing season, high water use efficiency and high stress tolerance. A number of promising biomass crops with special physiological characteristics and stress resistance and hence suitable for growth in different environments, have also been identified, including *Miscanthus*, switchgrass, poplar and even dessert succulent plants such as Agave. A number of additional features that can potentially be used to further improve productivity needs to be investigated, including changes in root:shoot ratio, respiratory rate, photosynthetic rates and also the biomass composition. The rapid advances in DNA sequencing technology combined with the natural variations of these different crops can be used to rapidly advance the future biomass crop breeding. So far the ideotype of biomass crops has not been well established. Development of crop systems models can be used to not only support identification of features for a biomass crop ideotype, but also support identification of optimal agronomic practices to gain higher productivity.

Acknowledgements The author's research has been supported by EU FP7 project GrassMargin (#EU 289461), CAS Strategic Project "Designer Breeding by Molecular Modules" (#XDA08020301), National Basic Science Research Programs (# 2015CB150104), National Science Foundation of China (#30970213), Ministry of Science and Technology of China (#2011DFA31070), Bill and Melinda Gates Foundation (#OPP1060461) and CAS International Collaboration Grants (#GJHZ1501).

References

Amthor JS (2010) From sunlight to phytomass: on the potential efficiency of converting solar radiation to phyto-energy. New Phytol 188:939–959

Asao S, Bedoya-Arrieta R, Ryan MG (2015) Variation in foliar respiration and wood CO2 efflux rates among species and canopy layers in a wet tropical forest. Tree Physiol 35:148–159

Ball JT, Woodrow IE, Berry JA (1987) A model predicting stomatal conductance and its contribution to the control of photosynthesis under different environmental conditions. In: Biggens J (ed) Progress in photosynthesis research, vol IV. Martinus Nijhoff, Dordrecht, pp 221–224

Berndes G, Hoogwijk M, van den Broek R (2003) The contribution of biomass in the future global energy supply: a review of 17 studies. Biomass Bioenergy 25:1–28

Bryngelsson DK, Lindgren K (2013) Why large-scale bioenergy production on marginal land is unfeasible: a conceptual partial equilibrium analysis. Energy Policy 55:454–466

Campbell JE et al (2008) The global potential of bioenergy on abandoned agriculture lands. Environ Sci Technol 42:5791–5794

Cassman K, Liska AJ (2007) Food and fuel for all: realistic or foolish? Biofuels, Bioproducts Biorefining 1:18–23

Claeys H, De Bodt S, Inze D (2014) Gibberellins and DELLAs: central nodes in growth regulatory networks. Trends Plant Sci 19:231–239

Clifton-Brown J, Lewandowski I (2000) Overwintering problems of newly established Miscanthus plantations can be overcome by identifying genotypes with improved rhizome cold tolerance. New Phytol 148:287–294

Dale VH et al (2010) Biofuels: implications for land use and biodiversity. Ecological Society of America, Washington, DC

Dauber J et al (2012) Bioenergy from “surplus” land: environmental and socio-economic implications. BioRisk 7:5–50

dePury DGG, Farquhar GD (1997) Simple scaling of photosynthesis from leaves to canopies without the errors of big-leaf models. Plant Cell Environ 20:537–557

Evans JR (1989a) Partitioning of nitrogen between and within leaves grown under different irradiances. Aust J Plant Physiol 16:533–548

Evans JR (1989b) Photosynthesis and nitrogen relationships in leaves of C3 plants. Oecologia 78:9–19

Farquhar GD, von Caemmerer S, Berry JA (1980) A biochemical model of photosynthetic CO_2 assimilation in leaves of C_3 species. Planta 149:78–90

Field CB, Campbell JE, Lobell DB (2008) Biomass energy: the scale of the potential resource. Trends Ecol Evol 23:65–72

Field JL et al (2013) Distributed biochar and bioenergy coproduction: a regionally specific case study of environmental benefits and economic impacts. Global Change Biology Bioenergy 5:177–191

Fischer G, Schrattenholzer L (2001) Global bioenergy potentials through 2050. Biomass Bioenergy 20:151–159

Fritz S et al (2013) Downgrading recent estimates of land available for biofuel production. Environ Sci Technol 47:1688–1694

GrassMargins (2015) Project consortium (KBBE-2011-5-289461)—final project report to EC, November 2015, www.grassmargins.com

Heaton E, Voigt T, Long SP (2004) A quantitative review comparing the yields of two candidate C-4 perennial biomass crops in relation to nitrogen, temperature and water. Biomass Bioenergy 27:21–30

Hoogwijk M et al (2003) Exploration of the ranges of global potential of biomass for bioenergy. Biomass Bioenergy 25:119–133

Jones HG, Jones MB (1989) Introduction: some terminology and common mechanisms. In: Jones HG, Flowers TJ, Jones MB (eds) Plants under stress. Cambridge university Press, Cambridge, pp 1–10

Lærke PE, Karki YK, Sorenson HB, Sorenson KK, Jorgenson U (2014) Screening genotypes of candidate energy grasses for flooding tolerance. Abstracts, Plant Biology Europe, FESPB/EPSO Congress, 2014, Dublin, Ireland

Lawson T, Kramer DM, Raines CA (2012) Improving yield by exploiting mechanisms underlying natural variation of photosynthesis. Curr Opin Biotechnol 23:215–220

Long SP, Marshall AM, Zhu XG (2015) Engineering crop photosynthesis and yield potential to meet global food demand of 2050. Cell 161:56–66

Lotze-Campen H et al (2010) Scenarios of global bioenergy production: the trade-offs between agricultural expansion, intensification and trade. Ecol Model 221:2188–2196

Maeda Y et al (2006) Variations in salt tolerance of reed canarygrass (Phalaris arundinacea L.) plants grown at sites with different degrees of cattle urine contamination. Soil Sci Plant Nutr 52:13–20

Marland G, Obersteiner M (2008) Large-scale biomass for energy, with considerations and cautions: an editorial comment. Climate Change 87:335–342

Meehan PG, Finnan JM, Mc Donnell KP (2013) The effect of harvest date and harvest method on the combustion characteristics of Miscanthus × giganteus. GCB Bioenergy 5:487–496

Naidu SL et al (2003) Cold tolerance of C-4 photosynthesis in Miscanthus × giganteus: adaptation in amounts and sequence of C-4 photosynthetic enzymes. Plant Physiol 132:1688–1697

Norman JM (1980) Interfacing leaf and canopy light interception models. In: Hesketh JD, Jones JW (eds) Predicting photosynthesis for ecosystem models, vol 2. CRC Press, Boca Raton, pp 49–67

Offermann R et al (2011) Assessment of global bioenergy potentials. Mitig Adapt Strat Glob Chang 16:103–115

Parton WJ, Stewart JWB, Cole CV (1988) Dynamics of C, N, P and S in grassland soils: a model. Biogeochemistry 5:109–131

Parton WJ et al (1993) Observations and modelling of biomass and soil organic matter dynamics for the grassland biome wordwide. Global Biogeochem Cycles 7:785–809

Peng S, Khush GS, Cassman KG (1994) Evaluation of a new plant ideotype for increased yield potential. In: Cassman KG (ed) Breaking the yield barrier: proceedings of a workshop on rice yield potential in favourable environments. International Rice Research Institute, Los Banos, pp 5–20

Peng SB et al (2004) Rice yields decline with higher night temperature from global warming. Proc Natl Acad Sci U S A 101:9971–9975

Peng S et al (2008) Progress in ideotype breeding to increase rice yield potential. Field Crop Res 108:32–38

Perez-Harguindeguy N et al (2013) New handbook for standardised measurement of plant functional traits worldwide. Aust J Bot 61:167–234

Popp A et al (2011) The economic potential of bioenergy for climate change mitigation with special attention given to implications for the land system. Environ Res Lett 6:034017

Richards RA (2006) Physiological traits used in the breeding of new cultivars for water-scarce environments. Agric Water Manag 80:197–211

Sellers PJ et al (1986) A simple biosphere model (SiB) for use within general circulation models. J Atmos Sci 43:505–531

Sellers PJ et al (1996) A revised land surface parameterization (SiB2) for atmospheric GCMs.1. Model formulation. J Climate 9:676–705

Slade R, Bauen A, Gross R (2014) Global bioenergy resources. Nature Climate Change 4:99–105

Song Q-F, Zhang G, Zhu X-G (2013) Optimal crop canopy architecture to maximise canopy photosynthetic CO_2 uptake under elevated CO_2—a theoretical study using a mechanistic model of canopy photosynthesis. Functional Plant Biol 40:108–124

Song Q-F et al (2016a) Genetics based dynamic systems model of canopy photosynthesis: the key to improved light and resource use efficiencies for crops. Food Energ Secur 5:18–25

Stewart JR (2015) Agave as a model CAM crop system for a warming and drying world. Front Plant Sci 6:684

Tilman D et al (2006) Carbon negative biofuels from low-input high-diversity grassland biomass. Science 247:699–702

von Caemmerer S (2000) Biochemical models of leaf photosynthesis. CSIRO, Canberra, ACT, Australia

Wang S, Zhu X-G (2008) Coupling cyberinfrastructure and geographic information systems to empower ecological and environmental research. Bioscience 58:6–7

Yan J et al (2012) Variability and adaptability of Miscanthus species evaluated for energy crop domestication. GCB Bioenergy 4:49–60

Yuan Z, Shen Y (2013) Estimation of agricultural water consumption from meteorological and yield data: a case study of Hebei, North China. PLoS One 8, e58685

Zheng B et al (2013) Quantification of the effects of VRN1 and Ppd-D1 to predict spring wheat (Triticum aestivum) heading time across diverse environments. J Exp Bot 64:3747–3761

Zhu X-G, Long SP, Ort DR (2010) Improving photosynthetic efficiency for greater yield. Annu Rev Plant Physiol Plant Mol Biol 61:235–261

Zhu B et al (2012a) Genome sequence of Enterobacter sp. strain SP1, an endophytic nitrogen-fixing bacterium isolated from sugarcane. J Bacteriol 194:6963–6964

Zhu X-G, Song Q-F, Ort DR (2012b) Elements of a dynamic systems model of canopy photosynthesis. Curr Opin Plant Biol 15:237–244

Zhuang D et al (2011) Assessment of bioenergy potential on marginal land in China. Renew Sustain Energy Rev 15:1050–1056

Chapter 19
Mycorrhiza-Based Biofertilizer Application to Improve the Quality of *Arundo donax* L., Plantlets

E. Baraza, M. Tauler, A. Romero-Munar, J. Cifre, and J. Gulias

Arundo donax as Energy Crop

Giant reed (*Arundo donax*) is an Asian plant considered one of the most promising species for energy, cellulose paste and second-generation biofuel production (Pilu et al. 2013; Lewandowski et al. 2003; Shatalov and Pereira 2002). This is due to its ability to adapt to a wide variety of environments, low input requirements after establishment (Cosentino et al. 2008; Pilu et al. 2013) and its high production of biomass, around 40 t year^{-1} ha^{-1} of dry biomass (Angelini et al. 2005, 2009; Mantineo et al. 2009). Another important advantage of this species is that it can be cultivated in marginal or sub-marginal lands due to its tolerance to environmental stresses, thus reducing competition with food crops, which generally require better-quality lands (Sims et al. 2010; o di Nasso et al. 2013). Once established, *A. donax* is highly drought tolerant (Perdue 1958; Lewandowski et al. 2003; Lambert et al. 2010). Individuals can survive and even grow well during extended periods of severe drought accompanied by low-pressure humidity or periods of excessive moisture (Perdue 1958). In addition, giant reed can tolerate excessive salinity levels (Perdue 1958; Lambert et al. 2010), inhabiting estuarine and coastal strand environments (Lambert et al. 2010). *A. donax* needs moderate nitrogen fertilization but application of K and P if the nutrient status of the soil is poor (Lewandowski et al. 2003).

Giant reed cannot produce viable seeds since its pollen is unviable (Perdue 1958). So, propagation is only possible by vegetative means using fragments of stems or rhizomes that present a high capacity to survive and to generate shoots and roots (Mann et al. 2013). Rhizomes are commonly the most used, but this method of propagation is time-consuming and involves considerable cost and effort to dig-up, break-apart and replant the rhizomes. In addition, this method of propagation

E. Baraza (✉) • M. Tauler • A. Romero-Munar • J. Cifre • J. Gulias
Balearic Islands University, Crt. Valldemossa Km 7.5, Palma, Baleares, Spain
e-mail: elena.baraza@uib.es

S. Barth et al. (eds.), *Perennial Biomass Crops for a Resource-Constrained World*, DOI 10.1007/978-3-319-44530-4_19

requires a large land surface to produce enough plants for use in energy production programs, as each plant produces only a few rhizomes each year (Pilu et al. 2013). Planting shoot cuttings directly in the field is a practical option when irrigation water is abundantly available to ensure high rates of rooting (Ceotto and Di Candilo 2010); however, this situation is not common in the Mediterranean area. According to Cavallaro et al. (2011), in vitro culture methods are the only ones offering the potential required for large-scale propagation of giant reed. In vitro culture methods facilitate the production of larger amounts of high-quality genetically homogeneous plants in modest time with the guarantee of being free of pathogens (Cavallaro et al. 2011). However, micropropagation is a complex procedure which can considerably increase the cost of crop establishment (Xue et al. 2015). On one hand, research efforts are required to reduce the material production costs. On the other hand, more research is needed to evaluate whether the increase in cost is compensated or not by increases in harvest.

Mycorrhiza-Based Biofertilizer Application on Plant Micropropagation

Most plant species, in their natural environment, are colonized both by external and internal microorganisms. Some of these microorganisms, particularly beneficial bacteria and mycorrhizal fungi, can improve plant performance under stressful environments and so enhance yield (Brown 1974; Levy et al. 1983; Creus et al. 1998). Arbuscular mycorrhizal fungi (AMF) establish symbioses with 80 % of plant species (Smith and Read 1996; Varma 2008). AMF increase root system surface, allowing plants to explore greater soil areas and enhancing water and nutrients access and uptake, while the fungus acquires a protected ecological niche and plant photosynthates (Varma 2008). The arbuscular mycorrhizae symbiosis is associated with a great range of benefits for the plant, including resistance to a wide variety of environmental stresses (Kapoor et al. 2008). Inoculation with AMF has been tested as a technique to improve plant production, increase crop field and enhance crop tolerance to a variety of environmental stresses in a wide range of plant species (Fester and Sawers 2011).

The use of AMF has been proposed as a solution to improve the micropropagation process of multiple plant species (Vestberg and Estaún 1994; Estrada-Luna et al. 2000; Kapoor et al. 2008). One of the phases of in vitro plant production that has generated mortality is the acclimatization or hardening phase (Kapoor et al. 2008). Directly after transfer from in vitro to ex vitro conditions, micropropagated plants are susceptible to various stresses because of the sudden shock of environmental changes and the incompleteness of certain physiological activities and anatomical developments such as photoheterotrophic considerations and the lack of cuticle or improper vascular connection (Alarcón and Ferrera-Cerrato 2000). Inoculation with AMF during hardening enhances the development of the root system, increases nutrient uptake, water conducting capacity and photosynthetic rates, and promotes defense against harmful pathogens and the synthesis of growth hormones (Rai 2001; Kapoor et al. 2008; Parkash et al. 2011; Singh et al. 2012). The benefits associated with the use of AMF inoculation for in vitro-raised plantlets have been reported in several horticultural, fruit, ornamental and forest species (Alarcón and Ferrera-Cerrato 2000; Rai 2001; Kapoor et al. 2008). On energy crops, da Silva Folli-Pereira et al. (2012) reported that arbuscular mycorrhization enhances the production system of micropropagated plantlets of *Jatropha curcas* increasing phosphorus uptake and growth during the acclimatization period.

Two experiments were conducted at the Universitat de les Illes Balears, Majorca (West Mediterranean basin, Spain) in order to evaluate the effect of AMF commercial inoculum on the acclimatization process of micropropagated plantlets of *A. donax*. The objective was to determine the potential of the use of mycorrhizal biotechnology to improve the plantlet production process of this promising energy crop, evaluating the effects of AMF on biomass accumulation of *A. donax* micropropagated plantlets.

Inoculation with Arbuscular Mycorrhizae Fungi (AMF) as a Method for Improving *A. donax* Propagation

Plantlets used were the Arundo-K12 clone, provided by BIOTHEK ECOLOGIC FUEL S.L. Micropropagated plants were received bare-root and immediately planted on trays with agricultural substrate (KEKKILÄ DSM 1W ©) which consisted of nutrient-rich (0.6 kg/m^3 N-P_2O_5-K_2O 14-16-18 + microelements) black peat (pH = 5.5). Plants were transplanted into 28 cm^3 (experiment one) or 50 cm^3 (experiment two) cell planting trays. In both experiments, at the moment of transplantation, 5 cm^3 of AMF inoculum was added to half of the plantlets, near to the root. The AMF commercial inoculum (AEGIS SYM ©) used contained 25 spores/g of *Rhizophagus intraradices* (before known as *Glomus intraradices*) and 25 spores/g of *Funneliformis mosseae* (before known as *Glomus mosseae*). To imitate the soil texture of inoculated cells, we added the same quantity of sterile zeolite to the others plantlets, which were used as a control. In both experiments, plantlets were maintained at a mean temperature of 18 °C and 86 % moisture in a greenhouse during the acclimatization period. The plantlets were watered according to their needs to always keep the substrate moist. In addition, we moved the tray periodically to avoid a position effect. See Tauler and Baraza (2015) for more details of experiment one. In experiment two, two different substrates, 100 % agricultural substrate (AS) and a mix of agricultural-substrate and sand (1:1, v/v) (MIX) were tested. The acclimatizing period was 46 days for experiment 1 and 100 days for experiment two.

During both assays inoculated plantlets presented significant positive increments in height, number of leaves, total dry biomass and chlorophyll content (Table 19.1) when compared to non-inoculated ones. Differences between mycorrhized and control plants were greatest on the MIX substrate (Table 19.1). Mycorrhizae helped the plant to overcome stress conditions (Krishna et al. 2005) and, therefore, we expected

Table 19.1 Percentages of variation (+ increases or – reductions) of the mean value of inoculated plantlets in comparison with control ones growing in Agricultural substrate (AS) or mixture of agricultural-substrate:sand (1:1, v/v) (MIX) after 46 days of acclimatization or 100 days of acclimatization

Variable	46 days AS	100 days AS	100 days MIX
Final height	+30.5*	+7.9	+65.7*
Total biomass	+43.7*	+5.5	+86.6*
Shoot:Root	+30.5*	+6.4	+100*
Leaf mass area	ND	−12.0*	−17.6*
Chlorophylls	+36.2*	+22.8*	+21.3*

ND not data
*Denote significant differences $P<0.05$ t-student for comparison among inoculated an control plants

differences between AM and control involving an advantage over the lack of nutrients and water generated by MIX substrate (less than 94.3 % of nitrogen content and 39.9 % of water holding capacity than the AS substrate). In fact higher leaf mass per unit area (LMA) is expected when plants grow under water stressful conditions (Liu and Stützel 2004) and a significant decrease of LMA in AM plants relative to control was found, mainly in MIX substrate (Table 19.1).

Singh et al. (2012) reported that the rate of infection doubled between 60 and 90 days after inoculation when infection levels rose above 80 %. In our case, we expected an increase of infection with time, which may be accompanied by an increase of the effect of fungi on plants, as we found that the percentage of AMF infection was positively correlated with heights and shoot:root ratios in experiment one. However, the mean percentage of infection during the first experiment (after 46 days) was even higher (36 %) than after 100 days in the same substrate (22.1 %), or in the MIX one (32.5 %). Moreover, beneficial effects of AMF were less significant in experiment two in the same substrate (Table 19.1). In both cases the percentage of AMF infection was low and highly variable among inoculated plants, which suggests the need to study other inoculums in order to find more infectious ones that could have greater beneficial effects.

Micropropagated plantlets of *A. donax* do not require AMF inoculation to reach high survival rates during the acclimatization period. However, inoculation with AMF improves the growth of the plantlets. *Arundo donax* can tolerate a large diversity of stress conditions, being tolerant to drought, salinity, flood and heavy metals (Lewandowski et al. 2003; Papazoglou 2005; o di Nasso et al. 2013). However, our study suggests its sensitivity to lack of water and nutrients during the early stages of growth. This is consistent with Perdue (1958), who reported that giant reed can be seriously retarded by lack of moisture during its first year. According to our results, inoculation with AMF attenuates the effects of stress, allowing plants to present similar growth to plants in optimal conditions, since AM plants in suboptimal substrate (MIX) had a similar development to plants in the optimal substrate (AS), while control plants in MIX were 50 % smaller.

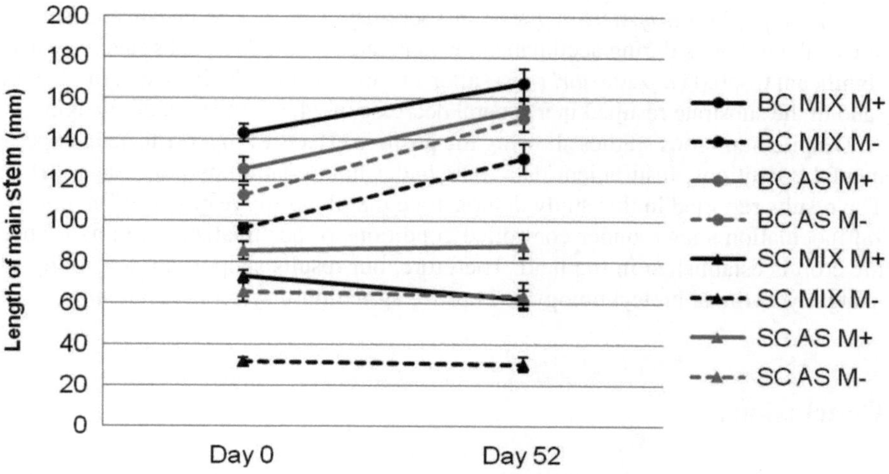

Fig. 19.1 Effect of the inoculation with mycorrhizae, substrate type and cell size during maintenance on the height of the plants before (Day 0) and after 52 days of transplantation to field. Mean and standard error values for AM plants (M+) and control plants (M−) in agricultural-substrate:sand (MIX) and agricultural substrate (AS) after being in small cells (SC) or bigger cells (BC) during maintenance

Is the Quality of Plantlets Important for Crop Establishment and Future Harvests?

Although most studies that have shown the positive effect of mycorrhizae on plant growth have been conducted under controlled conditions (Tawaraya 2014), several studies have described an increase in yields under field conditions (Díaz-Hernández et al. 2013; Gholamhoseini et al. 2013). However, the studies that evaluated the effect of mycorrhizae during hardening of micropropagated plants do not analyze the subsequent effect of these on the field crop. In order to estimate the mycorrhizal effect and the importance of size of plantlets on the survival and growth during crop establishment and the first year plant production, the plantlets obtained at experiment two were transferred to an open field. Previously 20 plants per treatment were transplanted to 300 cm^3 cells and 20 maintained on small cells (50 cm^3).

During establishment (first 52 days) in the field, plant size—obviously related to the size of the cell—became as important as AMF inoculation (Fig. 19.1). Larger sizes ensure a greater initial growth, which facilitates competition with weeds and ensures successful establishment. In fact only four plants died and all of them were from small cell treatments. In the same way mycorrhization seems to play an important role on plantlet survival since in experiment one only control plants died (Tauler and Baraza 2015). Moreover, in general inoculated plants were still higher than control ones in similar conditions (Fig. 19.1). After 11 months eight plants per treatment were cut at ground level and their dry weight calculated. The mycorrhizal effects were still present on dry biomass, yet inoculated plantlets showed a significant increase

(+19.75 % $p<0.05$ *a posteriori t*-test) in shoot dry weight. The negative effects of stressful conditions during acclimatization generated by the MIX substrate were also significant ($p<0.05$ *a posteriori t*-test) after 11 months since the inclusion of 50 % of sand in the substrate resulted in a general decrease of 24.8 % of shoot dry weight.

Despite numerous studies showing the positive effects of mycorrhizae in experimental conditions, insufficient data exist under field conditions (Tawaraya 2014). The results reported in this study demonstrate that the positive effects of mycorrhizal inoculation shown under controlled conditions of acclimatization remain once the crop is established in the field. Therefore, our results support the feasibility of using mycorrhizal biotechnology to improve agricultural systems sustainably.

Conclusions

The standardized use of AMF inoculum, early inoculation of giant reed plantlets with AMF can be very useful in improving plant tolerance to marginal lands, enhancing one of the characteristics that made this species especially interesting as an energy crop. Moreover, field results showed the importance of the quality of the plantlet for field establishment and first year crop production. However, more research is needed in order to find more efficient inoculums than the ones tested in these experiments, to determine the generality of the positive effect of this biotechnology on *A. donax* and to evaluate their economic viability.

Acknowledgments We thank Catina Alomar and Miquel Nolla for their help on laboratory work. Thanks to Biothek Ecologic Fuel for giving plants and give information about the conditions of acclimatization that are used on a commercial basis. We thank EMAYA for providing farmland and irrigation facilities. This study was supported by the "European Project Optimization of Perennial Grasses for Biomass Production" (FP7; 289642).

References

Alarcón A, Ferrera-Cerrato R (2000) Manejo de la micorriza arbuscular en sistemas de propagación de plantas frutícolas. Terra 17(3):179–191

Angelini LG, Ceccarini L, Bonari E (2005) Biomass yield and energy balance of giant reed (*Arundo donax* L.) cropped in central Italy as related to different management practices. Eur J Agron 22(4):375–389

Angelini LG, Ceccarini L, Nassi o Di Nasso N, Bonari E (2009) Comparison of *Arundo donax* L. and *Miscanthus* x *giganteus* in a long-term field experiment in Central Italy: analysis of productive characteristics and energy balance. Biomass Bioenerg 33(4):635–643

Brown ME (1974) Seed and root bacterization. Annu Rev Phytopathol 12(1):181–197

Cavallaro V, Tringali S, Patanè C (2011) Large-scale *in vitro* propagation of giant reed (*Arundo donax* L.), as a promising biomass species. J Hortic Sci Biotech 86:452–456

Ceotto E, Di Candilo M (2010) Shoot cuttings propagation of giant reed (*Arundo donax* L.) in water and moist soil. The path forward? Biomass Bioenerg 34(11):1614–1623

Cosentino SL, Copani V, Patanè C, Mantineo M, D'Agosta GM (2008) Agronomic, energetic and environmental aspects of biomass energy crops suitable for Italian environments. Ital J Agron 3(2):81–96

Creus CM, Sueldo RJ, Barassi CA (1998) Water relations in Azospirillum-inoculated wheat seedlings under osmotic stress. Can J Bot 76(2):238–244

da Silva Folli-Pereira M, Meira-Haddad LSA, Rasool N, Otoni WC, Kasuya MCM (2012) Development of mycorrhized vitroplants of *Jatropha curcas* L. at different rooting stages. Plant Biotech Report 6(4):355–362

Di Nasso NNO, Roncucci N, Bonari E (2013) Seasonal dynamics of aboveground and belowground biomass and nutrient accumulation and remobilization in giant reed (*Arundo donax* L.): A Three-Year Study on Marginal Land. BioEnergy Res 6(2):725–736

Díaz-Hernández BG, Aguirre-Medina JF, Díaz-Fuentes VH (2013) Yield of *Jatropha curcas* L. inoculated with mycorrhiza and fertilized with sugarcane compost. Revista Mexicana de Ciencias Agrícolas 4(4):599–610

Estrada-Luna AA, Davies FT, Egilla JN Jr (2000) Mycorrhizal fungi enhancement of growth and gas exchange of micropropagated guava plantlets (*Psidium guajava* L.) during ex vitro acclimatization and plant establishment. Mycorrhiza 10(1):1–8

Fester T, Sawers R (2011) Progress and challenges in agricultural applications of arbuscular mycorrhizal fungi. Crc Cr Rev Plant Sci 30(5):459–470

Gholamhoseini M, Ghalavand A, Dolatabadian A, Jamshidi E, Khodaei-Joghan A (2013) Effects of arbuscular mycorrhizal inoculation on growth, yield, nutrient uptake and irrigation water productivity of sunflowers grown under drought stress. Agr Water Manage 117:106–114

Kapoor R, Sharma D, Bhatnagar AK (2008) Arbuscular mycorrhizae in micropropagation systems and their potential applications. Sci Hortic 116(3):227–239

Krishna H, Singh SK, Sharma RR, Khawale RN, Grover M, Patel VB (2005) Biochemical changes in micropropagated grape (Vitis *vinifera* L.) plantlets due to arbuscular-mycorrhizal fungi (AMF) inoculation during ex vitro acclimatization. Sci Hortic 106(4):554–567

Lambert AM, Dudley TL, Saltonstall K (2010) Ecology and impacts of the large-statured invasive grasses *Arundo donax* and *Phragmites australis* in North America. Invas Plant Sci Mana 3(4):489–494

Levy Y, Dodd J, Krikun J (1983) Effect of irrigation, water salinity and rootstock on the vertical distribution of vesicular-arbuscular mycorrhiza in citrus roots. New Phytol 95(3):397–403

Lewandowski I, Scurlock JM, Lindvall E, Christou M (2003) The development and current status of perennial rhizomatous grasses as energy crops in the US and Europe. Biomass Bioenerg 25(4):335–361

Liu F, Stützel H (2004) Biomass partitioning, specific leaf area, and water use efficiency of vegetable amaranth (*Amaranthus* spp.) in response to drought stress. Sci Hortic 102(1):15–27

Mann JJ, Kyser GB, Barney JN, DiTomaso JM (2013) Assessment of aboveground and belowground vegetative fragments as propagules in the bioenergy crops *Arundo donax* and *Miscanthus × giganteus*. BioEnergy Res 6(2):688–698

Mantineo M, D'agosta GM, Copani V, Patanè C, Cosentino SL (2009) Biomass yield and energy balance of three perennial crops for energy use in the semi-arid Mediterranean environment. Field Crop Res 114(2):204–213

Papazoglou EG, Karantounias GA, Vemmos SN, Bouranis DL (2005) Photosynthesis and growth responses of giant reed (Arundo donax L.) to the heavy metals Cd and Ni. Environ Int 31(2):243–249

Parkash V, Sharma S, Aggarwal A (2011) Symbiotic and synergistic efficacy of endomycorrhizae with *Dendrocalamus strictus*. L Plant Soil Environ 57(10):447–452

Perdue RE (1958) *Arundo donax*, source of musical reeds and industrial cellulose. Econ Bot 12(4):368–404

Pilu R, Manca A, Landoni M, Barrière Y, Motto M (2013) *Arundo donax* as an energy crop: pros and cons of the utilization of this perennial plant. Maydica 58(2013):54–59

Rai MK (2001) Current advances in mycorrhization in microporopagation. Vitro Cell Dev-Pl 37(2):158–167

Shatalov AA, Pereira H (2002) Influence of stem morphology on pulp and paper properties of *Arundo donax* L. reed. Ind Crop Prod 15(1):77–83

Sims RE, Mabee W, Saddler JN, Taylor M (2010) An overview of second generation biofuel technologies. Bioresource Technol 101(6):1570–1580

Singh NV, Singh SK, Singh AK, Meshram DT, Suroshe SS, Mishra DC (2012) Arbuscular mycorrhizal fungi (AMF) induced hardening of micropropagated pomegranate (Punica *granatum* L.) plantlets. Sci Hortic 136:122–127

Smith SE, Read DJ (1996) Mycorrhizal symbiosis. Academic, London

Tauler M, Baraza E (2015) Improving the acclimatization and establishment of *Arundo donax* L. plantlets, a promising energy crop, using a mycorrhiza-based biofertilizer. Ind Crop Prod 66:299–304

Tawaraya K (2014) Use of arbuscular mycorrhizal fungal inocula for horticultural crop production. In: Mycorrhizal fungi: use in sustainable agriculture and land restoration. Springer, Berlin, pp 81–87

Varma A (ed) (2008) Mycorrhiza: state of the art, genetics and molecular biology, eco-function, biotechnology, eco-physiology, structure and systematics. Springer, New York

Vestberg M, Estaún V (1994) Micropropagated plants, an opportunity to positively manage mycorrhizal activities. In: Impact of arbuscular mycorrhizas on sustainable agriculture and natural ecosystems. Birkhäuser, Basel, pp 217–226

Xue S, Kalinina O, Lewandowski I (2015) Present and future options for *Miscanthus* propagation and establishment. Renew Sust Energ Rev 49:1233–1246

Chapter 20
Biomass Yield and N Uptake in Tall Fescue and Reed Canary Grass Depending on N and PK Fertilization on Two Marginal Sites in Denmark

S.U. Larsen, U. Jørgensen, and P.E. Lærke

Introduction

Agriculture faces a complicated challenge: On the one hand, there is need for a larger biomass production for a growing population and for displacing the use of fossil resources and, on the other hand, the production needs to be more sustainable with minimal negative environmental impact. Often, intensive agricultural production with large biomass yields have been associated with negative side effects such as increased leaching of nutrients and increased use of pesticides. However, various perennial crops such as grass species and willow have demonstrated the ability to produce higher yields and a better energy balance than cereals such as winter wheat and, at the same time, to have a lower leaching of nitrate (Pugesgaard et al. 2014). Therefore, a biomass production based on perennial grasses could contribute to the double-edged challenge.

It is estimated that there are about 10,000 grass species worldwide (Hubbard 1984). Hence, there is a considerable chance to find grass species adapted to a specific set of conditions. For temperate climates as found in Denmark combined with relatively moist sites, species such as tall fescue (TF) (*Festuca arundinacea* Schreb.) and reed canary grass (RCG) (*Phalaris arundinacea* L.) may be candidate crops. The

The original version of this chapter was revised. An erratum to this chapter can be found at DOI 10.1007/978-3-319-44530-4_25

S.U. Larsen (✉)
Department of Agroecology, Aarhus University, Blicher Allé 20, DK-8830 Tjele, Denmark

Division AgroTech, Danish Technological Institute,
Agro Food Park 15, DK-8200 Aarhus, Denmark
e-mail: slar@teknologisk.dk

U. Jørgensen • P.E. Lærke
Department of Agroecology, Aarhus University,
Blichers Allé 20, Postboks 50, DK-8830 Tjele, Denmark

S. Barth et al. (eds.), *Perennial Biomass Crops for a Resource-Constrained World*, DOI 10.1007/978-3-319-44530-4_20

biomass yield of grass species is highly dependent on the availability of N, often with a characteristic yield response curve envisaging the declining effect per kg of N at higher fertilization rates (e.g., Morrison et al. 1980). However, K is also needed in large amounts and is closely related to N nutrition (Kayser and Isselstein 2005), and K availability may be a limiting factor for yield within certain conditions (Larsen et al. 2012). In order to give recommendations regarding the productivity and sustainability of the grass species TF and RCG, it is necessary to know the relationship between fertilization and the biomass yield. In particular, it is relevant to know which quantities of N can be applied with limited and acceptable loss to the environment.

We conducted two field trials with TF and RCG. The aim was to study the effect of fertilization with N and PK on biomass yield and N uptake in the two species during the first 3 harvest years when grown on semi-marginal land, i.e., on coarse sandy soil with relatively high humidity.

Materials and Methods

Two field trials were established in 2011 on two sites near Sunds in Central Jutland, Denmark. The distance between the sites was approx. 4 km. The soil type was coarse sandy soil on both sites (75.6 % coarse sand, 14.8 % fine sand, 4.1 % silt, 4.7 % clay, and 6 % organic matter and corresponding values of 75.7 % coarse sand, 18.1 % fine sand, 2.6 % silt, 3.1 % clay, 5.4 % organic matter on the other site). The soil had relatively high humidity on both sites and, therefore, the sites had limited suitability for cereal production.

On both sites, a series of experimental plots were established on 9th of April 2011 by sowing either 35 kg ha^{-1} of TF cv. Barolex or 20 kg ha^{-1} of RCG cv. Bamse, both species being sown in pure stand and without a cover crop. In the year of establishment, all plots were treated identically, i.e., fertilized with 50, 20, and 105 kg ha^{-1} of N, P, and K, respectively, and two cuts were taken at the end of June and in September, respectively. No measurements of yield were performed in the establishment year.

Fertilization treatments were initiated in spring 2012, and the same treatments were performed each year from 2012 to 2014. Both trials were designed according to a split-split-plot design with four replicate blocks and three experimental factors: (1) The whole-plot factor was grass species with two levels of either TF or RCG. (2) The sub-plot factor was PK fertilization with two levels of either no PK fertilization or fertilization with 25 kg ha^{-1} y^{-1} of P and 257 kg ha^{-1} y^{-1} of K. (3) The sub-sub-plot factor was N fertilization with four rates of 0, 150, 300, or 450 kg ha^{-1} y^{-1} of N. Net plot size was 1.5 × 12 m (18 m^2) and unfertilized plots of the same size were located between net plots.

N was applied as NS fertilizer (24 % N, 6 % S) with a distribution of 47, 33, and 20 % of the N quantity applied prior to first, second, and third cut, respectively. PK fertilizer (4 % P, 21 % K, 7 % S, 1.2 % Mg) was applied prior to first cut and KCl fertilizer (49 % K) was applied prior to second and third cut. Hence, the full annual

P quantity was applied prior to first cut including micro nutrients whereas the K quantity was distributed as 51 %, 29 %, and 20 % to the three cuts, respectively.

Yield was measured at three cuts each year, and harvest was always performed on the same day in both trials. Harvest dates were 5th of June, 8th of August, and 24th of October 2012, 10th of June, 6th of August, and 7th of October 2013, and 4th of June, 14th of August, and 13th of October 2014. For each treatment combination, a sample was taken from all replicate blocks and mixed to one sample which was analyzed for DM content and N content in DM. In 2012, samples were analyzed for crude protein content and N content was calculated by dividing crude protein content by 6.25. In 2013 and 2014, total N content was analyzed directly.

Data were analyzed using SAS software (SAS Institute 2008). For each trial and cut, fresh matter yield was analyzed by analysis of variance taking into account the split-split-plot structure of the design. Least square mean values (LS means) for each cut and treatment combination were then used to calculate the DM yield and N removal (i.e., the quantity of N harvested with the crop) per cut. Subsequently, the total annual DM yield and N removal per year and treatment combination was calculated by summarizing yields from each cut. The resulting values per trial, year and treatment combination were subsequently used for further analyses of treatment and year effects.

For each individual year, data were analyzed using a model with either annual DM yield or N content or N removal as response variable and with the three experimental factors and their interactions as explanatory variables as well as field trial site as a systematic factor. The analyses were made both with N fertilization as class variable and as numerical variable with the latter providing response curves. For DM yield, an additional analysis was made across all 3 years with N fertilization as class variable and where the model was extended by the year factor as a class variable and with all possible interactions between the three treatment factors and year. For N removal, an additional analysis was performed across years with N quantity as numerical variable and grass species, PK fertilization and year as class variables and with all possible interactions between the three treatment factors and year. The analyses across years took into account that repeated measurements were performed over time. All models were reduced successively until only significant factors and interactions remained.

Results and Discussion

Biomass yield: Results from analyses with N fertilization as class variable are shown in Table 20.1, and results from analyses with N fertilization as numerical variable are shown in Figs. 20.1 and 20.2. DM yield increased significantly with increasing N fertilization in all 3 years (Table 20.1) and similarly in both grass species (Fig. 20.1). In accordance with other studies with N fertilization (e.g., Morrison et al. 1980), the yield response per kg N applied decreased gradually at high N rates (Fig. 20.1). DM yield was also significantly increased by PK fertilization in all 3 years and similarly in both species (Table 20.1). However, the effect of PK increased

Table 20.1 DM yield, N content, and N removal with the crop during 3 years in fertilization trials with tall fescue and reed canary grass

Grass species	Fertilization: PK fertilization	Fertilization: N fertilization (kg ha^{-1} y^{-1})	Yield (Mg ha^{-1} y^{-1} DM) 2012	2013	2014	2012–2014	N content (% in DM) 2012	2013	2014	N removal with crop (kg ha^{-1} y^{-1}) 2012	2013	2014
Tall fescue	–	0	3.2	2.4	3.4	3.0	1.6	1.5	1.4	53	37	48
		150	9.9	7.3	7.5	8.2	1.7	1.9	1.9	169	138	142
		300	13.0	8.4	7.7	9.7	2.0	2.6	2.7	259	214	206
		450	15.2	7.5	7.3	10.0	2.1	3.0	3.0	312	223	220
	+	0	3.3	2.8	3.4	3.2	1.9	1.6	1.3	62	43	44
		150	10.9	9.1	9.8	9.9	1.8	1.7	1.5	192	155	153
		300	15.1	11.9	13.1	13.4	1.9	2.2	1.9	280	263	249
		450	19.3	12.1	14.2	15.2	1.7	2.6	2.2	326	310	318
Reed canary grass	–	0	4.5	3.2	3.2	3.6	1.6	1.7	1.6	74	53	51
		150	11.5	7.6	5.2	8.1	1.7	2.1	2.6	194	162	133
		300	15.1	8.9	5.0	9.7	2.0	2.8	3.0	298	251	152
		450	16.3	8.7	4.8	9.9	2.3	3.2	3.2	382	275	150
	+	0	4.7	3.4	3.3	3.8	1.5	1.8	1.4	73	60	47
		150	12.3	9.7	9.9	10.6	1.6	1.7	1.5	195	162	144
		300	16.0	13.3	13.9	14.4	1.9	2.2	1.9	296	295	256
		450	17.3	14.4	14.3	15.3	2.4	2.5	2.4	409	352	340
Mean value, PK fertilization												
	÷		11.1	6.7	5.5	7.8	1.9	2.3	2.4	218	169	138
	+		12.3	9.6	10.2	10.6	1.8	2.0	1.8	229	205	194
Mean value, grass species												
Tall fescue			11.2	7.7	8.3	9.0	1.8	2.1	2.0	207	173	173
Reed canary grass			12.2	8.6	7.4	9.4	1.9	2.2	2.2	240	201	159

LSD, N fertilization			*1.2*	*0.9*	*1.0*	*0.7*	*0.2*	*0.2*	*0.2*	*23*	*15*	*14*
LSD, PK fertilization			*0.8*	*0.7*	*0.7*	*0.5*	*ns*	*0.1*	*0.1*	*ns*	*10*	*10*
LSD, grass species			*0.8*	*0.7*	*0.7*	*0.5*	*ns*	*0.1*	*0.1*	*16*	*10*	*10*
LSD, interaction N×PK			*ns*[a]	*1.3*	*1.4*	*1.0*	*ns*	*0.2*	*0.3*	*ns*	*21*	*20*
LSD, interaction N×grass species			*ns*	*ns*	*ns*	*ns*	*0.3*	*ns*	*ns*	*32*	*ns*	*ns*
LSD, interaction PK×grass species			*ns*	*ns*	*1.0*	*ns*	*ns*	*ns*	*0.2*	*ns*	*ns*	*14*
LSD, interaction N×PK×grass species			*ns*	*ns*	*ns*	*ns*	*ns*	*ns*	*ns*	*ns*	*ns*	*28*

LS mean values based on two trials

[a]*ns* indicates that the there was no significant effect ($P>0.05$)

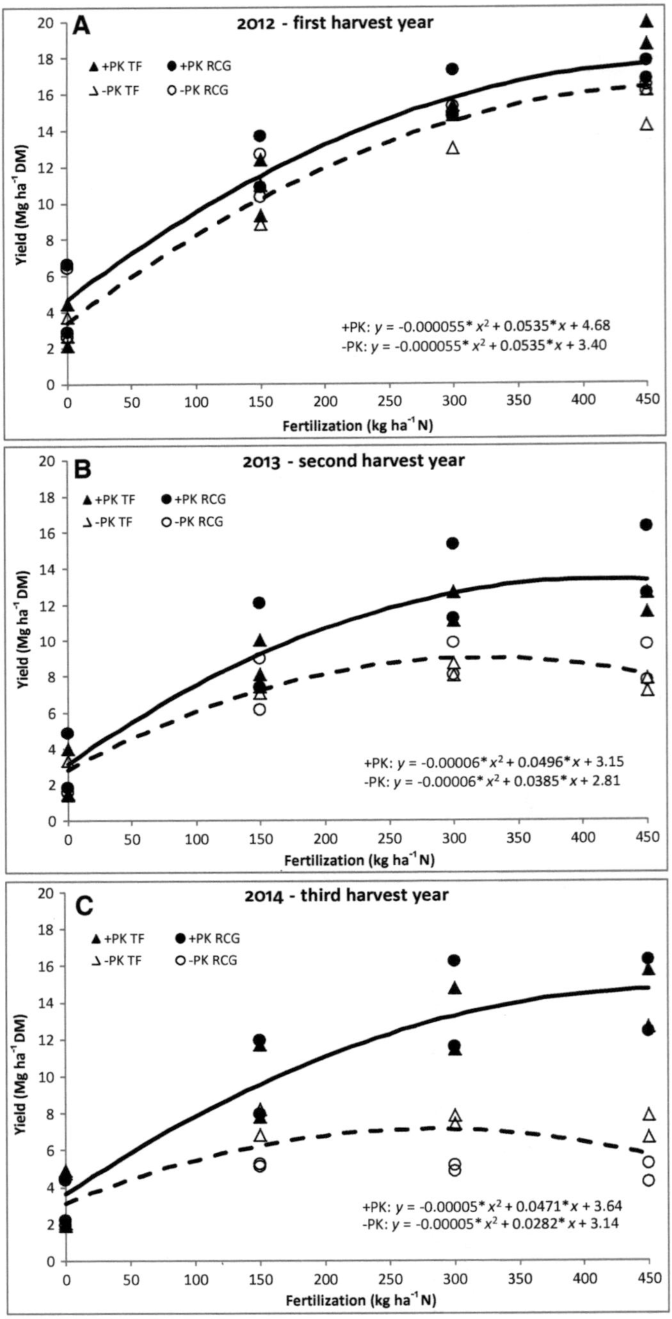

Fig. 20.1 DM yield of tall fescue (TF) and reed canary grass (RCF) at various levels of *N* fertilization and with (*full line*) and without (*dashed line*) fertilization with PK. Results from two field trials and each symbol represents the mean value from one treatment combination and one trial

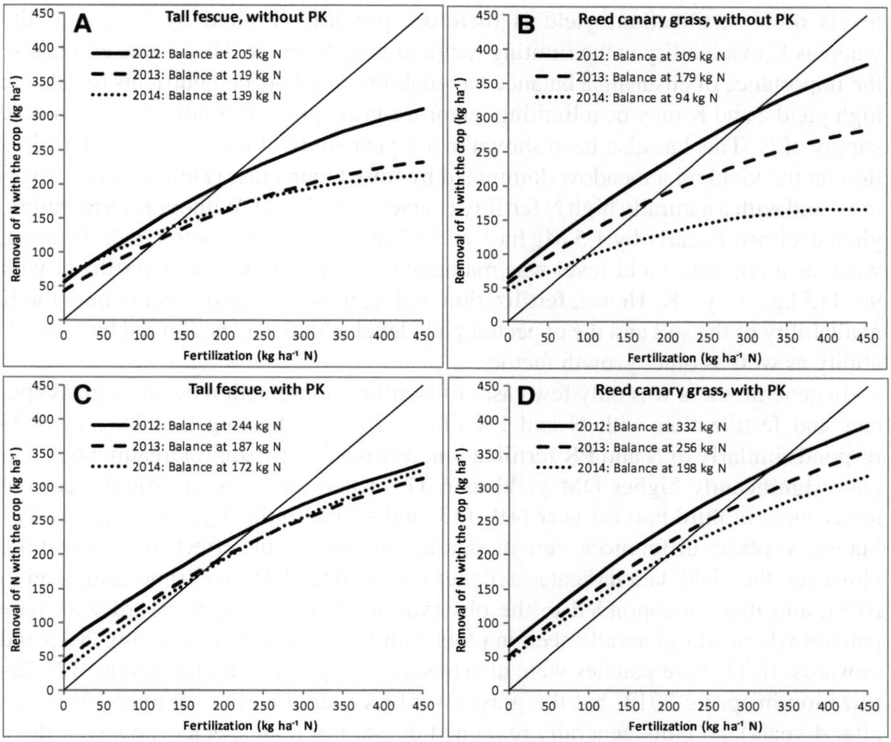

Fig. 20.2 Removal of *N* with the harvested crop depending on *N* fertilization. Predicted relationships shown for 3 harvest years for tall fescue and reed canary grass with and without fertilization with PK. Based on data from two trials. Balance point indicates the *N* fertilization rate at which *N* removal equals *N* applications

over time and, moreover, in second and third harvest year PK fertilization also interacted significantly with N fertilization (Table 20.1). Thus, the effect of N fertilization decreased over time when combined with no PK fertilization; in third harvest year, fertilization with 450 kg ha^{-1} N resulted in a very limited yield increase of 2.6 Mg ha^{-1} DM (82 %) across grass species (no significant effect of species, $P=0.177$) when not combined with PK fertilization compared to an effect of 11.1 Mg ha^{-1} DM (305 %) when combined with PK fertilization (Fig. 20.1).

The developing interaction between N fertilization and PK fertilization over time is most likely a consequence of a gradual depletion of particularly available K in the soil. This was supported by soil analyses in spring 2014, showing significantly lower K availability in plots without PK fertilization than in plots with PK fertilization whereas there was no significant difference in P availability (data not shown).

These results are fully in line with the well-known Liebig's law of the minimum, stating that growth is controlled by the scarcest resource (Liebig 1862). At low levels of N fertilization, yield is therefore primarily limited by N availability whereas K availability is the limiting factor at high N levels. The results emphasize the importance of ensuring a balanced availability of all needed nutrients to achieve high yields, and K may be a limiting factor for grass yield in conditions with ample supply of N. This has also been shown in a 4-year study of the effect of K fertilization on the yield on a meadow dominated by tufted hair-grass (*Deschampsia cespitosa*) and with a naturally high N fertility (Larsen et al. 2012). Without K fertilization, yield declined linearly by 1.5 Mg ha^{-1} y^{-1} DM over 4 years to approx. half the level, whereas a constant yield level was maintained when the grass was fertilized with 90–115 kg ha^{-1} y^{-1} K. Hence, fertilization with K must take into account both the K availability of the soil and the expected yield level which may be limited by N availability as well as other growth factors.

In general, there was only few cases of significant interaction between grass species and fertilization with N and PK (Table 20.1), indicating that TF and RCG respond similarly to N and PK fertilization. Across all fertilization treatments, RCG gave significantly higher DM yield than TF in first and second growth year but lower yield in third harvest year (+9, +13, and ÷10 % in the 3 years, respectively), but grass species differences were moderate compared to the effects of fertilization. However, the yield data indicates a higher persistence of TF over time compared to RCG, and this corresponded to the observation of a more open sward with bare patches which was generally seen in plots with RCG already from spring 2013 and onwards. In TF, bare patches were also observed in plots with high levels of N fertilization in spring 2013, but the grass sward recovered and in summer 2015 (i.e., after 4 years) TF plots generally remained dense. For both species, however, there was a decrease in maximum DM yield from first harvest year to second and third harvest year. This may be due to year-to-year variation in growth conditions or, alternatively or additionally, due to a general yield decline over time in perennial grass.

In general, the N content of the biomass increased significantly with increasing N fertilization (Table 20.1). Conversely, fertilization with PK resulted in significantly lower N content, possibly due to a higher biomass yield and, hence, a dilution of the N in the biomass. Moreover, the results indicate a higher N content in RCG than in TF which was significant in second and third harvest year.

N *removal*: The quantities of N removed with the biomass were significantly affected by grass species, N fertilization and PK fertilization (significant in second and third year), but there were also some interactions between the factors (Table 20.1, Fig. 20.2). The results also indicate a general reduction in N removal over the 3 years, particularly for RCG (Table 20.1). The predicted relationships between N fertilization and N removal clearly illustrate that more N is removed than applied at the lowest N fertilization levels (Fig. 20.2), and N is probably depleted from the pool of soil N. Conversely, at higher N fertilization levels there is a surplus of N which indicates a potential risk of leaching of nitrate from the root zone. If N fertilization levels are kept below the balancing point where N fertilization equals N removal, it is possible to ensure a sustainable production of grass biomass with very limited risk of loss of N. The predicted N balancing points, however, depend largely on PK fertilization but also on grass species and year. In combination with PK fertilization, the balancing points were 244, 187, and 172 kg ha^{-1} y^{-1} in the 3 years for TF and 332, 256, and 198 kg ha^{-1} y^{-1} for RCG. Without PK fertilization, the balancing point decreased to considerably lower levels over time, particularly for RCG (Fig. 20.2). The results indicate that the N balancing point may decrease over time, and this may in part be related to an initial buildup of root biomass. Provided that the availability of other nutrients such as K is ensured, it may still be possible to apply more than 150 kg ha^{-1} y^{-1} N without building up an N surplus and, therefore, with limited risk of nitrate leaching.

The use of mineral N fertilizer may affect the carbon footprint of grass crops (Haugaard-Nielsen et al. 2016). Since both fertilizer quantity and biomass yield are important factors for the greenhouse gas balance for a biomass production system, the results of the present study may be useful in the modeling of greenhouse gas costs and benefits of producing biomass by the grass species TF and RCG. In this way, the results may aid in selecting the most environmentally appropriate land use.

Conclusions

The results show that relatively high DM yields can be obtained over the first 3 harvest years in both TF and RCG on semi-marginal, moist sandy soil, provided sufficient and balanced availability of N and K as well as other nutrients. Moreover, the N balancing points show that both species may be fertilized with considerable quantities of N with limited risk of nitrate leaching. Thus, these species may provide a rather sustainable biomass production in conditions poorly suited for, for example, cereal production.

Acknowledgements The study was funded by the European Agricultural Fund for Rural Development and the Danish fund for agricultural production levy ("Promilleafgiftsfonden for landbrug").

References

Haugaard-Nielsen H, Lachouani P, Knudsen MT, Ambus P, Boelt B, Gislum R (2016) Productivity and carbon footprint of perennial grass–forage legume intercropping strategies with high or low nitrogen fertilizer input. Sci Total Environ 541:1339–1347

Hubbard CE (1984) Grasses, 3rd edn. Penguin, St. Ives, 476 pp

Kayser M, Isselstein J (2005) Potasium cycling and losses in grassland systems: a review. Grass For Sci 60(3):213–224

Larsen SU, Nielsen L, Hald AB, Lærke PE (2012) Græs på engarealer [Grass on meadows]. In: Pedersen JB (ed) Oversigten over Landsforsøg 2012 [Summary of field trials within Danish Agricultural Service 2012]. pp 214–219. Knowledge Centre for Agriculture, Aarhus, Denmark

Liebig J (1862) Die Naturgesetze des Feldbaues. Vieweg, Braunschweig, Germany

Morrison J, Jackson MV, Sparrow PE (1980) The response of perennial ryegrass to fertilizer nitrogen in relation to climate and soil. Technical Report 27:71, Grassland Research Institute, Hurley. 71p

Pugesgaard S, Schelde K, Larsen SU, Lærke PE, Jørgensen U (2014) Comparing annual and perennial crops for bioenergy production—influence on nitrate leaching and energy balance. GCB Bioenergy 7(5):1136–1149

SAS Institute (2008) SAS Institute SAS/STAT. Release 9.2. SAS Institute, Cary, NC

Chapter 21
Energy Balance of Cardoon (*Cynara cardunculus* L.) Cultivation and Pyrolysis

P. Bartocci, G.i Bidini, F. Cotana, and F. Fantozzi

Introduction

Cardoon is a species which belongs to the *Asteraceae* Family (*Compositae*), along with artichoke, sunflower, safflower, and Jerusalem artichoke. It is a perennial crop which sprouts in autumn. *Cynara cardunculus* L. is a promising crop for the production of biomass in Mediterranean areas, its main advantages are the possibility to grow in dry regimes and the low moisture content, because it dries at the end of the growing season reaching a moisture that is around 10 %; so when it is harvested the crop can be immediately used as a fuel. This crop has undergone an important work of selection and improvement carried out at the Politechnical University of Madrid (Curt et al. (2002), Gominho et al. (2011), Fernández et al. (2006)). In literature, few works are available on the production of energy through cardoon pyrolysis. Encinar et al. (2000) performed isothermal pyrolysis tests. Mass and energy balances of the pyrolysis of other perennial crops have been analyzed by McKendry (2002) and Gaunt and Lehman (2008). Energy Return on Investment (EROI) was never calculated for a pyrolysis plant. Most energy analysis use EROI (Energy Return On Investment) as a final index to focus on the efficiency of the proposed technology. Weißbach et al. (2013) analyze EROIs of electricity generating power plants based on renewable (wind, solar photovoltaic, biogas, and hydro) and nonrenewable energies (e.g., coal). EROI depends on the technology used for the energy conversion, pyrolysis process is not analyzed in the study. The only study that takes into account EROI on biomass CHP is that of Buonocuore et al. (2012) that obtained a value of 3.5 for a biogas CHP plant. Cherubini and Ulgiati (2010) analyze EROI of a bio refinery system on a life cycle basis. Sastre et al. (2014) analyze energy yields of electricity production from winter cereals. They consider in their analysis

P. Bartocci (✉) • G. Bidini • F. Cotana • F. Fantozzi
Department of Engineering, University of Perugia, Perugia, Italy
e-mail: bartocci@crbnet.it

S. Barth et al. (eds.), *Perennial Biomass Crops for a Resource-Constrained World*, DOI 10.1007/978-3-319-44530-4_21

also the amount of machinery used to perform specific agricultural operations, referring it to the functional unit of the life cycle. Goe and Gaustad (2014), focus their study on solar photovoltaics and consider as an important energy performance index EPBT (Energy Payback Time), that is an input in EROI calculation.

University of Perugia has managed about 1 ha of experimental fields, cultivated with: cardoon (*Cynara cardunculus* L.), giant reed (*Arundo donax* L.), Topinambour (*Helianthus tuberosus* L.), and Miscanthus (*Miscanthus giganteus* L.). The plantations were realized in 2006 and were continuously monitored. Data were used in previous studies to asses suitable herbaceous energy crops for Umbria region (Italy) cultivated areas (G. Bidini et al. (2007)). Besides University of Perugia has developed an Integrated Pyrolysis Regenerated Plant (IPRP) (see D'Alessandro et al. 2013), based on a gas turbine fuelled with pyrogas produced by the slow pyrolysis of biomass in an externally heated rotary kiln pyrolyser (D'Alessandro et al. (2011), Fantozzi et al. (2009)).

For these reasons the aim of the work is to combine experimental fields monitoring with pyrolysis tests in a lab scale batch reactor already described in Paethanom et al. (2013) and evaluate also nonisothermal conditions. In fact no data are available on cardoon nonisothermal pyroysis at the moment. The results of the batch pyrolysis tests are used to calculate the yields of electricity, heat, and biochar of an IPRP (Integrated Pyrolysis Regenerated Plant) fuelled with *Cynara cardunculus*. These data finally allow to calculate the Energy Return on Investment (EROI) of the technology, this is another innovative aspect of the presented work.

Materials and Methods

Experimental Fields

As previously mentioned, 1 ha experimental field cultivated with cardoon was realized in 2006 in the experimental farm (see Fig. 21.1), actually owned by the Foundation for Agricultural Instruction of the University of Perugia. The seeds were provided by prof. J. Fernandez, Universidad Politécnica de Madrid. The planting density was of one plant per square meter. At the end of the life cycle the energy crop was removed in 2015 by plowing and two times weeding (using a systemic herbicide).

Main cultivation processes have been: spreading organic and inorganic fertilizers, plowing, planting, weeding, irrigation, and harvest. While the above mentioned operations were performed only one time at the beginning of the plantation, harvest was performed every year.

Soil has a silty-clay-loam texture. The growth season of *Cynara cardunculus* L. is between March and the beginning of July. The plant sprouts in September–October, in November produces a winter rosette and stem grows between April and May, full blossom in June while the fruit ripens in July and the plant completely dries in August.

Fig. 21.1 Experimental fields, University of Perugia, August 2013

Pyrolysis Plant Layout and Experimental Methodology

Figure 21.2 shows the batch pyrolysis plant layout. Pyrolysis reactor consists of a 5 L cylindrical reactor. The upper part of the reactor hosts the connections to the tar sampling line, the sensors and the inert gas (nitrogen) inlet connection. So the connections of the batch reactor are the following: 1 thermocouple to measure reactor internal atmosphere temperature; 1 thermocouple to measure biomass temperature; 1 thermocouple connected to P.I.D. (Watlow EZ-ZONE PM) to control the heaters power; 1 pipe with fins linked to the pressure sensor, hosting also the inlet of inert gas (Nitrogen); 1 pipe for the connection to the tar sampling line. The pressure sensor used is an OMET TRIP/ZPT-510; while the flowmeter is a Tecora G1.6. Two types of pyrolysis tests have been performed: isothermal and nonisothermal. In isothermal experiments the temperature of the reactor is set to 600 °C and then biomass is inserted with a flow rate of about 3 g per minute. Nitrogen is injected only before the start of the test, to avoid the presence of oxygen inside the reactor. In the nonisothermal tests, 100 g of biomass are inserted in the reactor before it is closed and then heated up. So reactor and biomass reach gradually the final pyrolysis temperature of 600 °C. The heating rate is about 20 °C/min. Final pyrolysis temperature was chosen based on previous TGA (Thermogravimetric Analysis) experiments done on biomasses and wastes (Slopiecka et al. (2012), Bidini et al. (2015), Buratti et al. (2015)).

Volatiles coming out from the reactor condensed in two baths containing three Drechsel bottles each. One is the warm bath (see number 8 in Fig. 21.2) and it is kept at a temperature of about 35 °C, the other is the cold bath (number 9 in Fig. 21.2) and is kept at the temperature of –20 °C, maintained using an eutectic system, based on ice and NaCl. Drechsel bottles are filled with isopropanol, which adsorbs tars. At the end of the pyrolysis test the isopropanol-tar mixture is collected in a 250 ml vessel and heated, to evaporate isopropanol, as reported in CEN/TS

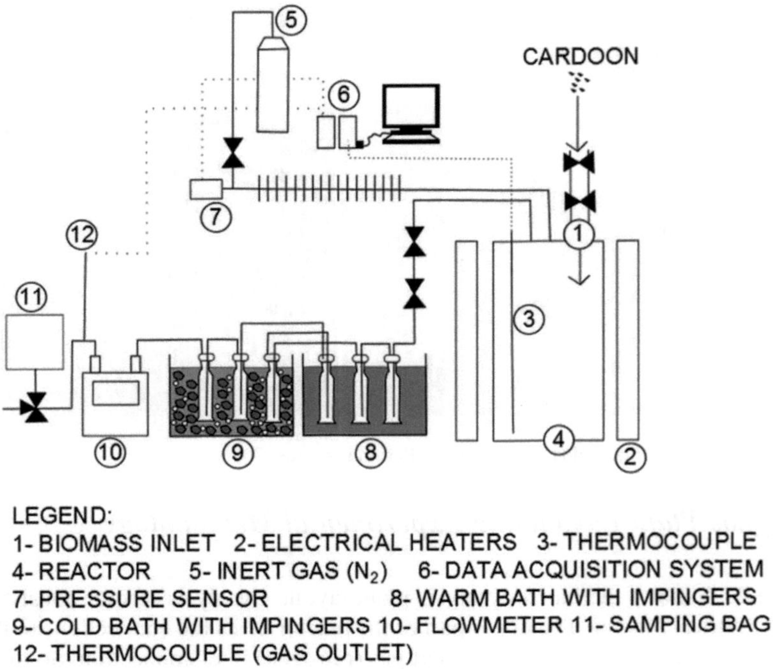

Fig. 21.2 Batch pyrolysis plant layout

15439:2006. Rotary evaporator working pressure is set to 100 mbar while temperature is set to 55 °C. Tar is weighted in a balance with precision equal to 1 mg. Evaporation is stopped when the time passed between one drop and another in the condenser of the evaporator is about 4 s. Then another evaporation step is performed for about 45 min, after that period the vessel is dried for 2 h. Pyrolysis tests were performed in triplicates. Pyrolysis gas was analyzed using the gas chromatograph Varian 490-GC PRO Micro-GC, to evaluate its composition and then calculating the heating value.

Biomass Characterization

Biomass proximate analysis and ultimate analysis were performed in the Biomass Research Centre Laboratory. Proximate analysis was performed according to the norms UNI EN 14774-2:2010, UNI EN 14775:2010 and UNI 15148:2009; ultimate analysis was performed according to the norm UNI EN 15104:2011, calorimetry was performed according to the norm UNI EN 14918:2010. Cardoon seeds oil content was determined through Soxhlet extraction using hexane as solvent; extraction sample was represented by 20 g of mashed cardoon seeds. Figure 21.3 shows the

Fig. 21.3 Cardoon oil extraction process and Higher Heating Value measurement

sample of seeds (Fig. 21.3a), that is mashed in a mortar (Fig. 21.3b), then it undergoes Soxhlet extraction with hexane (Fig. 21.3c), oil is separated by the solvent through a rotary evaporator (Fig. 21.3d), then it is inserted in a Mahler bomb to measure higher heating value (Fig. 21.3e).

Oil production from cardoon is very important; in fact this can be used for example for biodiesel production (Curt et al. (2002)) or can be used for gas scrubbing (Paethanom et al. 2013, Phuphuakrat et al. 2011, Paethanom and Yoshikawa 2012, Paethanom et al. 2012). The cake obtained from oil extraction can be burned to produce heat or may be used as a fuel in CHP plants (Fantozzi et al. 2010). Characterization analysis for biomass, pyrolysis products and oils were also performed in triplicates.

Theory and Calculations

Mass Balances

The consumption of diesel fuel during cultivation is reported in Table 21.1. Other materials employed during implantation phase are urea (50 kg/ha), seeds (5 kg/ha), and herbicide (0.2 kg/ha). During production years, that is the years following to

Table 21.1 Fuel consumption during cultivation of cardoon (referred to 1 ha of surface cultivated in the experimental fields of the Foundation for Agricultural Instruction of the University of Perugia, described in this chapter)

Process	Fuel input (l)
Implantation phase and plantation	
Solid manure loading	10
Fertilizing	14
Plowing	90
Sowing	10
Irrigating	50
Total	174
Production phase	
Harvesting	54
Transport	20
Total	74
Plantation removal phase	
Plowing	90
Herbicide application	14
Total	104

implantation, only harvest and transport of biomass are performed. The consumption of diesel fuel during implantation and plantation removal phases are also considered.

Cultivation mass balances were determined taking notes of the materials consumed and of biomass yield per hectare. Pyrolysis mass balances were realized measuring char left inside the reactor; pyrogas yield was calculated considering data recorded by the flowmeter and pyrogas density (calculated based on the exiting temperature and pressure, and gas composition, measured with a microGC). Tar production was derived by difference.

Energy Balances

Energy required for biomass cultivation is calculated considering diesel consumption, shown in Table 21.1. The total energy consumption to grow cardoon is about 8300 MJ/ha, that is lower than the amount obtained by Fernandez et al. (1997); this is due to the fact that cultivation in the experimental fields was very extensive and low quantities of energy and chemical inputs were employed. On the other hand, also biomass yields are lower in this case, if compared with other studies. In order to consider also the energy needed to produce agricultural machinery, an average weight and composition of the equipment used for cultivation was assumed, considering data shown in Sastre et al. (2014). The same procedure was considered for the IPRP plant. Dealing with pyrolysis energy balance, an important term is the heat required for pyrolysis process (Q_{pyr}), defined by D. E. Daugaard and R. C. Brown (2003) as "the energy required to raise biomass from room temperature to the reaction temperature and convert the solid biomass into the reaction products of gas, liquids and char". This was calculated with Eq. (21.1).

$$Q_{pyr} = H_c\left(T_{pyr}\right) + H_t\left(T_{pyr}\right) + H_g\left(T_{pyr}\right) - H_{B,0}\left(T_{298,15K}\right) \qquad (21.1)$$

Biomass composition in terms of carbon, hydrogen, nitrogen, and oxygen was obtained with ultimate analysis; biomass standard enthalpy of formation ($H_{B,0}$) was found in Baratieri et al. (2008). The enthalpies of char, tar and water are calculated according to Yang et al. (2013). Tar specific heat and standard enthalpy of formation have been calculated using formulas and methodology proposed by Yang et al. (2013). Experimental data on pyrolysis energy balance, obtained with the batch reactor, were used to model the IPRP (Integrated Pyrolysis Regenerated Plant) calculating the amount of electricity, heat and biochar produced. Also the weight of different materials composing the plant was measured and estimated, to obtain the energy necessary to realize the plant. The results of energy balance are shown in Fig. 21.4 by a Sankey diagram. Data produced with the energy balance were used to calculate EROI (Enery Return On Investment), through Eq. (21.2).

$$EROI = \left(OE_{el} + OE_{th}\right) / \left(PE_{cult} + PE_{tr} + PE_{conv} + PE_{mat}\right) \qquad (21.2)$$

Where OE_{el} is the electricity produced from pyrolysis gas, tar, and part of the produced biochar, OE_{th} is thermal energy produced by the IPRP plant. PE_{cult} is the energy required to cultivate cardoon, PE_{tr} is the energy necessary to transport cardoon to the IPRP plant, PE_{conv} is the energy necessary to pyrolyze cardoon and PE_{mat} is the energy necessary to produce and assembly the IPRP plant and the agricultural machinery (referred to the amount of machinery used to transform/produce the

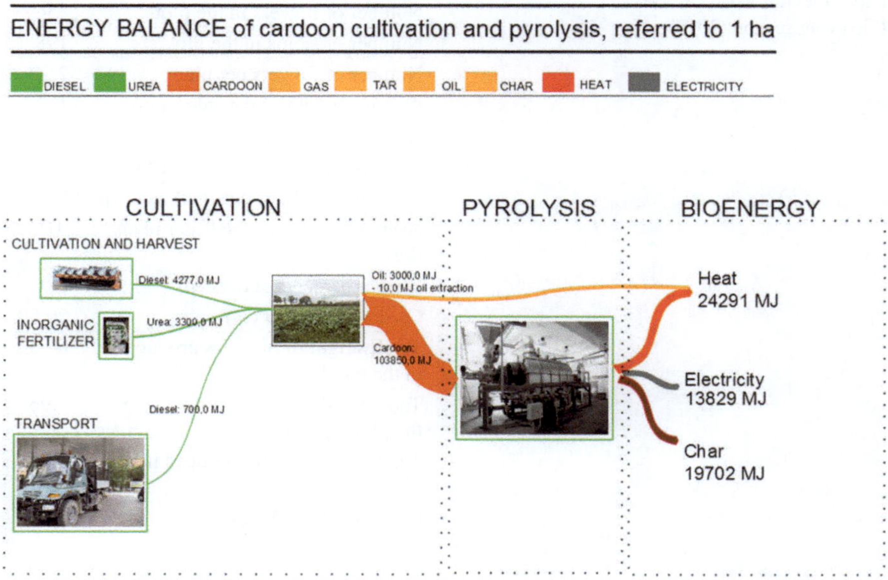

Fig. 21.4 Cardoon cultivation and pyrolysis energy balance

biomass obtained from 1 ha). All the terms contained in the EROI equation are referred to the functional unit of 1 ha.

Results and Discussion

Cultivation of Cardoon on Experimental Fields

The results of cardoon experimental fields monitoring activities are shown on Table 21.2. All measured data are referred to a plot of 100 m^2. The data presented are slightly lower than those obtained during other years of monitoring activity on the same field and with similar cultivation practices. This was due to the competition of weeds and the absence of fertilization and other treatments.

From the information proposed on Table 21.2 it can be calculated that 1 ha produced 463 kg of seeds, 97 kg of oil and about 6.7 t of dry biomass, considering also the stems. Comparing these data with Fernandez et al. (2006) and Fernandez et al. (1997), biomass yield and seeds production is low. This is due to the extensive cultivation of cardoon, that was performed with reduced cultivation inputs, in both fertilizing and weeding. Oil yield has a good correspondence with literature and the HHV of cardoon oil measured in this study is higher than that reported by other studies.

Table 21.2 Results of field experimentation on cardoon (Growing season: 2013)

Parameter	Value
Number of flowers in the plot	454
Number of stems in the plot	198
Number of flowers per plant	2.29
Average height of the stem (cm)	177
Average diameter of the stem (cm)	1.76
Average number of seeds per flower	257
Average weight of entire plant (kg)	0.34
Average weight of seeds per plant (kg)	0.02
Weight percentage of seeds (%)	6.0
Cardoon oil yield (%)	20.90
Total weight of biomass produced in the plot (kg d.m./100 m^2)	67.32
Total weight of seeds produced in the plot (kg/100 m^2)	3.96
Total weight of oil produced in the plot (kg/100 m^2)	0.8
Cardoon Oil HHV (MJ/kg)	36.5

Characterization of Pyrolysis Products

Table 21.3 shows the results of proximate analysis, ultimate analysis and calorimetry, made on cardoon and its pyrolysis products (tar and char).

As shown in Table 21.3 both LHV and HHV of char are quite low, because of the high ashes content. The average gas composition is reported in Table 21.4.

Table 21.4 shows that gas heating value increases with the increase of the temperature.

Pyrolysis Mass Balance

Table 21.5 shows the mass balances for biomass pyrolysis tests. Nonisothermal pyrolysis produces a higher char yield and a lower production of tar and pyrogas; these results depend basically by the heating rate of the reaction. Literature data show that a higher heating rate induces a higher production of gas against a lower production of char. These differences become less evident with the increase of pyrolysis temperature. Table 21.5 shows also a higher tar production in the isothermal tests.

Comparing the results shown in Table 21.5 with the only publication available at the moment (Encinar et al. (2000)) on cardoon pyrolysis, the most evident difference is on tar production, this can be explained with different reactors configurations and with a residence time of volatiles inside it. The heat of pyrolysis (Q_{pyr}), calculated with Eq (21.1), is also reported in Table 21.5 and it is equal to 1.7 MJ/kg of pyrolyzed cardoon.

Cardoon Pyrolysis and Cultivation Energy Balance

At the end of the experimental tests, in field and in laboratory, mass and energy balance of the whole cardoon to energy conversion chain was calculated (see Fig. 21.4). Cultivation energy is derived from the use of fuel (see Table 21.1), fertilizer (urea) and herbicide. The primary energy corresponding to fuel, fertilizer and herbicide is taken from SimaPro software databases, version 7.2. Mass balances of pyrolysis products, referred to 1 ha, have been calculated. Nonisothermal conditions are considered because they are closer to the process carried out in industrial plants. The energy required for the production of the agricultural machines and IPRP plant used in the bioenergy chain is equal to 180 MJ/ha.

Figure 21.4 shows that the heat required to sustain the pyrolysis process (Qpy) may be provided by the combustion of part of the produced char and/or tar. Pyrogas LHV in this case is supposed to be about 13 MJ/kg, because it is difficult to obtain in a real plant a gas with the characteristics measured in a lab scale reactor. For the

Table 21.3 Cardoon biomass and pyrolysis products characterization

		Isothermal pyrolysis		Nonisothermal pyrolysis	
Material	Cardoon	Cardoon tar	Cardoon Char	Cardoon tar	Cardoon char
Proximate analysis (%w_db)					
Moisture (UNI 14774-2:2010)	8.12±0.3	/	5.01±0.1	/	5.62±0.3
Volatiles (UNI 14775:2010)	86.9±2.3	100±0.1	34.47±5.4	100±0.1	17.43±2.9
Ashes (UNI 15148: 2009)	6.18±0.9	/	32.71±0.27	/	26.53±0.5
Fixed Carbon	6.69±0.1	/	32.82±5.6	/	56.05±2.7
Ultimate analysis (%w_wb)					
Carbon (UNI 15104:2011)	39.7±0.2	51.6±2.0	52.3±0.9	52.06±1.3	62.7±0.9
Hydrogen (UNI 15104:2011)	5.67±0.2	7.6±0.6	1.51±0.09	7.24±0.7	2.87±0.04
Nitrogen (UNI 15104:2011)	2.68±0.1	0.7±0.1	1.63±0.7	0.48±0.1	1.39±0.5
Heating Value (kJ/kg)					
HHV (UNI 14918:2010)	15,242±232	31,432±431	18,778±312	29,761±371	23,381±413
LHV (UNI 14918:2010)	15,521±331	29,866±530	19,562±374	28,262±410	24,284±639

Table 21.4 Average gas composition and LHV

	Isothermal pyrolysis	Nonisothermal pyrolysis			
Final Pyrolysis Temperature (°C)	600	400	450	550	600
H_2 (%v)	33±1.6	22±2.2	34±2.3	42±1.8	46±2.1
CH_4 (%v)	15±2.0	20±2.9	17±2.3	10±1.1	3±0.8
CO (%v)	38±0.8	25±1.2	27±1.3	33±1.6	46±2.3
CO_2 (%v)	14±0.5	33±1.7	22±1.3	15±0.9	5±0.5
Total	100	100	100	100	100
LHV (kJ/kg)	15,484±333	11,301±212	14,287±308	15,045±291	16,097±434

electricity and heat production, data already available from the Integrated Pyrolysis Regenerated Plant (IPRP) realized by the University of Perugia were considered, see D'Alessandro et al. (2013) and Fantozzi et al. (2007). The energy produced per hectare in this case would be about 60.8 GJ/ha. This value is about 59 % of the input biomass energy content. This is even higher than that reported by Gaunt and Lehman (2008), that is equal to about 50 %, due to pyrolysis technology optimization. So the gross electric efficiency of the plant is about 13 %. Table 21.6 shows the calculation of EROI, that is an energy index useful to compare the bioenergy chain presented in this study with those calculated in other works.

Table 21.5 Mass balances of pyrolysis tests

Isothermal pyrolysis tests	
Char (%w)	21 ± 0.7
Tar (%w)	18 ± 2.8
Gas (%w)	54 ± 4.5
Condensed water (%w)	7 ± 0.3
Total (%w)	100
Nonisothermal pyrolysis tests	
Char (%w)	29 ± 1.2
Tar (%w)	13 ± 3.5
Gas (%w)	51 ± 3.6
Condensed water (%w)	7 ± 0.5
Total (%w)	100

Table 21.6 Mass and energy balances of cardoon cultivation and pyrolysis (referred to 1 ha, totals for each phase—cultivation, transport, pyrolysis—and totals for the employed materials are indicated in italic)

	Mass	Energy
Input		
Diesel fuel	102 l/ha	4.4 GJ/ha
Inorganic fertilizer (urea)	50 kg/ha	3.3 GJ/ha
Herbicide	0.13 kg/ha	0.6 GJ/ha
Cultivation phase total (PE_{cult})	*8.3 GJ/ha*	
Diesel fuel used in transport	20 l/ha	0.9 GJ/ha
Transport phase total (PE_{tr})		0.9 *GJ/ha*
Electricity required to run the plant	/	0.12GJ/ha
Electricity required to extract cardoon oil	/	0.01 GJ/ha
Pyrolysis phase total (PE_{conv})	/	*0.13 GJ/ha*
Agricultural machinery materials	9 kg/ha	0.15×10^{-3}GJ/ha
IPRP plant materials	16 kg/ha	0.3×10^{-3} GJ/ha
Materials total (PE_{mat})	15 kg/ha	*0.18×10^{-3}GJ/ha*
Output		
Cardoon oil	97 kg/ha	3 GJ/ha
Ashes	0.4 t/ha	/
Output electric energy (OE_{el})	/	13.8 GJ/ha
Output thermal energy (*OEth*)	/	44.0 GJ/ha
Output Total energy (OE_{TOTAL})	/	*60.8 GJ/ha*
$EROI_{CHP}$	/	6.5

There are not so many studies performing EROI calculation on biomass. Buonocore et al. (2012) obtained a value of 3.5 for biogas, which is lower than the one obtained in this study (i.e., 6.5). The value obtained in this study would increase with the increase of cardoon production per hectare. In this case the bioenergy chain could be very challenging respect to other biomass-to-energy chains. Biomass could be challenging, if compared with solar photovoltaics that in Buonocore et al. (2012) has an EROI of about 3.9, while wind power has a higher EROI, equal to 16. The limits of the analysis presented in this chapter are the following: no economic feasibility is considered in this case; tar combustion may present problems caused by the high tar viscosity.

Discussion

Cardoon is a promising perennial energy crop. The chapter presents a bioenergy chain based on cardoon cultivation and pyrolysis. Cardoon is a herbaceous crop, which most promising energetic uses are the production of bioethanol (see Cotana et al. (2015)) or thermal degradation. Containing important concentration of chloride in its ashes the conversion of cardoon biomass through pyrolysis can be a promising technique, avoiding dangerous emissions (such as chlorine dioxide or dioxins), see Angelini et al. (2009).

The chapter indicates that with a yield of 6.7 td.m./ha about 24,291 MJ of heat can be produced, 13,829 MJ of electricity and 811 kg of biochar. Biochar can be used also as a soil conditioner, and application rates from 10 to 100 t/ha can be considered (see Jeffery et al. (2011)). This can be used also as a carbon sink in soil gaining carbon credits to the farmer and pyrolysis plant owner. In this sense the technology developed by the University of Perugia appears to have several advantages such as: the supply of the energy required from pyrolysis process from waste heat and partial processing of pyrolysis outputs, an efficient cleaning system based on oil scrubbing (see Paethanom et al. (2013)), the polygeneration of electricity–heat–biochar, avoiding negative emissions due to chlorine presence in the feedstock.

Taking into account literature tests on cardoon cultivation, Angelini et al. (2009) mostly confirm the results obtained in this study. In fact during a long-term evaluation of cardoon in Italy it was noticed already after the sixth year of plantation a decrease of biomass productivity in the range of 10 t/ha. In this study the data were collected in the year 2013, that is the seventh year after crop establishment.

Dealing with the energy balance of cardoon, Angelini et al. (2009) consider an energy input per hectare comprised between 11 GJ/ha and 15.3 GJ/ha, while in this study an average of 8.3 GJ/ha is indicated in Table 21.6. But it has been already said that the inputs for fertilization and weeding in the case presented in this paper were very low and this implies lower energy inputs, but also lower biomass output. It should be taken into account that energy crops like miscanthus can reach even higher yields (see Xue et al. (2016), Lewandowski (2015), Iqbal et al. (2015)) and so perform better by a point of view of EROI (Energy Return on Investment). EROI of seven can be reached with higher yielding crops (see Gaunt and Lehman (2008)).

While several studies are available on cardoon cultivation, its use in pyrolysis tests is very reduced. So the data presented in this chapter can help bioenergy operators to develop chains on cardoon pyrolysis.

Biochar production and use can be promoted through the mechanism of Kyoto Protocol, being it a carbon sink. Clean Development Mechanism project can promote its use to generate Certified Emission Reductions to be sold on the carbon market. The use of controlled pyrolysis as a strategy to avoid emissions from crop residues and stabilize carbon and the principle that avoided emissions associated with changes in agricultural practice can be monetized is established under the small scale CDM methodology AMS-III.L (21), as reported in Gaunt and Lehman (2008). Besides slow pyrolysis process can be usefully introduced in agro-energy districts as reported in Fantozzi et al. (2014) and Manos et al. (2014).

Conclusions

The entire bioenergy chain of cardoon (from cultivation to energy production through pyrolysis process) is analyzed in this chapter. Cardoon is an interesting energy crop, because it has a good heating value and through on field drying moisture content can be reached up to 6 %. Besides from the harvested seeds about 20 % of oil could be extracted. This energy crop has undergone an important work of selection and improvement made mainly in Spain. The University of Perugia has realized experimental fields to test Cynara cardunculus varieties. Biomass and oil yields were respectively 6.7 t d.m/ha and 97 kg seed/ha. The harvested biomass was used for pyrolysis tests in a batch reactor. The data obtained have been used to calculate mass and energy balances for isothermal and nonisothermal pyrolysis. Results have shown that in isothermal conditions about 21 ± 0.7 %w char, 18 ± 2.8 %w tar, 54 ± 4.5 %w gas and 7 ± 0.3 %w condensed water are obtained; while in nonisothermal conditions about 29 ± 1.2 %w char, 13 ± 3.5 %w tar, 51 ± 3.6 %w gas and 7 ± 0.5 %w condensed water are obtained. Using part of the produced tar and char to supply the energy required for pyrolysis, about 59 % of the energy contained in cardoon can be turned into thermal and electrical energy through cogeneration technology in an Integrated Pyrolysis Regenerated Plant. The EROI of this bioenergy chain is equal to 6.5. While several studies are available on cardoon cultivation, its use in pyrolysis tests is very reduced. So the data presented in this chapter can help bioenergy operators to develop chains on cardoon pyrolysis. Biochar production and use can be promoted through the mechanism of Kyoto Protocol, being it a carbon sink. Clean Development Mechanism project can promote its use to generate Certified Emission Reductions to be sold on the carbon market. Besides slow pyrolysis process can be usefully introduced in agro-energy districts. The limits of the analysis presented in this chapter are the following: no economic feasibility is considered in this case; tar combustion to supply the energy required by pyrolysis process may present problems caused by the high tar viscosity and generation of corrosive compounds.

Acknowledgements The authors acknowledge Prof. Jesus Fernandez of the Politechnical University of Madrid for the cardoon seeds he had selected. The authors would also like to thank the would also like to thank the Foundation for Agricultural Instruction of the University of Perugia; Dr. V. Mazza for the analysis of data obtained from experimental fields; Dr. R. Petesse for the analysis of data obtained by pyrolysis tests. Eng. Bruno D'Alessandro should be acknowledged for calculations on mass and energy balances of the IPRP pyrolysis plant. Dr.ess Pilar Ciria and dr. Luis Saul Esteban of CEDERCIEMAT (Soria, Spain) should be acknowledged for their explanations on cardoon crop productivity and cardoon combustion tests.

References

Angelini LG, Ceccarini L, Nassi o Di Nasso N, Bonari B (2009) Long-term evaluation of biomass production and quality of two cardoon (Cynara cardunculus L.) cultivars for energy use, Biomass and Bioenergy. 33(5):810–816

Baratieri M, Baggio P, Fiori L, Grigiante M (2008) Biomass as an energy source: thermodynamic constraints on the performance of the conversion process. Bioresour Technol 99:7063–7073

Bidini G, Fantozzi F, Buratti C, Bartocci P (2007) Most suitable areas for the cultivation of herbaceous crops in Umbria region (Italy) and biomass production evaluation. 15th European Biomass Conference and Exhibition, 7–11th May, Berlin, Germany

Bidini G, Fantozzi F, Bartocci P, D'Alessandro B, D'Amico M, Laranci P, Scozza E, Zagaroli M (2015) Recovery of precious metals from scrap printed circuit boards through pyrolysis. J Anal Appl Pyrol 111(1):140–147

Buonocore E, Franzese PP, Ulgiati S (2012) Assessing the environmental performance and sustainability of bioenergy production in Sweden: a life cycle assessment perspective. Energy 37:69–78

Buratti C, Barbanera M, Bartocci P, Fantozzi F (2015) Thermogravimetric analysis of the behavior of sub-bituminous coal and cellulosic ethanol residue during co-combustion. Bioresour Technol 186:154–162

CEN/TS 15439:2006. Biomass gasification. Tar and particles in product gases. Sampling and analysis

Cherubini F, Ulgiati S (2010) Crop residues as raw materials for biorefinery systems—a LCA case study. Appl Energy 87:47–57

Cotana F, Cavalaglio G, Gelosia M, Coccia V, Petrozzi A, Ingles D, Pompili E (2015) A comparison between SHF and SSSF processes from cardoon for ethanol production. Ind Crop Prod 69:424–432

Curt MD, Sanchez G, Fernandez J (2002) The potential of Cynara cardunculus L. for seed oil production in a perennial cultivation system. Biomass and Bioenergy 23:33–46

D'Alessandro B, D'Amico M, Desideri U, Fantozzi F (2013) The IPRP (Integrated Pyrolysis Regenerated Plant) technology: from concept to demonstration. Appl Energy 101:423–431

D'AlessandroB, Bartocci P, Fantozzi F (2011) Gas turbines CHP for bioethanol and biodiesel production without waste streams. Proceedings of the ASME Turbo Expo, ASME 2011 Turbo Expo: Turbine Technical Conference and Exposition, GT2011; Vancouver, BC; Canada; 6 June 2011 through 10 June 2011. 1:691–700

Daugaard DE, Brown RC (2003) Enthalpy for pyrolysis for several types of biomass. Energy & Fuels 17:934–939

Encinar JM, Gonzalez JF, Gonzalez J (2000) Fixed-bed pyrolysis of Cynara cardunculus L. Product yields and compositions. Fuel Process Technol 68:209–222

Fantozzi F, D'Alessandro B, Desideri U (2007) An IPRP (Integrated Pyrolysis Regenerated Plant) Microscale demonstrative unit in central Italy. ASME paper GT 2007-28000, Montreal, Canada ASME Turbo Expo 2007

Fantozzi F, D'Alessandro B, Bartocci P, Desideri U, Bidini G (2009) Performance evaluation of the IPRP technology when fuelled with biomass residuals and waste feedstocks. Proceedings of the ASME Turbo Expo, ASME Turbo Expo; Orlando, FL; United States; 8 June 2009 through 12 June 2009. 1:449–458

Fantozzi F, D'Alessandro B, Bartocci P, Desideri U, Bidini G (2010) Assessment of the energy conversion of whole oil fruits with a pyrolysis and gas turbine process. Proceedings of the ASME Turbo Expo. 1:685–693. ASME Turbo Expo 2010: Power for Land, Sea, and Air, GT 2010; Glasgow; United Kingdom; 14 June 2010 through 18 June 2010

Fantozzi F, Bartocci P, D'Alessandro B, Arampatzis S, Manos B (2014) Public-private partnerships value in bioenergy projects: economic feasibility analysis based on two case studies. Biomass and Bioenergy 66:387–397

Fernández J, Marquez L, Venturi P (1997) Technical and economic aspects of Cynara Carduncculus L.: an energy crop for the Mediterranean region. 1:48–51

Fernández J, Curt MD, Aguado PL (2006) Industrial applications of Cynara cardunculus L. for energy and other uses. Ind Crop Prod 24:222–229

Gaunt JL, Lehman J (2008) Energy balance and emissions associated with biochar sequestration and pyrolysis bioenergy production. Environ Sci Technol 42:4152–4158

Goe M, Gaustad G (2014) Strengthening the case for recycling photovoltaics: an energy payback analysis. Appl Energy 120:41–48

Gominho J, Lourenço A, Palma P, Lourenço ME, Curt MD, Fernández J, Pereira H (2011) Large scale cultivation of Cynara cardunculus L. for biomass production—a case study. Ind Crop Prod 33:1–6

Iqbal Y, Gauder M, Claupein W, Graeff-Hönninger S, Lewandowski I (2015) Yield and quality development comparison between miscanthus and switchgrass over a period of 10 years. Energy. 89(1):268–276

Jeffery S, Verheijena FGA, van der Velde M, Bastos AC (2011) A quantitative review of the effects of biochar application to soils on crop productivity using meta-analysis. Agric Ecosyst Environ 144(1):175–187

Lewandowski I (2015) Securing a sustainable biomass supply in a growing bioeconomy. Glob Food Sec 6:34–42

Manos B, Bartocci P, Partalidou M, Fantozzi F, Arampatzis S (2014) Review of public-private partnerships in agro-energy districts in Southern Europe: The cases of Greece and Italy. Renew Sust Energy Rev 39:667–678

McKendry P (2002) Energy production from biomass (part 2): conversion technologies. Bioresour Technol 83:47–54

Paethanom A, Yoshikawa K (2012) Influence of pyrolysis temperature on rice husk char characteristics and its tar adsorption capability. Energies 5:4941–4951

Paethanom A, Nakahara S, Kobayashi M, Prawisudha P, Yoshikawa K (2012) Performance of tar removal by absorption and adsorption for biomass gasification. Fuel Process Technol 104:144–154

Paethanom A, Bartocci P, D'Alessandro B, D'Amico M, Testarmata F, Moriconi N, Slopiecka K, Yoshikawa K, Fantozzi F (2013) A low-cost pyrogas cleaning system for power generation: scaling up from lab to pilot. Appl Energy 111:1080–1088

Phuphuakrat T, Namioka T, Yoshikawa K (2011) Absorptive removal of biomass tar using water and oily materials. Bioresour Technol 102:543–549

Sastre CM, Maletta E, González-Arechavala Y, Ciria P, Santos AM, del Val A, Pérez P, Carrasco J (2014) Centralised electricity production from winter cereals biomass grown under central-northern Spain conditions: global warming and energy yield assessments. Appl Energy 114:737–748

Slopiecka K, Bartocci P, Fantozzi F (2012) Thermogravimetric analysis and kinetic study of poplar wood pyrolysis. Appl Energy 97:491–497

UNI 14774-2:2010 Solid biofuels—determination of moisture content—oven dry method—part 2: total moisture—simplified method. Italian Organization for Standardization, Rome, Italy

UNI 14775:2010 Solid biofuels—determination of ash content. Italian Organization for Standardization, Rome, Italy

UNI 14918:2010 Solid biofuels—determination of calorific value. Italian Organization for Standardization, Rome, Italy
UNI 15104:2011 Solid biofuels. Determination of total content of carbon, hydrogen and nitrogen. Instrumental methods. Italian Organization for Standardization, Rome, Italy
UNI 15148: 2009 Solid biofuels. Determination of the content of volatile matter. Italian Organization for Standardization, Rome, Italy
Weißbach D, Ruprecht G, Huke A, Czerski K, Gottlieb S, Hussein A (2013) Energy intensities, EROIs (energy returned on invested), and energy payback times of electricity generating power plants. Energy 52:210–221
Xue S, Lewandowski I, Wang X, Yi Z (2016) Assessment of the production potentials of Miscanthus on marginal land in China. Renew Sust Energy Rev 54(1):932–943
Yang H, Kudo S, Kuo HP, Noriaga K, Mori A, Masek O (2013) Estimation of enthalpy of bio-oil vapor and heat required for pyrolysis of biomass. Energy & Fuels 27:2675–2686

Part V
Examples for End Use of Perennial Biomass Crops

Chapter 22
Harvesting and Separation of Different Plant Fractions in *Cynara cardunculus* L.

L. Pari, V. Alfano, A. Acampora, A. Del Giudice, A. Scarfone, and E. Sanzone

Introduction

In the last few years the Italian agro-scenario has changed. In fact, with the increasing cost of labor, marginal agricultural lands have been almost abandoned to make room for industrial crops in areas devoted to mechanization and extensive agriculture (Stijker 2005).

However, scientists stress that the utility of marginal areas could be revalued in favor of energetic cultivations requiring low agronomic inputs (Vasilakoglou and Dhima 2014). This may ensure a rational use of the lands, promoting at the same time the production of clean energy and determining new revenues for farmers.

Concerning the use of energetic biomasses in the EU, the European strategy on bio-economy furthers the hierarchical use of biomass fractions (including agricultural and agro-industrial residues), with the highest economic value. In this regard, the promotion of multipurpose energy crops has to be taken into account.

The plant species *Cynara cardunculus* L. or "cardoon" has recently gained interest as multipurpose energetic crop in the Mediterranean areas since it fits very well to xerothermic conditions of Southern Europe than other energetic species (Raccuia et al. 2004a, b). Cardoon seeds, capitula, stem, and leaves can be used as raw material for the production of solid and liquid biofuels. Moreover, the dispersal mechanism of the seeds called pappus finds application as high quality pulp for the paper industry (Fernandez et al. 2006; Gominho et al. 2009).

From cardoon seeds it is possible to extract edible oil, which can be appositely converted into biofuels through chemical processes (Curt et al. 2005; Fernandez et al. 2007). The unsaturated fatty acid composition of the oil is similar to that of the sunflower. Oil can be also processed into biopolymers for producing bio-plastics,

L. Pari (✉) • V. Alfano • A. Acampora • A. Del Giudice • A. Scarfone • E. Sanzone
Consiglio per la ricerca in agricoltura e l'analisi dell'economia agraria, Unità di ricerca per l'ingegneria agraria (CREA-ING), Monterotondo, Rome, Italy
e-mail: luigi.pari@entecra.it

S. Barth et al. (eds.), *Perennial Biomass Crops for a Resource-Constrained World*, DOI 10.1007/978-3-319-44530-4_22

while the extraction flours can be utilized in the feed industry. Furthermore, the active ingredients such as inulin, silymarin, and cynarina can be used in pharmacology for different purposes (Raccuia and Melilli 2004; Curt et al. 2005).

The growing period of the crop is between fall and winter when rainfall increases, while seed ripening occurs during summer. In Italy, the seed harvest takes place from August to mid-September. In this period the moisture content of the seeds is lower than 10%; this facilitates remarkably their conservation during storage. Research conducted in Mediterranean environments pointed out that the aboveground biomass production after the first year planting can reach up to 20.0 t ha^{-1} of fresh matter, while the seed yield is around 1.0 t ha^{-1} (Raccuia and Melilli 2007).

In the framework of the National Operative Programme (PON) Enerbiochem, this work aims at testing the performance of a combine equipped with three harvesting systems (traditional and modified head for wheat and CREA—ING cardoon head) to harvest cardoon in Mediterranean Italian Environment.

The operative capacity of the combine was assessed by studying the working times, while the work quality was evaluated quantifying the harvesting losses of both capitula and aboveground biomass. In the view of designing new machines for the selective harvesting of the different plant fractions, a characterization and a statistical analysis of the cardoon capitula components was carried out as well.

Material and Methods

Experimental Field

The harvesting trials on *Cynara cardunculus* L. (cv. *Altilis* D.C.) were carried in a marginal plain area of Porto Torres (Sardinia) during summer 2013. The cultivation was at its second growing season. It was extended for 1.1 ha, with a theoretical density of 20.833 plants per hectare (planting system 0.60 × 0.80 m). Data concerning air temperature and precipitation for the year 2013 were collected from a meteorological station provided by the ISPRA-SCIA (Italy's Institute for Environmental Protection and Research), located in the vicinity of the experimental field.

Combine Performance and Work Quality

The combine CS540 New Holland was tested on cardoon using three different harvesting systems which involved the use of a traditional head for wheat (width 6100 mm, weight 1700 kg), a modified head for wheat (width 6100 mm, weight 1784 kg), and the CREA-ING head prototype (width 4940 mm, weight 3342 kg) for cardoon.

The modification on the head for wheat was applied in correspondence with the cutting bar and consisted of a device formed by 21 triangular plates with rectangular

Fig. 22.1 Particular of the triangular plates applied at the head for wheat

base (larger side = 24.6 cm) spaced 5.6 cm from each other. This device was designed to prevent the capitula losses during harvest due to front tipping after cut (Fig. 22.1).

The cardoon head realized by CREA-ING was a prototype matching a six rows maize harvesting head (upper device) with a tradition head for wheat (lower device). The upper device provides the cut of the capitula and their conveying toward the threshing system, whereas the lower one performs the cut, the conditioning, and the windrowing of the remaining aboveground biomass. Specifications of the prototype are reported in Pari et al. (2009), Pari et al. (2008), and Pari et al. (2011).

The performance of the combine was evaluated according to the official methodology proposed by the American Society of Agricultural and Biological Engineers (ASABE). To this aim, the combine was assessed in 12 plots of 24 m^2 each. The cutting height of each head, the impurities, the seed damage, and the losses due to unthreshed capitula and unmowed stalks were collected. The biomass harvesting losses (excluding capitula) were calculated by studying the dry matter (DM) of the stem per linear centimeter through a logarithmic function ($y = a + b \ln ø$) where (ø) is the stalk diameter. Successively, the trends of biomass losses were put in relation to the cutting height that was utilized.

Morpho-biometrical Traits and Characterization of the Product

The morpho-biometrical traits of the plantation were evaluated in five representative plots (20 m^2 each) considering: plantation density, lodging percentage, average height of the plants, average stalk diameter (measured at 10 cm from the ground level), average height of the first branching point, number of capitula per hectare, and weight of the aboveground biomass (stalks and capitula) per hectare.

Seed production were characterized calculating the seed density and the moisture content according to EN 15103:2010 and EN 14774-1:2009, respectively.

Cardoon capitula of different diameters were studied separating the main fractions (seeds, hairs, bracts, receptacle, and stamens) by hand and weighing them using an analytical scale. Moreover, the diameters of the capitula present in the five plots used for the morph-biometrical analysis (1250 in total) were measured using a digital caliber. Subsequently, the values obtained were divided into seven size representative classes included in a range between 3.0 and 9.0 cm.

The study of capitula portioning and diameter frequency distribution allowed determining the frequency of the capitula for each diametric class, and the weight percentage of seeds and pappus that can be associated to this frequency. The frequency analysis was adjusted to the gaussian function $y = a \exp[-0.5 (x - x_0)/b^2]$, where a, b and x_0 were obtained through iterative method using the software SigmaPlot version 10.0.

The working speed, the machine work capacity, the unthreshed capitula and the capitula losses were studied through the analysis of variance (ANOVA) using the software package for data analysis CoStat Version 6.003 (CoHort Software, Monterey, California, USA). Before the ANOVA analysis, the percentage values of capitula losses were transformed as square root of the arcsine. Where significant differences were displayed, the Student–Newman–Keuls (SNK) method for means separation was applied.

Results and Discussion

Climatic Data and Crop Description

During the year 2013 the average maximum daily temperatures reached about 28 °C in July and August, while the minimal occurred in February with 7 °C (Fig. 22.2). Average daily temperatures at the beginning of the springtime were already over 14 °C. The rainfall was almost absent between July and August and concentrated in winter and autumn.

At the harvest time, the crop showed an effective density of 12,700 plants ha^{-1}. The average height of the plants was 2.3 m, while that of the main bunching point was 1.6 m. The base diameter observed had a mean value of 35.2 mm, while the average diameter of the capitula was 53.7 mm. The mean number of the capitula per each plant was 10.1 corresponding to a total of 127,800 heads ha^{-1} (Table 22.1).

Technical Aspects and Working Time

The different systems utilized for cardoon harvesting influenced significantly the performance of the combine. The combine with traditional head for wheat operated at a working speed of 2.77 km h^{-1} displaying a working capacity of

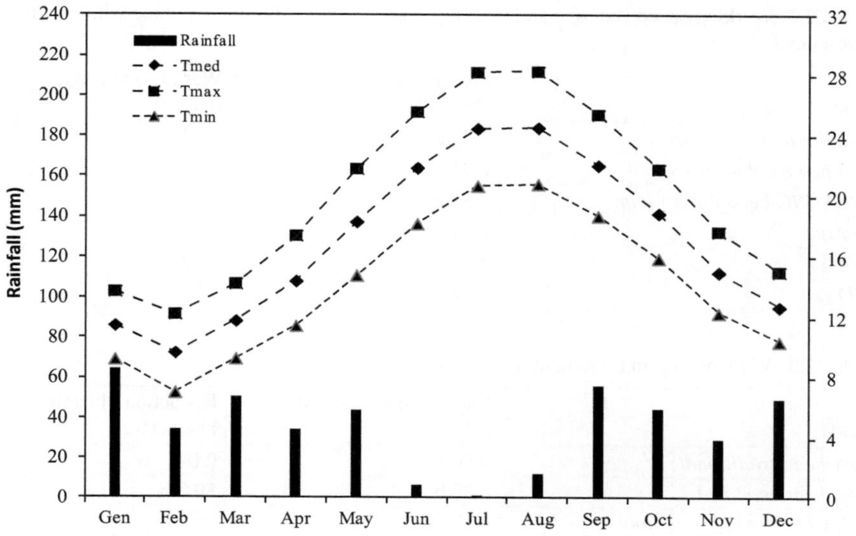

Fig. 22.2 Meteorological data recorded during the year 2013

Table 22.1 Average morphological traits of the plants

Sowing plant density	*plant ha⁻¹*	20,833
Plant density at the harvest	*plant ha⁻¹*	12,700
Height	*m*	2.3
Ø at 10 cm	*mm*	35.2
Intersection of the main branch	*m*	1.6
Stalks per plant	*n°*	1.8
Capitula per plant	*n°*	10.1
Capitula per hectare	*n°*	127,800
Diameter of the capitula	*mm*	53.7
Weight of the stalks (fresh basis)	*t ha⁻¹*	7.5
Weight of the capitula (fresh basis)	*t ha⁻¹*	3.7
Weight of 1000 seeds	*g*	38.2
Moisture content of seeds	*%*	9.2
Bulk density (seeds)	*kg m⁻³*	627.0

1.67 ha h^{-1}. With the modified head, the working speed of the machine was 2.92 km h^{-1} (working capacity 1.74 ha h^{-1}).

On the other hand, the speed of the combine equipped with the CREA-ING was 2.84 km h^{-1} and the working capacity 1.27 ha h^{-1} (Table 22.2). Although the combine equipped with different heads operated at similar working speeds, the lower performances recorded with the CREA-ING head were due to its reduced working width.

Table 22.2 Working speed (km h^{-1}) and working capacity (ha h^{-1}) of the combine with different harvest heads

Heads	Working speed *(km h^{-1})*	Working capacity *(ha h^{-1})*
Traditional head for wheat	2.77 a	1.67 b
Modified head for wheat	2.92 a	1.74 a
CREA-ING head for cardoon	2.84 a	1.27 c
Average	2.84	1.56
CV (%)	4.16	1.69
$LSD_{0.05}$	0.27	0.06

Table 22.3 Work quality in relation to the different harvesting systems

Heads	Capitula not threshed *(t ha^{-1})*	Reduction of capitula losses *(%)*
For wheat, traditional	0.86 a	0.0 c
For wheat, modified	0.35 b	59.2 b
CREA-ING head for cardoon	0.12 c	86.0 a
Average	0.44	48.60
CV (%)	6.25	3.48
$LSD_{0.05}$	0.06	3.84

Work Quality

The use of the traditional head for wheat determined the highest capitula losses (0.86 t ha^{-1}). In fact, the modified head for wheat and the CREA-ING prototype have showed a reduction of losses respectively of 59.2 % (0.35 t ha^{-1}) and 86 % (0.12 t ha^{-1}) compared to the traditional head (Table 22.3). So the modification applied to the traditional head for wheat demonstrated efficient in recovering the capitula. The efficiency of this device depended by the configuration and the shape of the separating elements fixed to the lower teeth of the cutting bar. The elements, penetrating into the crop before cutting the stems, prevented the loss of the capitula.

The harvesting losses occurred with the CREA-ING head were 86 % lower than those displayed by the traditional head for wheat. The highest efficiency of this head was due to the fact that it performed a more efficient separation of stems and capitula with larger size, the ones having a very high seed production.

Maintaining a cutting height of 10–20 cm, as in the case of the CREA-ING head, the biomass losses (excluding capitula) ranged from 0.4 to 0.8 t ha^{-1}, corresponding respectively to 5.6 and 11.1 % of the total DM ha^{-1} (Table 22.4). Cutting heights equal to 1.20 m, as those utilized with the other two systems, led to DM losses even higher than 50 % (>4.0 t ha^{-1}).

The cutting system of the CREA-ING head for cardoon allowed harvesting the crop almost at ground level, with a consequent reduction of aboveground biomass losses (below one ton per hectare) in field. The adoption of such cutting height help the drying process of the windrows since residual stems will keep the biomass slightly above the ground facilitating air circulation.

Table 22.4 Aboveground biomass losses (t ha^{-1}; % total DM) in relation cutting height (cm)

Cutting height	Aboveground biomass losses	
(cm)	*(t ha^{-1})*	*(%)*
10	0.4	5.6
20	0.8	11.1
30	1.3	16.4
40	1.7	21.8
50	2.1	27.0
60	2.5	32.1
70	2.8	37.1
80	3.2	42.0
90	3.6	46.8
100	4.0	51.5
110	4.3	56.1
120	4.6	60.5
130	5.0	64.8
140	5.3	69.0
150	5.6	73.1

Product Classification

The seed impurities such as weed seeds and inert materials constituted the 4.1 % of the total seeds harvested, while the seeds damaged were the 1.5 % of the total. The average weight of 1000 seeds was 38.2 g, while the bulk seed density was 627.0 kg m^{-3}. The average moisture content of the seeds at harvesting was 9.2 %.

Biomass Capitula Partitioning and Distribution According to Size

The dry biomass fractions of the capitula were put in relation with the capitula size according to their diametrical dimensional classes. The distribution showed significant differences according to the class considered.

In particular, the highest value of seed weight compared to the total dry matter content was identified in capitula with diameter between 5 and 9 cm, while a total absence of seeds was observed in the capitula with diameter ranging from 2 to 4 cm. The fraction of hairs was significantly higher in the two higher size classes (Table 22.5).

Figure 22.3 shows the trend of the capitula frequency put in relation to their diameter as according to Gaussian function: $y = 30.23\ exp[-0.5(x - 5.40)/1.32)^2]$

Considering the frequency of each dimensional class, the distribution of the two much interesting fractions (seeds and hairs) was determined (Fig. 22.4). In this case, the capitula having the greatest impact on the yield of hairs were those with a diameter

Table 22.5 Dry matter partitioning (g) into capitula in relation to different size classes

Classes	Seeds	Hairs	Bracts	Other	Total
2–3	0.0 d	0.6 f	4.7 d	4.5 b	9.9 g
3–4	0.0 d	1.2 f	5.5 cd	8.1 b	13.8 f
4–5	2.8 d	1.9 e	6.5 cd	6.9 b	19.2 e
5–6	6.6 c	2.8 d	9.9 bc	13.5 a	32.8 d
6–7	11.5 b	4.2 c	13.1 b	18.9 a	47.6 c
7–8	16.7 a	4.9 b	19.3 a	18.6 a	59.5 b
8–9	17.9 a	5.8 a	20.6 a	19.9 a	64.2 a
Average	*7.9*	*3.1*	*11.4*	*12.9*	*35.3*
C.V. (%)	*20.6*	*10.1*	*18.4*	*21.6*	*5.4*
$LSD_{0.05}$	*2.9*	*0.6*	*3.7*	*4.9*	*3.4*

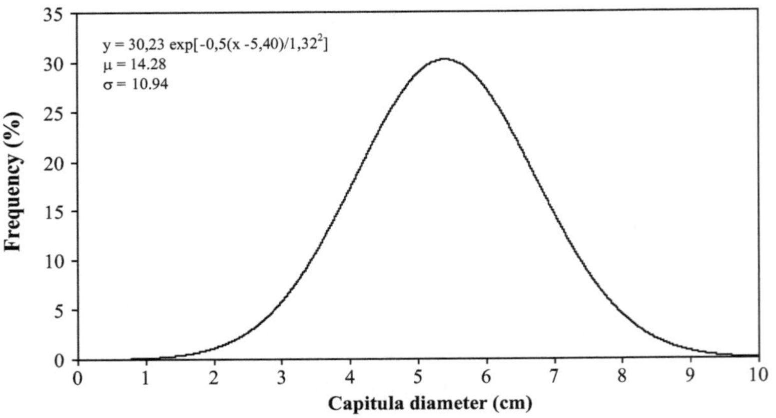

Fig. 22.3 Trend of the capitula frequency in relation to their size

between 5 and 7 cm (from 25 to 31.1 % of the total production). On the other hand, the class of capitula 6–7 greatly affected the seed production, with percentages ranging from 30 to 35.8 % compared to the total yield.

Discussion

In agriculture, the highest costs for production are given by the costs for mechanization (Buckmaster 2003). Studies show they can represent between the 15 % and the 50 % of the total production costs (mean data related to field crops, Anderson 1988; E. U. FADN 2007). These aspects are very important for assessing the economic viability of the traditional yards used for harvesting cardoon.

The study presented showed that the effective speed of the machine did not change significantly during the trials. This indicates that the lower working capacity

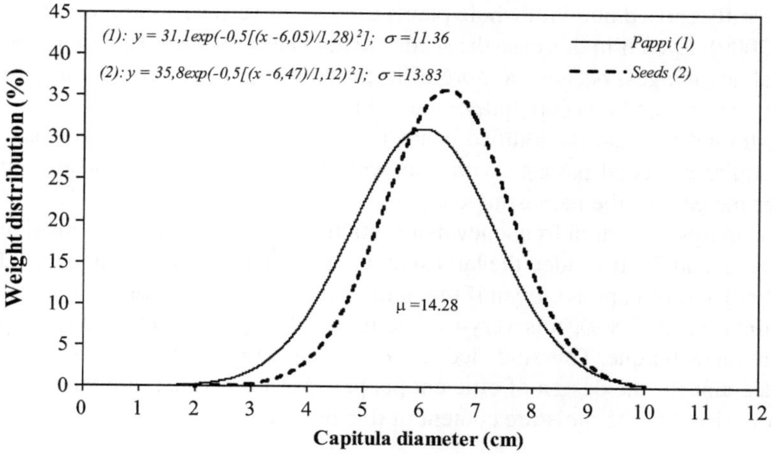

Fig. 22.4 Weight distribution (%) of the two main fractions (seeds and hairs) in relation to their frequencies

displayed by the combine equipped with the CREA-ING head was mainly due to the width of the mower, which was tighter compared to the other heads.

The modification applied to the traditional head for wheat has significantly reduced the capitula losses, which corresponded approximately to 60 % compared to the head without modification. The efficiency of this device depends by the configuration and the shape of the separating elements fixed to the lower teeth of the cutting bar. The elements penetrated into the crop before cutting the stems preventing the loss of the capitula. The harvesting losses were due to the fact that after cut some of the capitula fell forward and thus out of the outlet of the harvest head. Adjustments of the reel would aim to prevent the falling out of the capitula, reducing biomass losses (Assirelli 2011).

The harvesting losses occurred with the CREA-ING head were 86 % lower than those displayed by the traditional head for wheat. The higher efficiency of this head was due to the fact that it performs a more efficient separation of stems and capitula with larger size, which have a very high seed production (Pari et al. 2009).

The cutting system of the CREA-ING head for cardoon allowed harvesting the crop almost at ground level, with a consequent reduction of aboveground biomass losses in field. In fact, in the case of cardoon, this study indicates that losses can be maintained under one ton per hectare if the cutting height is set below 20 cm. Adopting such a cutting height will facilitate the drying process of the windrows since residual stems will keep the biomass slightly above the ground facilitating air circulation. Studies show that adopting this method would be very effective to decrease the moisture content of the biomass in percentages even lower than 20 % before baling (Vogel et al. 2011).

The dry stems of cardoon can be used to produce pellets for combustion in domestic boilers for heating (Aho et al. 2008). However, the production of energy is generally carried out with whole plants used as pulverized solid biofuel (Pallares et al. 2009). Even if in this case the seeds are not valorized, this destination presents several advantages, because a more efficient conventional mechanization for harvesting can be applied (Fernandez et al. 2007).

Both modified and unmodified heads for wheat experience high biomass losses and require a second passage with a mower-windrower. This obviously implies a further increase of the harvesting costs.

The analysis of pappi frequency distribution showed that capitula with diameters between 5 and 7 cm yielded the largest amount of pappi (between 25 and 31.1 % of the total yield of pappus). Even if the industrial interest for pappi is already known (Gominho et al. 2009), this very-rich cellulosic fraction would require a specific separation technique. This study led us to understand that the parameters to be taken into account for the design of efficient pappi separation system are the weight, the bulk density, and the moisture content of this product.

Conclusions

The results achieved showed the good performance of the CREA-ING head in terms of work quality, but at the same time they displayed its lower working capacity compared to the other two systems. The capitula characterization revealed that the capitula with a diameter between 5 and 7 cm had the largest impact on both seed and hair production.

In order to encourage the exploitation of marginal lands with multipurpose crops, the design of new mechanization systems should be promoted. In the specific case of cardoon, the diversification of the crop productions has interesting perspectives on both economic affordability and environmental sustainability. In fact, beside the recovery of hairs and other cardoon capitula fractions, the interest is also addressed to other plant components such as stems and leaves. For this reason, the study and the design of new prototypes able to separate the different plant fractions should be encouraged.

References

Aho M, Gil A, Taipale R, Vainikka P, Vesala H (2008) A pilot-scale fireside deposit study of co-firing Cynara with two coals in a fluidised bed. Fuel 87:58–69

Anderson W (1988) Factors affecting machinery costs in grain production. ASAE 88:1057

Assirelli A (2011) Girasole: valutazione delle soluzioni commerciali disponibili per la raccolta. In Luigi Pari (ed) Lo sviluppo delle colture energetiche in Italia. Il contributo dei progetti di ricerca SUSCACE e FAESI. Ed. Nuova Cultura. pp 429–437

Buckmaster DR (2003) Benchmarking tractor costs. Appl Eng Agric. 19:151–154

Curt MD, Sanchez G, Fernandez J (2005) Cynara cardunculus a source of silymarin. In: Bianco VV, Calabrese N, Rubatzky V (eds) Proceedings of the 4th International Congress on Artichoke, Acta Hort. 681:461–467

EN 14774-1:2009. Solid biofuels. Determination of moisture content—oven dry method
EN 15103:2010. Solid biofuels. Determination of bulk density
European Union (2007) EU farm economics overview FADN. http://ec.europa.eu/agriculture/rica/pdf/report_2007.pdf
Fernandez J, Curt MD, Aguado PL (2006) Industrial applications of Cynara cardunculus L. for energy and other uses. Ind Crop Prod 24:222–229
Fernandez J, Pari L, Muller MG, Marquez L, Fedrizzi M, Curt MD (2007) Strategies for the mechanical harvesting of cynara. In: Maniatis K (ed) 15th European biomass conference. pp 657–668
Gominho J, Lourenço A, Curt MD, Fernandez J, Pereira H (2009) Characterization of hairs and pappi from Cynara cardunculus capitula and their suitability for paper production. Ind Crop Prod 29:116–125
Pallares J, Gil A, Cortes C, Herce C (2009) Numerical study of cofiring coal and Cynara cardunculus in a 350 MWe utility boiler. Fuel Process Technol 90:1207–1213
Pari L, Fedrizzi M, Gallucci F (2008) Cynara cardunculus exploitation for energy applications: development of a combine head for threshing and concurrent residues collecting and utilization. 16th European Biomass Conference & Exhibition. pp 504–508
Pari L, Civitarese V, Assirelli A, Del Giudice A (2009) Il prototipo che abbatte i costi della raccolta del cardo. Agroenergie dall'impianto alla raccolta, L'Informatore Agrario, 29(1):11–14
Pari L, Fedrizzi M, Pansini L, Gallucci F (2011) Prototype development for capitula harvesting of Cynara cardunculus. In: Maniatis K (ed) 15th European Biomass Conference. pp 494–497
Raccuia SA, Melilli MG (2004) Cynara cardunculus L., a potential source of inulin in Mediterranean environment: screening of genetic variability. Aust J Agric Res 55:693–698
Raccuia SA, Melilli MG (2007) Biomass and grain oil yields in Cynara cardunculus L. genotypes grown in a Mediterranean environment. Field Crop Res 101:187–197
Raccuia SA, Cavallaro V, Melilli MG (2004a) Intraspecific variability in Cynara cardunculus L. var. sylvestris Lam. Sicilian populations: seed germination under salt and moisture stresses. J Arid Environ 56:107–116
Raccuia SA, Mainolfi A, Mandolino G, Melilli MG (2004b) Genetic diversity in Cynara cardunculus revealed by AFLP markers: comparison between cultivars and wild types from Sicily. Plant Breeding 123:280–284
Stijker D (2005) Marginal lands in Europe-causes of decline. Basic Appl Ecol 6:99–106
Vasilakoglou I, Dhima K (2014) Potential of two cardoon varieties to produce biomass and oil under reduced irrigation and weed control. Biomass and Bioenergy 63:177–186
Vogel KP, Sarath G, Saathoff A, Mitchell R (2011) Switchgrass. In: Halford N, Karp A (eds) Energy crops. The Royal Society of Chemistry, Cambridge, pp 341–380

Chapter 23
Masonry Blocks from Lightweight Concrete on the Basis of Miscanthus as Aggregates

D. Waldmann, V. Thapa, F. Dahm, and C. Faltz

Introduction

In recent years, the political and social awareness about sustainability in different aspects of private and public environment has considerably increased. This trend has become a highly discussed topic in all kinds of branches, from chemical engineering and environmental science to resource management and construction engineering. One of the most important issues to solve in order to provide sustainability is the reduction of CO_2 emissions. CO_2 is a greenhouse gas that is emitted in enormous amounts into the atmosphere by human activities. Due to the resulting high concentration, the natural carbon cycle and the natural balance are influenced, as a considerable portion remains in the atmosphere for a fairly long time and retards the cooling of the Earth. This is known to increase atmospheric temperatures, resulting in the so-called "global warming" and, locally, "climate change".

In order to counteract this phenomenon, in the field of construction and building materials different measures for reduction of heating energy have been promoted and new requirements and directives have been successively issued by the authorities. Especially the requirements for thermal insulation of building envelopes have been intensively tightened, which has mainly driven the masonry construction sector to investigate new technical innovations and new insulation products in order to meet the growing requirements.

In Europe, the European Parliament issued the directive 2010/31/EU of 19 May 2010 on the energy performance of buildings. This directive sets as the objective for all European Member States that, until 2020, they shall take all appropriate measures

D. Waldmann (✉) • V. Thapa
Laboratory for Solid Structures, University of Luxembourg, Esch-sur-Alzette, Luxembourg
e-mail: daniele.waldmann@uni.lu

F. Dahm • C. Faltz
University of Luxembourg, Esch-sur-Alzette, Luxembourg

S. Barth et al. (eds.), *Perennial Biomass Crops for a Resource-Constrained World*, DOI 10.1007/978-3-319-44530-4_23

for the transition of all new buildings and any transformation of existing buildings into nearly zero-energy buildings.

"Member States shall take the necessary measures to ensure that minimum energy performance requirements are set for building elements that form part of the building envelope and that have a significant impact on the energy performance of the building envelope when they are replaced or retrofitted, with a view to achieving cost-optimal levels."[1]

In order to act on this directive, the Luxembourgish government has established new objectives and requirements which demand that from 2017 on, every construction of new residential buildings should be built according the passive house standards.

These requirements force the manufacturers of masonry building materials to launch new and innovative products into the market in order to meet the required thermal insulation properties. Additionally, the attractiveness of using "green" material as insulation has become more and more important as the market for organically sourced insulation material has experienced strong growth.

More precisely, one very promising organic material which could be used in order to meet the requirements is the pretreated "Miscanthus Giganteus" plant. In the following, the potential of masonry blocks from lightweight concrete on the basis of Miscanthus as aggregates is going to be discussed. Before initiating the topic of Miscanthus as aggregates in lightweight concrete, it is necessary to explain the main characteristics of this organic material.

Miscanthus is a natural, reed-like grass which can be found mostly in humid areas like African or South-East Asian regions. Originally, it is an energy crop from Japan and primarily was used as a sustainable and energy-rich raw material for biomass cogeneration plants due to its high calorific value and its favourable carbon balance during growth. Additionally, it was found that Miscanthus presents advantages in conjunction with concrete as it is a C-4 plant and therefore shows a higher photosynthetic efficiency and lower water use requirements than other plants. Furthermore, this plant needs 2–3 times less water for growth than other plants and also can be grown on barren grounds.

These characteristics make Miscanthus very interesting for its use in lightweight concrete mixtures. In the following, the material characteristics of concrete based on Miscanthus aggregates for use within structural elements are analysed. It is discussed if the material properties of this concrete, especially its compressive strength and its heat conductivity, will fulfil the requirements necessary to use it within structural insulating masonry blocks.

[1] Excerpt from Article 4 of directive 2010/31/EU of the European Parliament and of the council of 19 May 2010 on the energy performance of buildings [1]

State of the Art

In general, water, cement and organic aggregates do not bond strongly together due to stability issues of the organic plant, as its behaviour in the bond is usually unpredictable. However, under specific conditions, Miscanthus becomes a useful aggregate for concrete mixtures. The outer shell of Miscanthus is covered with a tiny silicon layer which normally stabilises the plants. This characteristic, respecting certain specific conditions, allows a concrete mixture using Miscanthus aggregates to be prepared. Concrete made out of Miscanthus has the following advantages when using it within concrete masonry blocks:

Firstly, such mixtures enable better insulation characteristics leading to a reduction of the thickness of the concrete masonry blocks. Secondly, this building material provides due to its natural insulation characteristics an improved protection against summer overheating. The durability of this material is given due to the mineral bonding of Miscanthus which thus increasingly hardens. Furthermore, a comfortable sound insulation is given by the numerous pores in the concrete containing Miscanthus aggregates. Fire protection is increased due to the fact that more energy has to be supplied to the material as it can be extracted from it, which means that Miscanthus is not self-combustible. Finally, it constitutes a sustainable alternative to natural resources like sand and gravel in a concrete mixture.

These properties are naturally given by the characteristics of Miscanthus, but in order to use this building material in structural load-bearing elements, the objective is to provide reasonable compression strengths combined with good thermal resistance.

The combination of concrete with organic materials is long known but the main issue is that the organic material has to be pre-mineralized in order to reduce its water absorption characteristics in favour of the actual concrete reactions. This process is very costly and time-consuming as the material has to be mixed with the mineraliser in a separate work step and, in addition, the mineraliser itself is very expensive and often harmful to the environment. In order to prevent such an expensive and inefficient process, there are alternative processes where the mineraliser is incorporated during mixing, thus resulting in lower compression strengths and durability due to weaker bonding between the organic material and the binder. However, this process remains economically interesting as it skips the step of drying after the mineralisation process, saving time and money.

Different processes already exist for concrete mixtures using organic, plant-based additives. In these guidelines, the composition of concrete and mortar made out of organic admixtures, a hydraulic binder, water and a mineraliser is described. Generally, in these processes, the main intention is to prevent the costly and time-consuming procedure of direct pre-mineralisation of organic materials before combination with the hydraulic binder. Most often, the solution is to use calcium carbonate ($CaCO_3$) during the mixture of the cement with the organic additives. It works like a mineraliser on the surface of the organic additives during mixing, and the better and finer $CaCO_3$ is distributed, the more efficient the cement bonding becomes. As a consequence, better compression and bending tensile strengths can be achieved. However, if the calcium carbonate does not react well, the strength values may strongly vary and the required final strengths would not be fulfilled.

This has been tested in the following (see Sect. 23.4) and different types of mineralisers like calcium carbonate and calcium hydroxide has been used as mineraliser. It came out that the mixtures presented varying strengths and thermal resistances after 56 days.

Description of Insulation Concepts and Concrete Masonry Block Types

Before explaining the different concepts, it is necessary to mention that lightweight concrete blocks themselves have insulation properties as they are made with lightweight aggregates that contain more air spaces than other aggregates and therefore conduct heat more slowly than normal or heavyweight concrete blocks.

Insulation Concepts

As the intention is to develop masonry blocks from Miscanthus concrete, an overview of different concepts of insulating masonry walls is given and their characteristics are explained.

The first concept is to place the insulation on the exterior face of masonry walls. As masonry has more mass, it is able to absorb and store more heat and therefore this concept would allow the masonry to absorb heat during day time and release it during night. The insulation material is attached with adhesive or with mechanical fasteners to the masonry. Normally, it is covered with a weather-resistant finishing coating in order to guarantee additional protection, resistance and longer durability.

The second concept is to incorporate insulation in the cores of the concrete block. The insulation material, such as granular materials, polystyrene inserts or foam, must be chosen accordingly to the use of the wall, as its amount is limited by the open core area. The advantage is that the total thickness of the exterior wall is reduced; however, due to its limited protection, this kind of structure is highly susceptible to thermal bridges.

The third concept is to attach a suitable, rigid insulation material to the interior side of masonry walls. They are usually combined with panelling or gypsum boards for finishing. This method is mostly used in older buildings for simple retrofitting as, due to the position of the insulation, the wall's thermal mass is located outside the insulation envelope.

The forth concept uses the air space in a so-called cavity wall as an insulation layer. This will increase the thermal resistance of the wall as granular materials, rigid foam boards and pumping foam can be placed inside the cavity. The advantages are that this concept maintains the existing wall thickness, easy to instal, cheaper than exterior or interior wall insulation and reduces condensation. In this case, the insulation fills the gap, keeping the warmth in, so that the water vapour in the air cannot condense on the cold wall.

Concrete Insulation Masonry Block Types

Concrete insulation masonry blocks are lightweight structural elements used in building construction which combine, in a single section, load-bearing as well as insulating functions. These blocks usually are rectangular and can be solid or have one or several hollow spaces. In the following, different block types are presented and later the main issues are discussed.

Table 23.1 lists the different types of concrete bricks including a short description illustrated in Fig. 23.1:

The main issue with concrete bricks is that currently it is only possible to increase either the insulation characteristics or the load bearing capacity of the concrete bricks. The combination of both characteristics is always related to the choice between

Table 23.1 Types of masonry concrete block

A	Full brick
	Lightweight concrete with open structure (expanded clay)
B	Full brick
	Lightweight concrete with open structure (natural pumice)
C	Full brick
	Lightweight concrete with open structure (natural pumice + expanded clay)
D	Full brick
	Lightweight concrete with open structure (natural pumice + expanded clay)
E	Brick to be filled with concrete C 16/20
	EPS—exterior insulation, interior shell of lightweight concrete with lost formwork of concrete (expanded clay)
F	Hollow brick with an cork insulation at the inner side
G	Full brick
	Lightweight concrete with open structure (natural pumice)
H	Full brick
	Lightweight concrete with open structure (expanded clay) and integrated insulation
I	Aerated concrete—planar brick
J	Full brick
	Lightweight concrete with open structure (natural pumice + expanded clay)
K	Aerated concrete—planar brick (different producer from I)
L	Insulating sandwich brick with interior insulation
	Lightweight concrete with open structure (expanded clay)

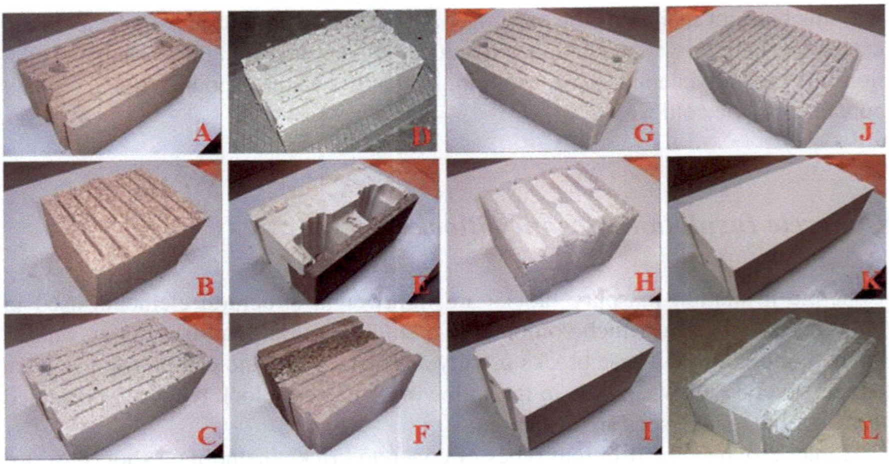

Fig. 23.1 Different types of concrete bricks

increasing the insulation properties by accepting losses in carrying capacities or, on the contrary, accepting a reduction of thermal properties in favour of more load capacity.

More precisely, if we consider one cross section of a concrete brick in terms of heat losses, three major thermal effects are significant: conduction, convection and radiation. Radiation can only be reduced by incorporating reflective insulation, which is an uneconomical and costly method, or by increasing the air space which raises block sizes. Therefore radiation is often reduced using other constructive elements.

The problem of conduction and convection when using concrete bricks can be reduced by applying lightweight concrete with lower thermal conductivity or by increasing the heat conduction path using several successive air spaces.

For example, the type A concrete brick (Fig. 23.1) has small, unfilled airspaces and therefore is exposed to lower convection effects, but higher conduction effects than for example the concrete section of the block type H (Fig. 23.1) which is exposed to higher convection, but lower conduction. Therefore in order to reduce convection effects, the air spaces are filled with insulation material.

As it can be seen, new concepts and technologies are needed to have an overall well-performing block composition in terms of combining insulation and load bearing capacities. In recent years, several studies and investigations on new materials and new compositions have been launched and consistently developed further (Leufgens 2010; Sagmeister 1999; Kvande 2001; Voß 2009). A promising conception would be the use of lightweight concrete made of Miscanthus aggregates.

In the following, different masonry block concepts and characteristics are discussed in order to find the most suitable concepts in terms of insulation performances using Miscanthus lightweight concrete blocks.

Experimental Investigations

The aim of this practical experiment was to develop new concrete formulations based on the renewable resource "Miscanthus" and lead primary investigations relating to the development of masonry bricks on the basis of this concrete. It is intended to develop a Miscanthus concrete with a certain load-bearing capacity and a low thermal conductivity. In order to cover a maximum of possibilities, different masonry block types were investigated and calculated. Finally, a conclusion is drawn on whether it is possible to develop masonry blocks incorporating Miscanthus concrete, which has a heat transition coefficient k equal or lower than 0.14 W/m^2K. This value is the minimal requirement for masonry wall insulations of passive house standards in Luxembourg as of January 2017.

Description of the Methodology for the Development of the Concrete Mixtures

In a first phase, different concrete mixtures were investigated to determinate the applicable relation between compressive strength and thermal conductivity. In general, load-bearing building materials possess high densities and low to no porosity which results in high thermal conductivity characteristics. On the other hand, insulation materials have low thermal conductivity but are not suitable to support high loads. Miscanthus has the characteristic of generating air spaces also in the compressed state due to its coarse and irregular form.

To be able to describe each fraction from the initial Miscanthus aggregate material, several grain-size distribution curves are established. Figure 23.2 shows the grain size distribution of three independent series by representing the through

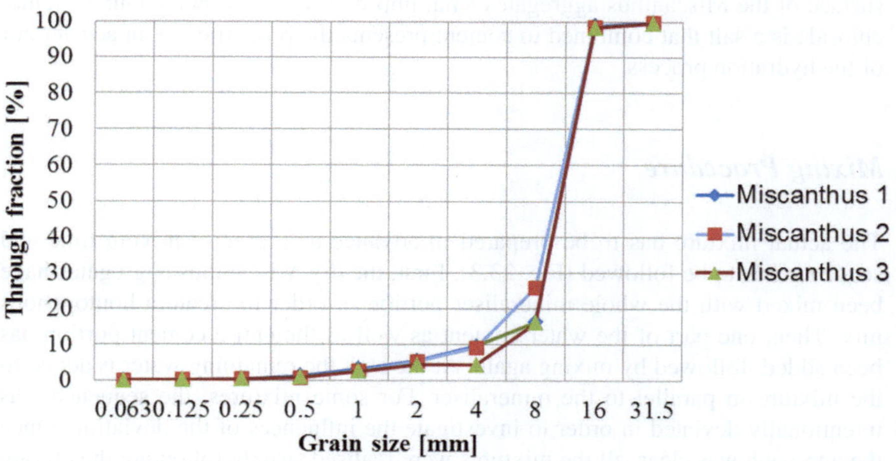

Fig. 23.2 Grain size distribution of three different Miscanthus series

fraction for the different sieves. It can be seen that the grain size distribution of this natural material, which is cut at the moment of the harvest, shows some variations in the main fractions 2/4 and 4/8. The effect of these variations must be further investigated and have not been considered within the current study.

Mineralisation of the Aggregates

The Miscanthus aggregates have been treated by a mineraliser. The goal of this treatment is to seal the pores of the Miscanthus thereby mineralising them. With this treatment, the Miscanthus does not absorb too much water during the mixing process and after, and more water remains in the matrix to react with the cement particles. This will improve the bond characteristics between the Miscanthus aggregates and the cement, which has a positive effect on the strength development of the concrete. A mineraliser must have a very fine grain size distribution to be able to seal the pores.

As mineraliser, different substances like magnesite (magnesium carbonate, $MgCO_3$), calcium hydroxide ($Ca(OH)_2$), chalk and calcium chloride ($CaCl_2$) have been used. Besides its effect of creating a better bond between the binder and organic substances and thus of improving compressive strength, magnesite improves the fire resistance and requires a relatively short hardening time. Calcium hydroxide is a colourless or white powder that is produced by the calcination of limestone at over 890 °C out of the raw material calcium carbonate. To obtain calcium hydroxide, the calcium oxide is mixed with water, where it reacts highly exothermically. If calcium hydroxide is dissolved in water, it forms an alkaline solution. In the construction domain, calcium hydroxide is mainly used for mortar mixtures as well as mixed with sand to produce lime plasters. Chalk is a building material that acts as a binder and is mostly mixed with mortars. As a mineraliser, it forms a film on the surface of the Miscanthus aggregates what improves the bond behaviour. Calcium chloride is a salt that combined to cement presents the properties of an accelerator of the hydration process.

Mixing Procedure

The actual mixture has to be prepared in advance as a precise mixing time and sequence has to be followed (Fig. 23.3). First, the dry Miscanthus aggregates have been mixed with the whole mineraliser portion in order to create a homogenous mix. Then, one part of the water content as well as the entire cement portion has been added, followed by mixing again. At the end, the remaining water is added to the mixture in parallel to the mineraliser. For some mixtures, the sequence was intentionally deviated in order to investigate the influences of the deviation. Once the approach was clear, all the mixtures were realised strictly following the chosen mixing sequence.

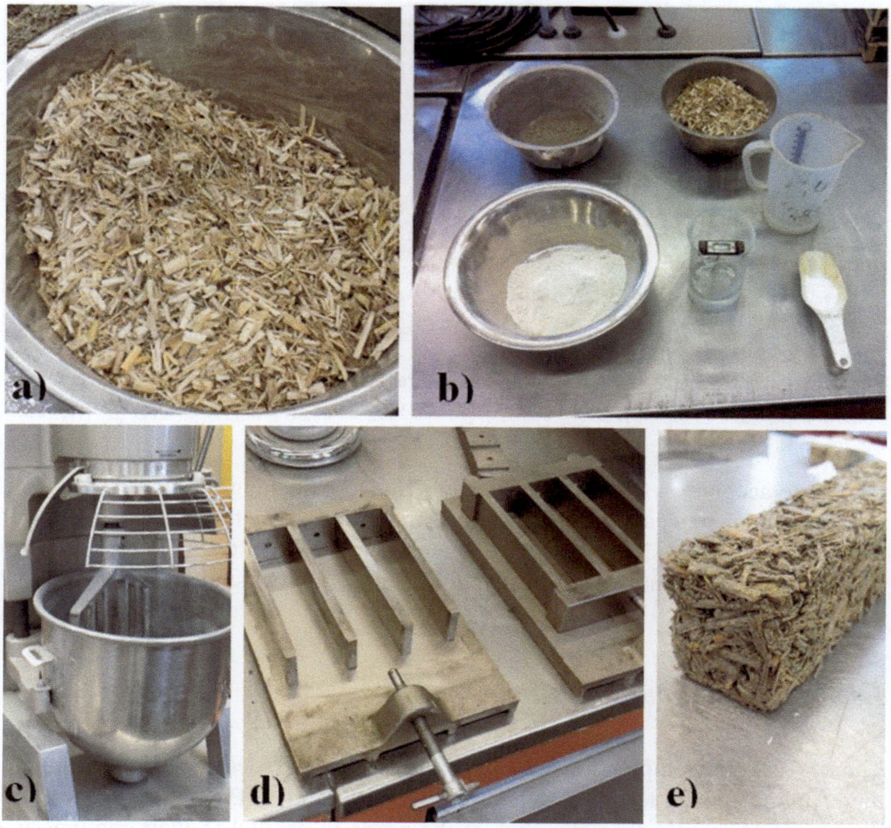

Fig. 23.3 (**a**) The loose Miscanthus in a bowl; (**b**) prepared components for the mixture; (**c**) the mixer; (**d**) standard prism-forms with the dimensions 4/4/16 cm; (**e**) example of a hardened prism sample

The mixtures are filled in standard prim-forms with the dimensions 4/4/16 cm (Fig. 23.3) in order to have standard samples. These samples are stored for about 1 week and then retracted from the forms. Afterwards they are packed in wrapping film in order to keep the humidity in the concrete as there will be more water for the cement reaction. After 14 days, the samples are unpacked from the cellophane and dried by air.

Description of the Methodology for Determination of the Material Properties

Apparent Density

The apparent density (kg/m^3) is determined by weighting the prisms and by evaluating their volume by measuring at different sections precisely the dimensions of the elements. This has been done on 14-day-old prisms. Hereby shrinkage has not been

Fig. 23.4 Strength tests—*Left side*: setup for flexural strength test—*Centre*: setup for compression strength test—*Right side*: displacement transducer

taken into account as well as the water loss due to hardening of the elements with increasing age. So it can be assumed that with further ageing the apparent density of the prisms will still slightly change. This effect will also still have to be further investigated and has not been considered within the current study.

Load-Deformation Behaviour

The flexural and compression strength of the prims were determined according to the common cement testing standard DIN EN 196 Part 1 (Fig. 23.4). To evaluate the load-deformation characteristics of the prims in parallel to their compressive strength, an inductive displacement transducer measured simultaneously to the recorded loading the respective vertical deformation Δh. By relating this deformation to the initial height h of the element the elongation $\varepsilon = \Delta h/h$ was calculated. The compressive strength has been determined as the mean value out of three tests on identical prims that have hardened within one mould.

Comparison of the Test Results of the Compressive Strength as a Function of Various Mineralisers, Water–Cement Ratios and Cement Content

Firstly, the compression strengths of different concrete prisms treated with different mineralisers in the mixture were investigated.

In Fig. 23.5, the impacts of different pretreatment substances/mineralisers on the compressive strength are illustrated. As apart from adding various mineralisers the mixture remained unchanged, the differences of the strength-elongation behaviours

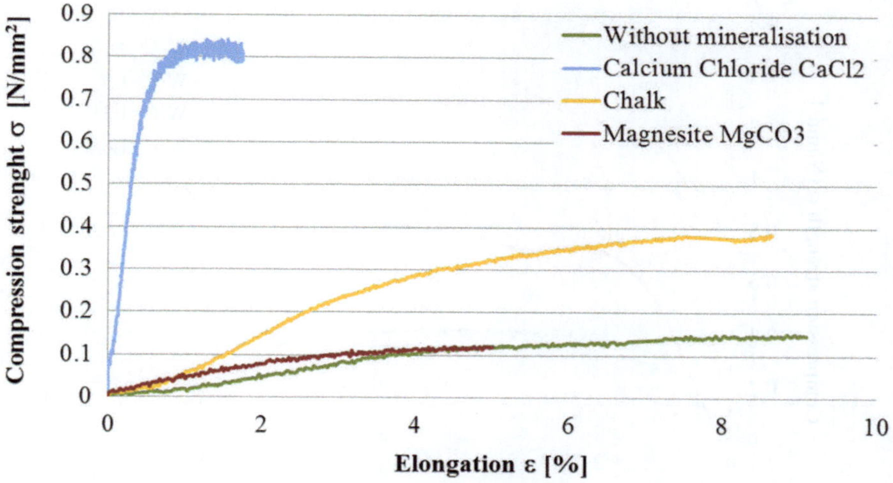

Fig. 23.5 Variation of mineralisation—Evolution of the compression strengths after 14 days of hardening

are considerable, especially for the mixtures with calcium chloride. So it can be concluded that the choice of mineraliser plays an important role in relation with the compression strength. Additionally, it can also be stated that calcium chloride ($CaCl_2$) is by far the most effective mineraliser as it has not only the highest compression strength but also presents the best bond behaviour between cement and Miscanthus aggregates as its strength-deformation curve shows the highest stiffness.

Variation of Water–Cement Ratio

In a second analysis, the strength-deformation behaviour of different concrete prisms with varying water–cement ratio in the mixture has been investigated.

In the first samplings, a low water–cement ratio has been chosen which led to the fact that these concrete mixtures achieved in general low compression strengths. Therefore different concrete mixtures with varying water–cement ratios were mixed and analysed. The most decisive factor is that there is enough water for the reaction of cement as Miscanthus aggregates are highly water absorbent.

As it can be considered in Fig. 23.6, also the water–cement ratio is, as expected, determinant for the resulting strength-deformation behaviour: a higher water–cement ratio is favourable for higher compression strengths which goes along with an increase of the stiffness as the curves become steeper.

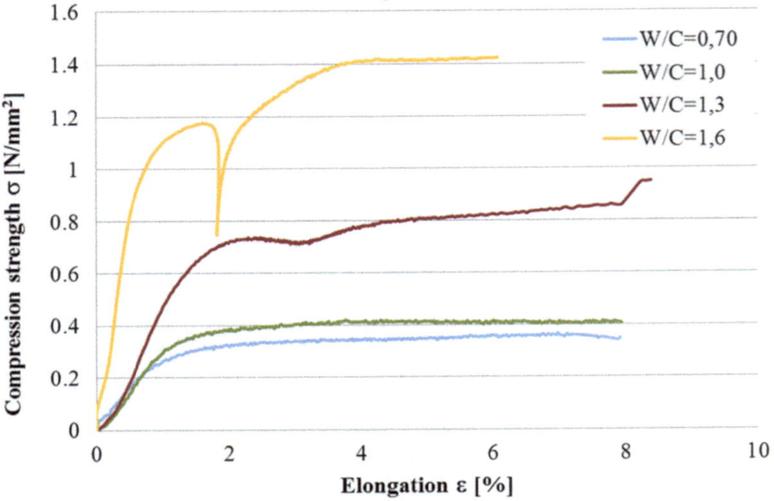

Fig. 23.6 Variation of water–cement ratio—Evolution of the compression strengths after 14 days of hardening

Investigation of an Increase of the Cement Content

In a third phase, the impacts of the cement content on the compression strength of different Miscanthus concrete prisms were investigated.

In Fig. 23.7, some interesting results can be observed. Predictably, with increasing cement content, the compression strengths increase, as generally a better bond can be expected because there is more cement present for the chemical reaction with water during the hydration process. In addition, it is remarkable that at same cement content, the concrete mixture with the highest water–cement ratio achieves higher compression strengths.

Investigation of Micro-silica Addition

Finally, micro-silica has been added to a mixture with a water–cement ratio of 1.0 containing 450 kg/m^3 cement and where calcium chloride has been used as mineraliser. Again, concrete prisms were investigated and the result can be seen in Fig. 23.8.

The compressive resistance as well as the stiffness of the mixtures with micro-silica are similar to the mixtures containing 450 kg/m^3 cement of Fig. 23.7. An even higher strength development has been observed for mixtures without micro-silica with a higher water–cement ratio, which again proves the importance of this parameter.

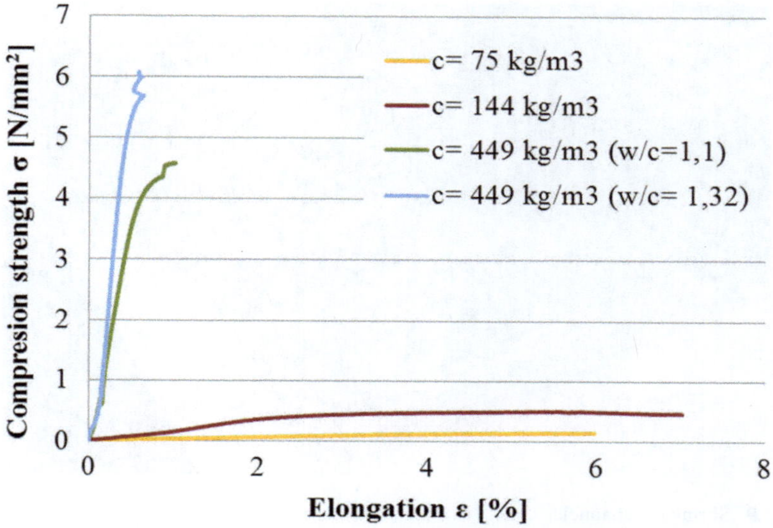

Fig. 23.7 Variation of cement content and water–cement ratio—Evolution of the compression strengths after 14 days of hardening

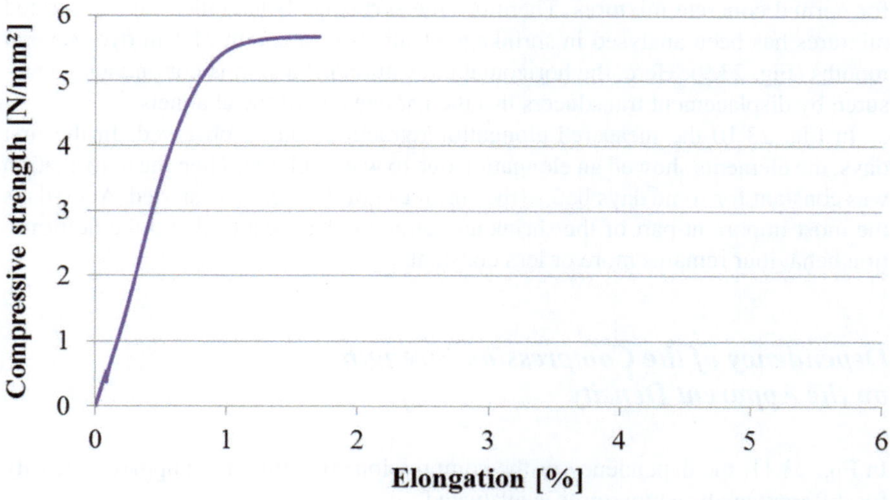

Fig. 23.8 Impact of an addition of micro-silica—Evolution of the compression strengths after 14 days of hardening

Fig. 23.9 Shrinkage channels

Investigation of Shrinkage

All concrete is subjected to shrinkage. This phenomenon is even more pronounced for lightweight mixtures on the basis of renewable water absorbing aggregates than for normal concrete mixtures. Therefore the shrinkage behaviour of the described mixtures has been analysed in shrinkage channels of a length of 1 m over several months (Fig. 23.9). Here, the horizontal elongation/retraction is continuously measured by displacement transducers installed at one end of the channels.

In Fig. 23.10 the measured elongation/retraction can be observed. In the first days, the elements showed an elongation due to water release. Then the deformation was constant for some days before the shrinkage of the elements started. At 60 days the most important part of the shrinkage retraction has occurred and the deformation behaviour remains more or less constant.

Dependency of the Compression Strength on the Apparent Density

In Fig. 23.11, the dependency of the compression strength on the apparent density for different analysed mixtures is evaluated.

The evaluation is carried out to examine if, like for other construction materials, a clear relation between the apparent density of the sample and the expected compressive strength is given. As expected this has been confirmed as the greater the density, the higher the compression strength of the sample. The determined values

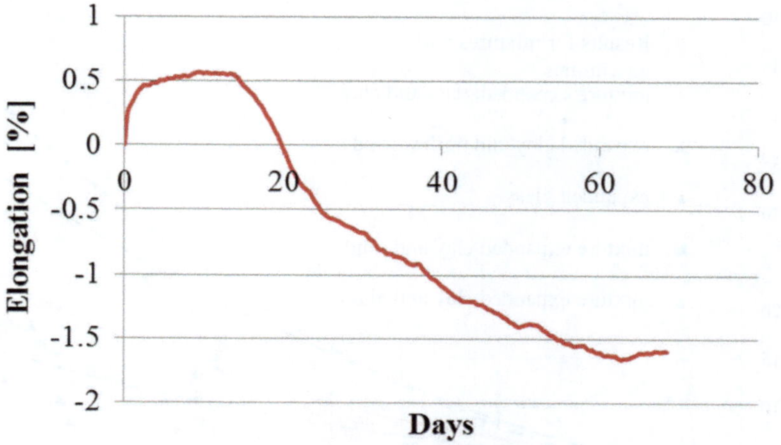

Fig. 23.10 Elongation/retraction behaviour of a mixture on the basis of Miscanthus aggregates

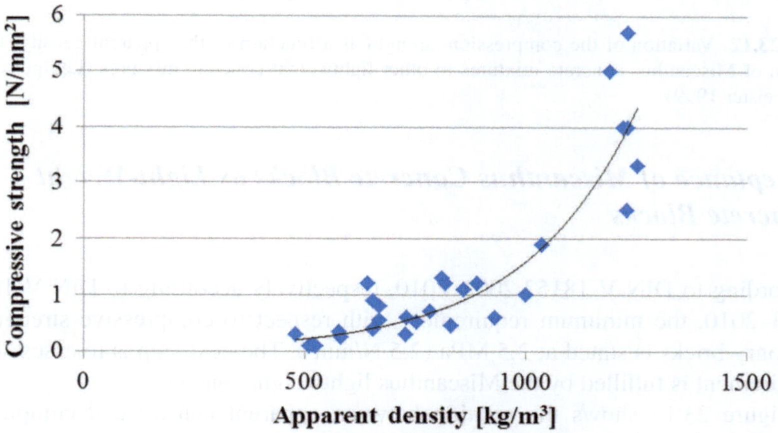

Fig. 23.11 Variation of compression strength in function of the apparent density—Evolution of the compression strengths related to the apparent density after 14 days of hardening (Dahm and Faltz, 2015)

can be placed along an exponential curve, but in order to confirm this behaviour it is necessary to evaluate more data in the future. However, this relation can be explained by the fact that a steady rise of the cement content leads to a gain in weight which conducts to higher strengths. Additionally, the water–cement ratio also has a considerable impact on the compressive strength as well as on the weight as the more water is available to the cement particles, the more hardened cement stone results, thus affecting the apparent density.

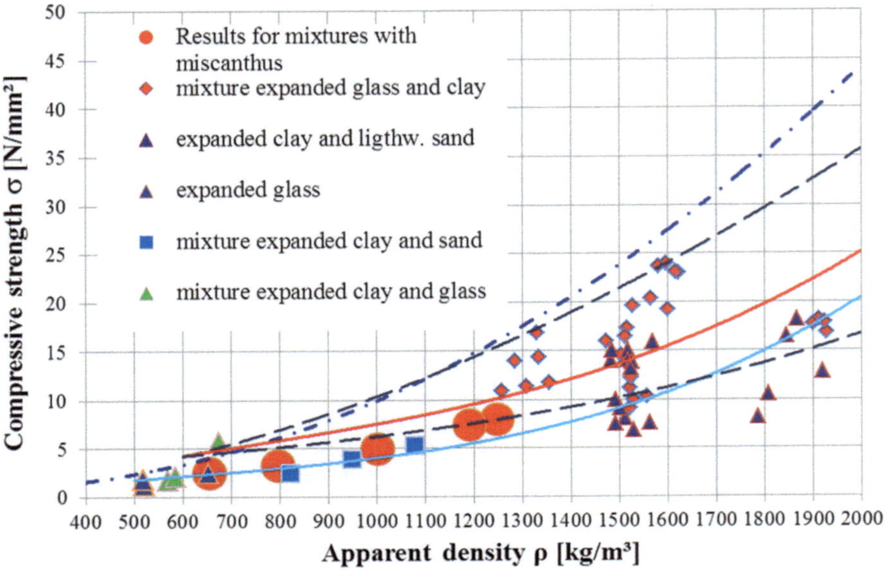

Fig. 23.12 Variation of the compression strength as a function of the apparent density: Comparison of Miscanthus concrete mixtures to other lightweight concrete mixtures (Leufgens 2010; Sagmeister 1999)

Acceptance of Miscanthus Concrete Blocks as Light-Weight Concrete Blocks

According to DIN V 18152:2003–2010, respectively according to DIN V 18152: 2003–2010, the minimum requirement with respect to compressive strength for masonry bricks is stated at 2.5 MPa (2.5 N/mm^2). The next step is to assess if this requirement is fulfilled by the Miscanthus lightweight concrete.

Figure 23.12 shows the relation between apparent density and compression strength of other lightweight mixtures in order to have some known references and state the position of Miscanthus concrete in the spectrum of existing similar options.

The previously evaluated relation points of the Miscanthus concrete are projected as red data points into Fig. 23.12, which allows us to make some conclusive observations about the aptness of Miscanthus concrete as lightweight concrete applicable for block fabrication.

Additionally to the red data points, a straight horizontal red line is drawn which represents the required minimal value of 2.5 MPa in compression strength. It can be considered that from an apparent density of about 800 kg/m^3 on, the analysed Miscanthus concrete fulfils this requirement. In addition, it can be noticed that the Miscanthus is close to the laboratory test range of other lightweight concrete mixtures. The fabrication of Miscanthus in production lines could even result in better performance values in terms of compressive strength in relation to the apparent

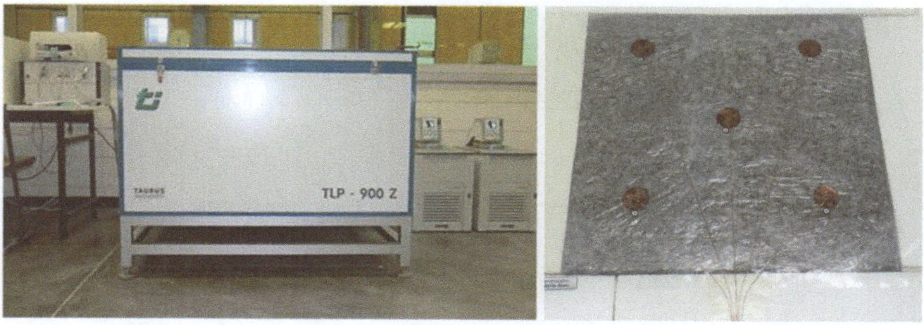

Fig. 23.13 Miscanthus board installed in a guarded hot plate (two-plate device from Taurus instruments)

density due to the industrial production process as the introduced compression and vibration energy during casting in production lines for masonry blocks is much higher.

Description of the Measurement of the Thermal Conductivity of Miscanthus Concrete

For selected Miscanthus concrete mixtures, two plates of dimensions 50/50/7 cm were produced and the thermal conductivity was determined using the Two-Plate Method according to DIN 52612, DIN EN 122664 and DIN EN 1946-1. This method allows determining the thermal conductivity of homogeneous and inhomogeneous test specimens (Fig. 23.13).

In this method, two samples of the same concrete mixture are placed and analysed such that the direction of the heat flow in the samples corresponds to those in the masonry. The samples are arranged between one heating plate and one cooling plate according to the sequence shown in Fig. 23.14 and a heat flow is established.

Once a stationary temperature state is installed during the measurement, a constant heat flow circulates through the sample plates and a one-dimensional temperature field in the measurement zone is given. The actual heat conductivity is measured by the determination of the electrical power of the mean temperature difference between the sample surface and the physical dimension of the specimen.

This method was applied to several Miscanthus concrete mixtures with different apparent densities and the results were projected in a diagram (Fig. 23.15) showing the relation of the apparent density and the thermal conductivity. Additionally, an approximated line representing the relation is drawn by assuming a linear relation.

As from previous figures, the minimal required apparent density is about 800 kg/m^3, the value for the corresponding thermal conductivity is determined by following the relation line. The associated thermal conductivity value is about 0.17 W/mK

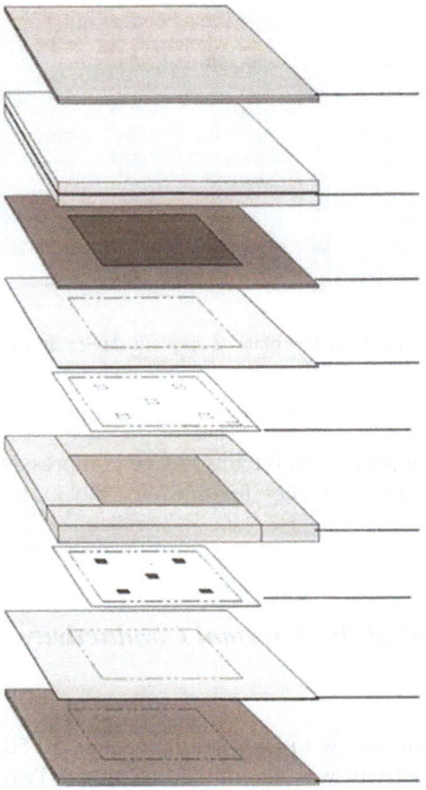

Fig. 23.14 Experimental setup for the measurement of thermal conductivity according to the Two-Plate Method (Taurus Instruments, user's guide 2005)

(Fig. 23.15). Now applying this value of 17 W/mK for the thermal conductivity of Miscanthus concrete, the thermal transmission coefficients for varying concrete block configurations can be calculated.

Analytical and Numerical Analysis of the Overall Heat Transfer Resistance for a Given Block Configuration

In Fig. 23.16 an example of an insulating masonry block configuration with its three layers is given by a top view: the inner load carrying part which is oriented towards the inside of a building, the middle insulation part and the outer leaf forming a resistant outer finish of the wall. The apparent densities with the according heat conductivities are also indicated in Fig. 23.16. The outer shell is realised with the same concrete as the inner load carrying part of the block.

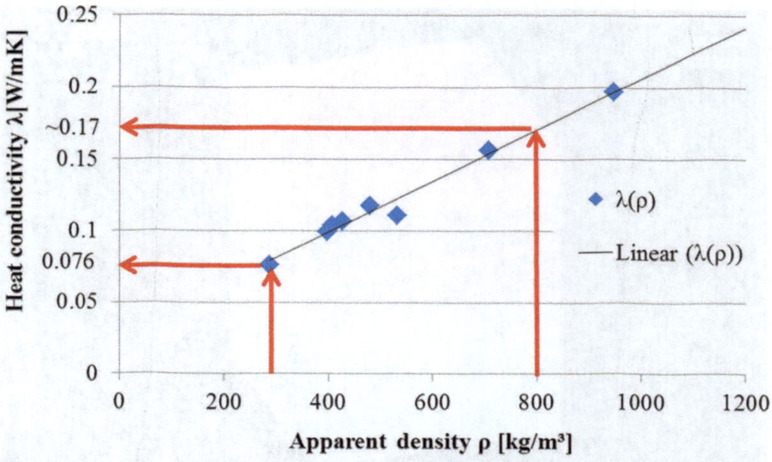

Fig. 23.15 Variation of the thermal conductivity as a function of the apparent density

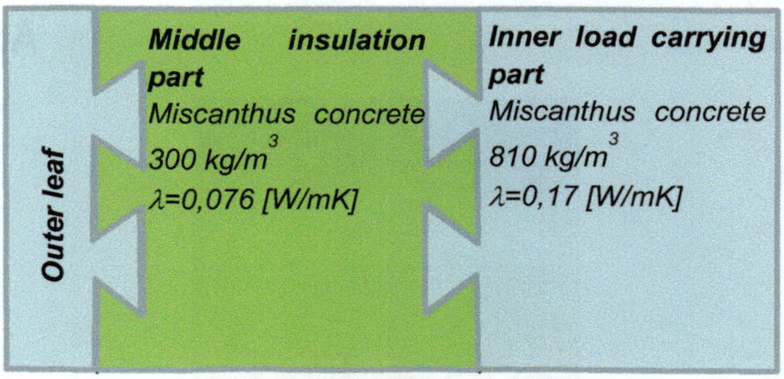

Fig. 23.16 Top view of a possible block configuration

The analytical calculation of the overall heat transfer resistance was performed using a two-dimensional model for the insulation block. The heat transfer resistance is composed of two parts:

1. RT′ = the upper extreme: analysis of the heat flow perpendicular to the element surface
2. RT″ = the lower extreme: analysis of the heat flow parallel to the element surface

The overall heat transfer resistance is then calculated as follows:
$RT = (RT' + RT'')/2$

These calculated values were used as reference values as additionally a numerical FE calculation is executed in order to check the acceptance of the primary test calculation. This has been realised by using an idealised 3D model in the software

Fig. 23.17 Finite Element Model of an insulating masonry block

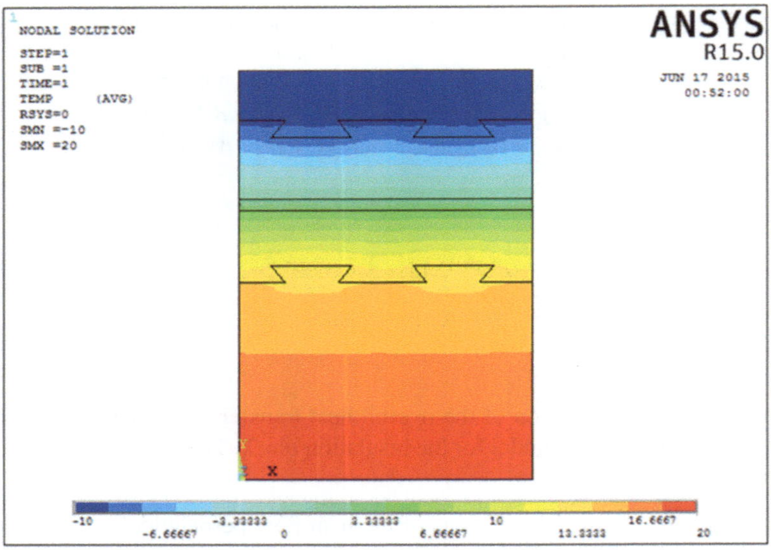

Fig. 23.18 Heat flow through an insulating masonry block

ANSYS on the basis of the Solid90 elements within a steady state analysis (Figs. 23.17 and 23.18). The following parameters have been used:

- Units: [m]; [°C], [W], [W/m^{2}K]
- Internal temperature: +20 °C
- External temperature: −10 °C
- Internal heat transfer coefficient: 8 W/m^{2}K
- External heat transfer coefficient: 20 W/m^{2}K

Table 23.2 Heat transfer resistance of different block configurations

Block configuration		Heat conductivity λ in [W/mK]	Heat transfer resistance [W/(m²K)]
The inner load carrying block	17.5 cm	Miscanthus concrete 810 kg/m³	**0.312**
		λ=0.17 W/mK	
Insulation	**15** cm	**Miscanthus concrete 300 kg/m³**	
		λ=0.076 W/mK	
Outer leaf	4 cm	Miscanthus concrete 810 kg/m³	
Total block width	**36.5 cm**	λ=0.17 W/mK	
The inner load carrying block	17.5 cm	Miscanthus concrete 810 kg/m³	**0.192**
		λ=0.17 W/mK	
Insulation	**15** cm	**Standard polystyrene**	
		λ=0.035 W/mK	
Outer leaf	4 cm	Miscanthus concrete 810 kg/m³	
Total block width	**36.5 cm**	λ=0.17 W/mK	
The inner load carrying block	17.5 cm	Miscanthus concrete 810 kg/m³	**0.259**
		λ=0.17 W/mK	
Insulation	**20** cm	**Miscanthus concrete 300 kg/m³**	
		λ=0.076 W/mK	
Outer leaf	4 cm	Miscanthus concrete 810 kg/m³	
Total block width	**41.5 cm**	λ=0.17 W/mK	
The inner load carrying block	17.5 cm	Miscanthus concrete 810 kg/m³	**0.150**
		λ=0.17 W/mK	
Insulation	**20** cm	**Standard polystyrene**	
		λ=0.035 W/mK	
Outer leaf	4 cm	Miscanthus concrete 810 kg/m³	
Total block width	**41.5 cm**	λ=0.17 W/mK	

Table 23.2 finally shows the results for the heat transfer of the analysed masonry block as a function of varying insulation thickness (15 and 20 cm) and insulation performance (λ=0.076 W/mK for Miscanthus concrete and λ=0.035 W/mK for standard polystyrene). The corresponding heat transfer resistance is indicated in the last column. The variations show heat transfer resistances from 0.312 to 0.15 W/(m²K) which constitutes an interesting performance.

Conclusions

Different masonry insulation block concepts and characteristics have been discussed in order to find the most suitable concepts in terms of insulation performances using Miscanthus lightweight concrete blocks. It has been proven that the compression strength that could be reached by lightweight concrete based on Miscanthus aggregates is comparable to other lightweight concrete mixtures. Furthermore, the porous character of this aggregate permits to generate interesting thermal conductivity performances. Thus, this new material is appropriate to be used as basis for masonry insulation blocks.

References

Achtziger J, Irmschler H.-J (2000) Bestimmung der Rechenwerte der Wärmeleitfähigkeit für Mauerwerk durch Messungen und durch Berechnungen Erschienen in: Mauwerkskalender, Abschnitt D, Teil IV, S. 635 ff

Anton H (1992) Bestimmung des Wärmedurchlasswiderstandes von Außenwänden aus Mauerwerk-Vergleich zwischen experimenteller Untersuchung und Berechnung nach der Finiten Differenzen Methode, Final report, Forschungsinstitut für Wärmeschutz München (FIW), DIBt Az.:IV 1-5-453/85

Brauer N, Heinecke S (2007) Neues Bemessungskonzept und seine Anwendung in der Baupraxis, Bemessung im Mauerwerksbau, Erschienen in: Deutsches-Ingenieurblatt

Dehn F (1998) Particular features on the production of lightweight aggregate concrete (LAC) with open structure. Lacer No. 03, Institut für Massivbau und Baustofftechnologie, University of Leipzig

DIBt-Richtlinie zur Messung der Wärmeleitfähigkeit $\lambda_{10,tr}$ von Mauersteinproben, Fassung Oktober 2002, Erschienen in: DIBt-Mitteilungen 2/2003

EuroLightCon-Report No. 1; Definition and International Consensus Report EuroLightCon; Economic Design and Construction with Light Weight Aggregate Concrete; Document BE96-3942/R1, April 1998

EuroLightCon-Report No. 2; LWAC Material Properties; State-of-the-Art; EuroLightCon; Economic Design and Construction with Light Weight Aggregate Concrete; Document BE96-3942/R2, December 1998

EuroLightCon-Report No. 8; Properties of lightweight concretes containing Lytag and Liapor, EuroLightCon; Economic Design and Construction with Light Weight Aggregate Concrete; Document BE96-3942/R8, March 2000

European Directive 2010/31/EU of the European Parliament and of the council of 19 May 2010 on the energy performance of buildings

Goltermann P (2008) Load-carrying capacity of lightly reinforced, prefabricated walls of lightweight aggregate concrete with open structure, Materials and Structures, Rilem

Kvande T (2001) Investigations of some Material Properties for Structural Analysis of LECA Masonry, Norwegian University of Science and Technology, Faculty of Civil Engineering, Dissertation

Leufgens N (2010) PhD Thesis Mechanische und thermische Untersuchungen zur Entwicklung eines Wärmedämmsteines aus Leichtbeton. PhD-FSTC-11_2010

Leufgens N, Waldmann D, Maas S, Zürbes A (2007) Development of hybrid-light concrete bricks with improved heat-insulating properties. Third Scottish Conference of Postgraduate Researchers of the Built and Natural Environment, PRoBE 2007, Conference Proceedings, pp. 277–289

Leufgens N, Waldmann D, Maas S, Zürbes A (2009) Thermische und Mechanische Eigenschaftswerte von haufwerksporigen Leichtbetonen (LAC) zur Optimierung von Wärmedämmsteinen, Tagungsband, ibausil -Internationale Baustofftagung, Bauhaus-Universität Weimar

Sagmeister B (1999) Mauersteine- Wärmedämmend mit haufwerksporigen Leichtbeton, Erschienen in: Betonwerk + Fertigteiltechnik, Heft 7, Vol 65, Bauverlag GmbH Wiesbaden, S. 48–55

Spitzner J (1975) Haufwerksporige Leichtbetone aus Blähton; Erschienen in: Beton, Heft 3, S. 92–94

Taurus instruments (2005) GmbH, Benutzerhandbuch, TLP 900 Z, Version Juni 2005

Dahm, Faltz (2015) Bachelor Thesis "Machbarkeitsstudie - Miscanthus", Bachelor Professional of Engineering, University of Luxembourg, not published

Voß K-U (2009) Target optimization of high thermal insulation lightweight aggregate concrete with open structure. Proceedings of the 53rd BetonTage- Building the Future- Ulm (Vol. 75, S. 112–113). Concrete Plant + Precast Technology

Standards

DIN EN 1946-1, Thermal performance of building products and components—Specific criteria for the assessment of laboratories measuring heat transfer properties—Part 1: Common criteria; German version EN 1946-1:1999

DIN 52612: Tesing of thermal insulating materials; determination of thermal conductivity by means of the guarded hot plate apparatus; conversion of the measured values for building applications

DIN EN 12664:2001, Thermal performance of building materials and products—Determination of thermal resistance by means of guarded hot plate and heat flow meter methods—Dry and moist products with medium and low thermal resistance; Beuth Verlag GmbH, May 2001

DIN EN 196, Methods of testing cement—Part 1: Determination of strength; German version EN 196-1:2005

DIN V 18151;4 Lightweight concrete hollow blocks; Deutsches Institut für Normung e.V., Beuth Verlag Berlin, Oktober 2003

DIN V 18152; Lightweight concrete solid blocks; Deutsches Institut für Normung e.V., Beuth Verlag Berlin, Oktober 2003

Chapter 24
Miscanthus Horse Bedding Compares Well to Alternatives

B. Rauscher and I. Lewandowski

Introduction

Wheat or rye straw is the most frequently used bedding material for horses. It can be easily handled, its visual appearance and smell are positively perceived by horse owners and it can be purchased at lower prices than alternative bedding materials (Kusch 2013). For example, straw bedding costs about 290 €, but wooden bedding materials about 510 € per year and horse (Beck 2005). Horses feel comfortable on straw, play with it and use it as additional coarse feed. However, cereal straw also has some disadvantages as bedding material, such as:

- Low water-holding capacity/ammonia (NH_3) absorption (Fleming et al. 2008a)
- High dust emissions, which can lead to allergic reactions (Garlipp et al. 2011)
- Pathogens (mostly fungi) (Fleming et al. 2008b)
- High manure volumes (Airaksinen et al. 2001)
- Daily mucking out required
- Continuous ingestion

Air quality is an important factor for the health of horses. Long periods of exposure to dust and noxious gases (NH_3, N_2O, CO_2) are seen as the main cause of several equine respiratory diseases (Fleming et al. 2008a; Gerber et al. 2003). In addition to inanimate particles (dust), animate particles such as bacteria, fungi, viruses etc. also have a significant impact on horses' health (Fleming et al. 2008b). All of these factors are strongly influenced by bedding material (Dunlea and Dodd 1999; Clarke et al. 1987). Therefore, an analysis of alternative bedding materials should take the importance of these factors and the above-mentioned disadvantages of straw into consideration.

B. Rauscher • I. Lewandowski (✉)
Department of Biobased Products and Energy Crops (340b), Institute of Crop Science, University of Hohenheim, Fruwirthstr. 23, Stuttgart 70599, Germany
e-mail: iris_lewandowski@uni-hohenheim.de

S. Barth et al. (eds.), *Perennial Biomass Crops for a Resource-Constrained World*, DOI 10.1007/978-3-319-44530-4_24

Alternative bedding materials used in practice are, for example, wood shavings, wood chips, paper cuttings, and straw, wood or miscanthus pellets. Miscanthus bedding material especially has recently become of interest for two reasons. First, it has been reported that horses are less likely to have allergic reactions to miscanthus than to wheat straw. Secondly, less material and time for mucking out is required (Redel 2015; Thoma 2015). It is this high time requirement in particular that is a problem for most horse owners. The overall annual workload can be up to 145 h per horse. About 60 % of this time is required for littering and mucking out (Hoffmann et al. 2007). However, there is little known about the competitive performance of these alternative bedding materials and whether their use could overcome the drawbacks of straw bedding mentioned above. Therefore, the objective of this study is to compare different alternative bedding materials, in particular miscanthus, with straw and to elaborate recommendations for their optimal uses. For this purpose a field trial was performed in a horse stable with four different bedding materials, namely wood chips, miscanthus chips, miscanthus pellets and straw. Operation time and costs were assessed. The water-holding and ammonium-binding capacity of the bedding material was measured in laboratory analysis.

Material and Methods

Trial Site

The field trial was performed on the holiday horse farm Reckerstal in Igersheim, south Germany. Sixteen boxes were available, and each box represented one "plot" of a fully randomised block trial design with four replications (see Fig. 24.1).

Bedding Material and Application

The Following Variants Were Used as Bedding Material

1. Straw: Straw from wheat, with smaller portions of rye and barley, was produced on the horse farm. A mixture of short and long straw was used. Short straw was produced by chopping the long straw (30 cm) to a length of 12 cm and then baling.
2. Wood shavings: 'Span Classic' from the company German Horse Pellets GmbH was used. This is a soft wood product, mainly from spruce and pine, which is dried, sieved, de-dusted and baled.
3. Miscanthus chips: These were supplied by the company Novabiom, France. *Miscanthus x giganteus* was harvested at 15 % moisture in April. A field chopper was used to cut miscanthus into chips of 3-cm length, which were de-dusted and baled.
4. Miscanthus pellets: These were supplied by the company Novabiom, France. *Miscanthus x giganteus* was harvested at 15 % moisture in April, pelleted and delivered in plastic bags. The pellet diameter was 0.8 mm.

Fig. 24.1 Arial view of the horse farm Reckerstal. Boxes are indicated in *yellow* with the following abbreviations for bedding material: *ST* straw, *HS* wood shavings, *H* miscanthus chips, *P* miscanthus pellets; *Green area* storage for bedding material; *Red area* manure storage. Source of photo: maps.google.com

Application of the Bedding Material

Straw: At the beginning of the 4-week investigation period, 40–50 kg of straw were applied per 12 m^2 box and spread evenly, to give a 10–15 cm deep litter. Horse droppings and wet spots were removed with a pitchfork and the removed material replaced on a daily basis. In sum, about 65 kg of straw was replaced per week per box.

Wood shavings: Three to four foiled bales of wood shavings of 20 kg weight each were applied per 12 m^2 box at the beginning of the 4-week investigation period, to give a litter of 10–15 cm depth. Horse droppings and wet spots were removed daily using a fine-pronged pitchfork and replaced. Around 1–2 bales were added per week during the investigation period, the amount being adjusted according to the amount of bedding material removed. The application followed recommendations by Levold (2015).

Miscanthus chips: The 12 m^2 boxes were filled with four to five foiled bales of miscanthus chips, each bale weighing 25 kg, to give a 10-cm-deep litter. Horse droppings were removed on a daily basis and wet spots every 2–3 days with a fine-pronged pitchfork. One bale was added per week during the 4-week investigation. The application followed recommendations by Pichon (2015).

Miscanthus pellets: 150–170 kg of miscanthus pellets were applied per 12 m^2 box at the beginning of the 4-week investigation period, to give a 5-cm-deep litter.

Horse droppings were removed daily and wet spots every 2–3 days with a fine-pronged pitchfork. One 15 kg bag was added per week in this way. The application followed recommendations by Pichon (2015).

Performance of Field Trial

Mucking out was performed daily by the same person at 6–8 am over a 4-week period.

The following parameters were assessed:

- Amount of manure (mass/bulk density/volume) produced
- Time needed to muck out ($t_{Mucking}$) and transport material and manure to storage places ($t_{Overall}$)
- Observation of animal behaviour

Labour and material costs for the different bedding alternatives were calculated based on an hourly wage of 8.50 €. When miscanthus and wood chips are used, horses need to be fed an additional source of roughage to support their natural feeding behaviour. An extra feed portion of 5 kg of hay was given daily. These costs were added to the overall costs.

Laboratory Analysis

The water-holding capacity was measured according to DIN 51718. For this purpose, 15 g of bedding material was weighed into water-permeable plastic bags. These were soaked in water for 24 h, hung up to drip for 10 min and then dried at 105 °C.

Ammonia absorption capacity was measured with Dräger diffusion tubes (NH3 20/a-D) placed in a plastic bucket containing 100 mL bedding material (loose bedding) and 400 mL fresh horse urine. Measurements were taken after 2 and 3 h of soaking at 25–27 °C. Ammonia absorption capacity was calculated using the following equation:

$$A = \frac{\frac{c_T - c_S}{t}}{\frac{c_T}{t}} * 100\%$$

A = ammonia absorption [%]
c_T = ammonia concentration in sample with pure horse urine [ppm*h]
c_S = ammonia concentration in sample with bedding material [ppm*h]
t = duration of soaking [h]

Statistical Analysis

Analysis of variance and comparison by Tukey test were performed using the statistics program SAS 9.4 at a confidence interval of 95 %.

Results

Field Trial Results

The lowest amount of manure per box was produced with miscanthus pellets and chips as bedding material. The lowest overall operation time for mucking out, transporting and storing the manure was also measured for these materials (Table 24.1). This is due to the fact that less material needs to be transported from the material stores and to the manure storage. However, the shortest time required for mucking out alone was found with straw bedding (Table 24.1).

The use of miscanthus pellets significantly reduces the manure volume compared to straw, followed by miscanthus chips and wood shavings (Fig. 24.2).

Water- and Ammonia-Binding Capacity

Miscanthus pellets and wood shavings had the highest water-holding capacity (Table 24.2). The water-holding capacity of straw was higher than expected. This could be due to the fact that the straw had to be cut to a length under 10 cm to get it into beakers for the experiment. Miscanthus chips showed a high ammonia absorption capacity, similar to wood shavings. Miscanthus pellets had by far the highest and straw the lowest ammonia-binding capacity of all the bedding materials (Table 24.2).

Table 24.1 Manure production and time (shown in minutes:seconds) required for mucking out and overall cleaning operations for different bedding materials

Bedding material	Manure (kg/day and box)	$t_{Mucking}$ (per day and box)	$t_{Overall}$ (per day and box)
Straw	34.93 ± 3.40	3:39 ± 00:17	5:49 ± 00:16
Wood shavings	30.28 ± 3.38	5:47 ± 00:08	6:16 ± 00:06
Miscanthus chips	28.44 ± 3.72	4:52 ± 00:34	5:26 ± 00:38
Miscanthus pellets	27.79 ± 8.51	5:05 ± 00:32	5:17 ± 00:27

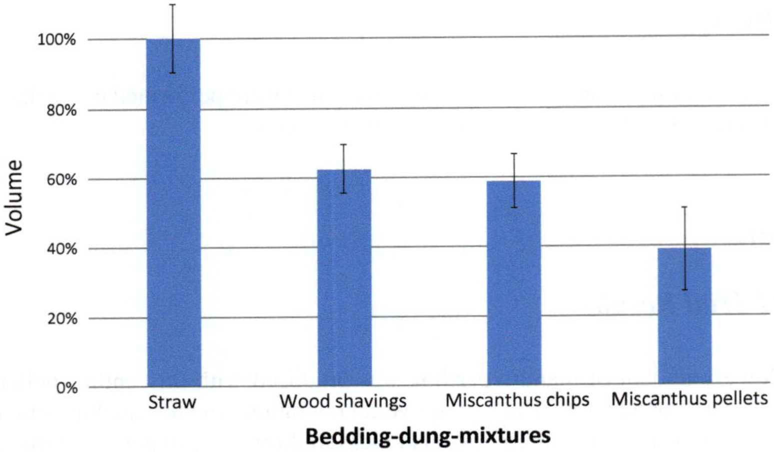

Fig. 24.2 Relative volume of manure derived from different bedding/dung mixtures compared to straw

Table 24.2 Water-holding capacity (WHC), ammonia emissions and ammonia absorption (binding capacity) of different bedding materials

Sample	WHC* [g water/g DM]	Ammonia emission [ppm]	Ammonia-binding capacity [%]
Horse urine (pure)	–	20.4±6.5	0
Straw	4.60±0.39	20.0±3.5	2
Wood shavings	5.52±0.24	9.6±0.6	53
Miscanthus chips	3.28±0.08	10.0±1.2	51
Miscanthus pellets	5.93±0.01	1.7±2.4	86

Observation of Animal Behaviour

Horses bedded on straw showed the most frequent recreational activities, such as grubbing or wallowing. Horses bedded on wood shavings and miscanthus chips showed less of these activities and were less occupied by the bedding material. Almost no activities were observed on miscanthus pellets. Despite less recreational activities, no behavioural disorders (aggressiveness, gnawing wood) were noticed on the alternative bedding materials.

Cost Comparison

Overall costs are lowest for straw bedding as it is a cheap bedding material and no additional feed is required (Table 24.3). Miscanthus bedding is the most expensive alternative. Purchasing costs per box are lower for miscanthus pellets than for chips.

Table 24.3 Costs for different bedding material alternatives, calculated for a 4-week time period per box

Costs for a 4-week period	Straw	Wood shavings	Miscanthus chips	Miscanthus pellets
Bedding material	98.68 €	192.40 €	337.17 €	216.00 €
Labour	92.32 €	99.43 €	86.17 €	83.78 €
Additional feed	0.00 €	58.10 €	58.10 €	58.10 €
Overall costs	**191.00 €**	**349.93 €**	**481.44 €**	**357.88 €**

Miscanthus pellets have a higher water-absorption capacity and therefore less material is required to bed horses on miscanthus pellets than on chips.

Discussion and Conclusions

The investigations showed that alternative bedding materials produce less manure mass and volume as well as higher bulk densities. This results in lower manure storage volume requirements for alternative bedding. Less time is also needed for mucking out and adding new bedding material to the boxes. In addition, they have a good hygiene status. Straw performed best in the parameters costs, handling, horse behaviour, appearance and smell (Table 24.4). The evaluation of smell and appearance in this case does not take staining from droppings and urine into account. If this was considered, the evaluation would change slightly in favour of the miscanthus products. Straw and Miscanthus pellets show the best results. These show much less discolouration than wood shavings and miscanthus chips. Miscanthus bedding is more expensive than straw, despite the fact that less material and time are required.

As shown in the comparative evaluation of the bedding materials (Table 24.4), miscanthus and wood show good field performance for those parameters where straw has the weakest performance and vice versa. Therefore, it is concluded that miscanthus could be applied in fields where straw is not used due to its specific disadvantages.

This leads to the following conclusions and recommendations for practical application of miscanthus horse bedding in the various fields of application:

1. Professional equestrian sport, on account of:
    - Short time required for mucking out
    - Ingestion can be adjusted to the individual performance of the horses
    - High health standards
2. Horse farms that need to reduce manure volumes, because:
    - Storage and disposal costs can be reduced

Table 24.4 Comparative evaluation of different horse bedding materials

Parameter		Straw	Wood shavings	Miscanthus chips	Miscanthus pellets
Manure	Mass	–	+	++	++
	Volume	––	+	+	++
Bulk density		–	+	+	++
Time	$t_{Mucking}$	++	––	–	0
	t_{Ges}	0	–	+	+
Costs		++	0	––	0
Handling		++	–	+	+
Hygiene	WHC	+	++	–	++
	NH_3	––	+	+	++
	Dust	––	+	+	–
Behaviour		++	+	0	––
Optics	Original	++	++	+	––
	With faecal pollution	+	–	–	0
Smell		++	++	+	0

- In urban areas manure application on fields is not possible

3. Owners of allergic horses, on account of:

- Low dust emissions
- High water-holding capacity
- High ammonia absorption
- Consequent reduction of respiratory diseases and allergic reactions

References

Airaksinen S, Heinonen-Tanski H, Heiskanen M (2001) Quality of different bedding materials and their influence on the compostability of horse manure. J Equine Vet Sci 21:125–130

Beck J (2005) Pferdemist—Problemlösung durch mechanische Aufbereitung, Kompostierung und thermische Verwertung. Landtechnik 60(1):40–41

Clarke AF, Madelin TM, Allpress RG (1987) The relationship of air hygiene in stables to lower airway disease and pharyngeal lymphoid hyperplasia in two groups of Thoroughbred horses. Equine Vet J 19:524–530

Dunlea AP, Dodd VA (1999) A mechanical ventilation system for horse stables to control respirable dust. Irish Vet J 52:257–262

Fleming K, Hessel EF, Van den Weghe HFA (2008a) Evaluation of factors influencing the generation of ammonia in different bedding materials used for horse keeping. J Equine Vet Sci 28: 223–231

Fleming K, Hessel EF, Van den Weghe HFA (2008b) Generation of airborne particles from different bedding materials used for horse keeping. J Equine Vet Sci 28:408–418

Garlipp F, Hessel EF, van den Weghe HFA (2011) Characteristics of gas generation (NH3, CH4, N2O, CO2, H2O) from horse manure added to different bedding materials used in deep litter bedding systems. J Equine Vet Sci 31:383–395

Gerber V, Robinson NE, Luethi S, Marti E, Wampfler B, Straub R (2003) Airway inflammation and mucus in two age groups of asymptomatic well-performing sport horses. Equine Vet J 35:491–495

Hoffmann G, Bockisch F, Kreimeier P (2007) Bessere Haltungsqualität bei weniger Arbeit—Verfahrenstechnische Lösungsansätze in der Pferdehaltung. Landtechnik 62(3):172–173

Kusch S (2013) Horse stall waste: amounts, management, bioenergy generation. Proceedings EIIC (Proceedings in EIIC—The 2nd Electronic International Interdisciplinary Conference), 2–6 September 2013 (pp. 423–428)

Levold R (2015) Usage of wood shavings "Span Classic". German Pellets GmbH. Pers. comm.

Pichon M (2015) Usage of miscanthus bedding. Novabiom. Pers. comm.

Redel R (2015) Miscanthus-Häcksel als Pferdeeinstreu. http://www.redel-miscanthus.de/einstreu.html. Accessed 26 Aug 2015

Thoma U (2015) Miscanthus-Einstreu. www.pferdeeinstreu-thoma.de/html/miscanthus_-_elephantengras__e. Accessed 26 Aug 2015

ERRATUM TO

Perennial Biomass Crops for a Resource-Constrained World

Susanne Barth, Donal Murphy-Bokern, Olena Kalinina, Gail Taylor, and Michael Jones

S. Barth et al. (eds.), *Perennial Biomass Crops for a Resource-Constrained World*,
DOI 10.1007/978-3-319-44530-4

DOI 10.1007/978-3-319-44530-4_25

In Chapter 3 titled "Perennial Grasses for Sustainable European Protein Production" the affiliation details of the authors' Drs. Jørgensen and Lærkeare are incorrect. The correct affiliation details should read as follows:

Department of Agroecology
Aarhus University
Blichers Allé 20, Postboks 50
DK-8830 Tjele

The above affiliation details are also updated in chapter 13 titled "Creation and Evaluation of Novel Cold Tolerant *Miscanthus* Hybrids" and chapter 20 "Biomass Yield and N Uptake in Tall Fescue and Reed Canary Grass Depending on N and PK Fertilization on Two Marginal Sites in Denmark" and also in the contributors list in the F.M.

The updated online version of the original book can be found under
DOI 10.1007/978-3-319-44530-4

S. Barth et al. (eds.), *Perennial Biomass Crops for a Resource-Constrained World*,
DOI 10.1007/978-3-319-44530-4_25

Index

S. Barth et al. (eds.), *Perennial Biomass Crops for a Resource-Constrained World*, DOI 10.1007/978-3-319-44530-4

MIX
Papier aus verantwortungsvollen Quellen
Paper from responsible sources
FSC® C105338

If you have any concerns about our products,
you can contact us on
ProductSafety@springernature.com

In case Publisher is established outside the EU,
the EU authorized representative is:
Springer Nature Customer Service Center GmbH
Europaplatz 3, 69115 Heidelberg, Germany

Printed by Libri Plureos GmbH
in Hamburg, Germany